Submarine Landslides and Tsunamis

NATO Science Series

A Series presenting the results of scientific meetings supported under the NATO Science Programme.

The Series is published by IOS Press, Amsterdam, and Kluwer Academic Publishers in conjunction with the NATO Scientific Affairs Division

Sub-Series

I. **Life and Behavioural Sciences**	IOS Press
II. **Mathematics, Physics and Chemistry**	Kluwer Academic Publishers
III. **Computer and Systems Science**	IOS Press
IV. **Earth and Environmental Sciences**	Kluwer Academic Publishers
V. **Science and Technology Policy**	IOS Press

The NATO Science Series continues the series of books published formerly as the NATO ASI Series.

The NATO Science Programme offers support for collaboration in civil science between scientists of countries of the Euro-Atlantic Partnership Council. The types of scientific meeting generally supported are "Advanced Study Institutes" and "Advanced Research Workshops", although other types of meeting are supported from time to time. The NATO Science Series collects together the results of these meetings. The meetings are co-organized bij scientists from NATO countries and scientists from NATO's Partner countries – countries of the CIS and Central and Eastern Europe.

Advanced Study Institutes are high-level tutorial courses offering in-depth study of latest advances in a field.
Advanced Research Workshops are expert meetings aimed at critical assessment of a field, and identification of directions for future action.

As a consequence of the restructuring of the NATO Science Programme in 1999, the NATO Science Series has been re-organised and there are currently five sub-series as noted above. Please consult the following web sites for information on previous volumes published in the Series, as well as details of earlier sub-series.

http://www.nato.int/science
http://www.wkap.nl
http://www.iospress.nl
http://www.wtv-books.de/nato-pco.htm

Submarine Landslides and Tsunamis

edited by

Ahmet C. Yalçiner
Middle East Technical University, Department of Civil Engineering,
Ocean Engineering Research Center,
Ankara, Turkey

Efim N. Pelinovsky
Laboratory of Hydrophysics and Nonlinear Acoustics,
Institute of Applied Physics,
Nizhny Novgorod, Russia

Emile Okal
Department of Geological Sciences,
Northwestern University,
Evanston, IL, U.S.A.

and

Costas E. Synolakis
Department of Civil and Aerospace Engineering,
University of Southern California,
Los Angeles, CA, U.S.A.

Springer Science+Business Media, B. V.

Proceedings of the NATO Advanced Research Workshop on
Underwater Ground Failures on Tsunami Generation, Modeling, Risk and Mitigation
Istanbul, Turkey
May 23–26, 2001

Additional material to this book can be doenloaded from http://extra.springer.com.

ISBN 978-1-4020-1349-2   ISBN 978-94-010-0205-9 (eBook)
DOI 10.1007/978-94-010-0205-9

Printed on acid-free paper

TABLE OF CONTENTS

Part 1. General Aspects of Tsunami Researches

1-1. Tsunamis of Seismic Origin; Science, Disasters and Mitigation

1-2. Needs and Perspectives of Tsunami Research in Europe

Part 2. Historical Tsunamis

2-1. Identification of Slide-generated Tsunamis in the Historical Catalogues

2-2. Updating and Revision of the European Tsunami Catalogue

2-3. Tsunami of Şarköy-Mürefte 1912 Earthquake: Western Marmara, Turkey

2-4. Spatial and Temporal Periodicity in the Pacific Tsunami Occurrence

Part 3. Submarine Landslides and Tsunami Generation

3-1. Submarine Landslide Generated Waves Modeled Using Depth-Integrated Equations.

3-2. Near Field Amplitudes of Tsunami from Submarine Slumps and Slides

3-3. Numerical Modeling of Tsunami Generation by Submarine and Subaerial landslides

3-4. Tsunami Simulation Taking Into Account Seismically Induced Dynamic Seabed
Displacement and Acoustic Effects of Water

vi

Part 5. Mitigation

PREFACE

Tsunamis are water waves   triggered by impulsive geologic events such as sea floor deformation, landslides, slumps, subsidence, volcanic eruptions and bolide impacts. Tsunamis can inflict significant damage and casualties both nearfield and after evolving over long propagation distances and impacting distant coastlines. Tsunamis can also effect geomorphologic changes along the coast. Understanding tsunami generation and evolution is of paramount importance for protecting coastal population at risk, coastal structures and the natural environment.

Accurately and reliably predicting the initial waveform and the associated coastal effects of tsunamis remains one of the most vexing problems in geophysics, and -with few exceptions- has resisted routine numerical computation or data collection solutions. While ten years ago, it was believed that the generation problem was adequately understood for useful predictions, it is now clear that it is not, especially nearfield. By contrast, the runup problem earlier believed intractable is now well understood for all but the most extreme breaking wave events.

Tsunami investigations have become highly interdisciplinary with contributions ranging from basic and applied science to coastal zone management. Most comprehensive current studies involve any number of marine surveys on tsunamigenic underwater faults, of landslide and slump geology, of sediment stability and underwater ground failures, seismic considerations, theoretical and computational approaches and hydromechanics. At present, even applied engineering design studies often extend to studying historical events, database management, instrumentation, risk, warning, preparedness, and hazards mitigation. This highly interdisciplinary culture has allowed for better identification of current limits in understanding mitigating hazard. For example, in the near future, it seems unlikely that it will be possible to assess the generating mechanism in real time with sufficient precision for timely warning of nearfield coastal residents. However, it is quite likely that within the next five years methodology will be developed for better identifying causative mechanisms from farfield measurements. It is also likely that most coastlines at risk in the US and Japan will have inundation maps for emergency preparedness while computational tools based on the shallow water wave equations will be readily available for developing maps for any coastline in the world.

Protection from natural disasters is now undisputable one of the societal priorities all over the world. Due to steady growth in coastal development in the last fifty years, future tsunamis could be even more catastrophic than historic ones. Rapid progress in tsunami science and engineering is needed for mitigating this deadly hazard. Following Plafker's classic analyses of the 1964 tsunami, recent tsunamis, including the 1979 Nice France, 1983 Sea of Japan, 1992 Nicaragua, 1992 Flores, 1993, Okushiri, Japan, 1994 Skagway, Alaska, 1994 East Java, Indonesia, 1994 Mindoro, Philippines, 1994 Kuril Islands, Russia, 1995 Manzanillo, Mexico, 1996, Biak Indonesia, 1996, Chimbote, Peru 1998 Sissano, Papua New Guinea, 1999 Izmit, Turkey, 1999 Pentecost, Vanuatu, 2001 Camana, Peru, and 2002 Wewak, Papua New Guinea, 2002 Stromboli, Italy, events, have all suggested the importance of considering multiple ground failure tsunamigenic sources for the same event. Tsunamis generated by slope failures may cause higher runup along shorter lengths of coastlines and often cause very strong, near-shore currents. Tsunamis generated by tectonic displacements often cause more widespread damage than those from slope failures, even though locally the inundation heights might be less extreme. Many events involve both types of tsunamigenesis.

Currently, sufficient quantitative comparison of different methods of tsunami generation is lacking, so that adequate discrimination of their sources from coastal inundation data is difficult and inaccurate. Even for the case of the 1998 Papua New Guinea tsunami, despite numerous extensive bathymetric surveys almost immediately following the event, only recently is consensus emerging about the crucial role of the landslide in the devastating tsunami. For many other events, necessary information on underwater ground failures is lacking, due to insufficient marine geologic data, or its unavailability for proprietary reasons. Even when suggestive evidence of earlier events are identified on the seafloor, the timing and the sequence of failures remain very difficult to predict, thus allowing only for the calculation of worst-case scenarios, at least for now. Clearly the best opportunity for real time tsunami warnings is the application of deep ocean tsunami measurements to forecast coastal impacts, regardless of the elusive generating mechanism. The technology to supply deep ocean tsunami data in real time is now available and in use in the North Pacific.

ix

The NATO Advanced Research Workshop on "Underwater Ground Failures on Tsunami Generation Modeling Risk and Mitigation" was organized in Istanbul on May 2001 through the support of the NATO Science Committee, Division of Scientific and Environmental Affairs. The objective was to better understand tsunamigenesis in the light of recent results, in a forum encouraging frank and uninhibited discussion in the most classic collegial manner of academia. The deliberations were covered in the national and international press, notably in the August 17, 2001 issue of SCIENCE magazine. The workshop discussed existing theoretical and experimental studies and examined new ideas and techniques to investigate and model the source mechanisms of tsunamis and their impact. In several instances, provocative ideas not readily available to English-only readers were discussed and debated. Several mitigation strategies were discussed with consensus emerging as to the need for further deployment of deep sea measuring transducers and for developing maps for all populations at risk. Advanced research papers selected from the presentations are presented in this book after peer review.

The CD containing full papers with colorful figures is attached to this book for providing illustrative and useful tool to the users for benefiting more from the information contained in the book.

The four editors hope that the new (and not so new) approaches, the comparisons among existing models and the development of newer ones, the coastal zone management and risk reduction considerations, will all help tsunami scientists, planners and policy makers to better mitigate present and future tsunami hazards.

Acknowledgements

We gratefully acknowledge NATO Public Diplomacy (formerly Scientific Affairs) Division, Environmental and Earth Science and Technology program for significant support to the NATO Advanced Research Workshop organization and preparation of this book.

The National Science Foundation (NSF) of the United States, Scientific and Technical Research Council (TUBITAK) of Turkey, Greek Ministry of Industry General Secretariat of Research&Technology International Organizations Department, Russian Fund for Basic Research, Middle East Technical University, University of Southern California, Northwestern University, Tohoku University, Yüksel Proje International Co, Kiska Komandit Co. are cheerfully acknowledged for their various and valuable forms of support to the workshop, either through travel grants or organizational funds.

We thank all participants and reviewers and contributors of this volume whose outstanding support made this book what it is. Many did so ignoring some unpleasant last minute nonacademic issues and deserve mention for their courage. Thanks are extended to Prof. Ayşen Ergin for her valuable guidance and help at all steps in the organization process of the workshop. We acknowledge Doğan Kısacık for his productive administrative assistance during the communications with reviewers and authors, and for his qualified technical assistance and contributions throughout the long term preparation process of this book. Cevher Coşanöz, Nilhan Çiftçi, Sibel Kızıltaş, Irep Güner and Nihal Yılmaz are also acknowledged for their efforts before, during and after the workshop organization and the preparation of this book.

Our sincere thanks are also extended to Dr. Alain H. Jubier for his kind guidance and understanding. His perseverance in the less illustrious moments was rewarded with the showcase of the workshop in the SCIENCE Magazine, and the entire tsunami mitigation community gained.

February 2003

Ahmet, Efim, Costas, Emile

In Memoriam:

Prof.Dr.Aykut Barka
1952-2002

AYKUT BARKA

This book is dedicated to the memory of Professor Aykut Barka, 1952-2002, an outstanding expert on earthquake geology and active tectonics, a unique example of success, productivity, enthusiasm and a unique character of friendship, lovable personality.

His great enthusiasm, warm friendship and lovable personality will be greatly missed not only by his Turkish colleagues and students, but also internationally.

LIST OF CONTRIBUTORS

C. AKIN
Yüksel Proje Int. Co. Birlik Mah. 9 Cad. No:41
Çankaya, Ankara, Turkey

B. ALPAR
University of Istanbul, Institute of Marine
Sciences and Management, Muskile Sok., 34470
Vefa, Istanbul, Turkey.

Y. ALTINOK
University of Istanbul, Engineering Faculty,
Department of Geophysics, 34850 Avcilar,
Istanbul, Turkey.

E. ALTUNEL
Osmangazi University, Engineering Faculty,
Eskisehir, Turkey

R. A. ALVAREZ
International Hurricane Centre, Centre for
Engineering and Applied Sciences, Florida
International University, 10555 West Flagler
Street, Miami, 33174, USA

N. N. AMBRASEYS
Imperial College of Science and Technology
London SW7 2BU UK

R. L. ANDERSON
California Seismic Safety Commission
1755 Creekside Oaks Drive, Suite 100
Sacramento, California 95833, USA

C.E. BALAS
Gazi University, ,Faculty of Engineering and
Architecture, Civil Engineering Department,
Ankara, Turkey.

A. BARKA
ITU, Eurasian Earth Science Institute,
Ayazaga, Istanbul, Turkey

E. BERNARD
NOAA/Pacific Marine Environmental
Laboratory, Seattle, Washington, USA

J. BORRERO
University of Southern California, Civil
Engineering Department, Los Angeles, CA
90089-2531, USA

J. BOURGEOIS
Dept. of Earth and Space Sciences
University of Washington
Seattle, WA 98195-1310

S. CAMERANI
Dept. di Fisica, University di Roma, "La
Spienza" Italy

Y. B. CHIRKOV
Joint Inst. Physics of the Earth, Russian
Academy of Sciences, Moscow, Russia

L. B. CHUBAROV
Institute of Computational Technologies
Lavrentyev ave. 6
630090, Novosibirsk, Russia

T. DAHLGREN
Ben C. Gerwick, Inc., 20 California Street
San Francisco, CA 94111 USA

T. DERMENTZOPOULOS
Ministry of Environment, Trikalon 36,Athens
GR-115 26,Greece

I. DIDENKULOV
Institute of Applied Physics of the Russian
Academy of Sciences, 46 Ulyanov Street, 603600
Nizhniy Novgorod, Russia

F. DIAS
Prof., CMLA, ENS-Cachan, 61 avenue du
President Wilson, 94235 Cachan Cedex, France

A. ERGİN
Middle East Technical University, Engineering
Faculty, Civil Engineering Department, Ankara,
Turkey.

M. L. ESKIJIAN
California State Lands Commission
330 Golden Shore, Suite 210, Long Beach, CA
90802 USA

Z. I. FEDOTOVA
Institute of Computational Technologies
Lavrentyev ave. 6, 630090, Novosibirsk, Russia

I. V. FINE
Heat and Mass Transfer Institute, Minsk, Belarus

M. FRANCIUS
Laboratoire de Mecanique et d'Acoustique, 31
Chemin Joseph Aiguier,13402 Marseille, Cedex
20, France

A. GANAS
Institute of Geodynamics, National Observatory
of Athens, P.O. Box 20048, Athens, Greece

P. GEORGOPOULOU
Institut of Geological and Mineral Exploration,
70 Mesogeion Av., Athens, Greece

F. I. GONZÁLEZ
NOAA/Pacific Marine Environmental
Laboratory, Seattle, Washington, USA

N. GORUR
President of TUBITAK Marmara Research
Center, PK: 21 Gebze 41470 Kocaeli Turkey

L. GRAZIANI
Istituto Nazionale di Geofisica e Vulcanologia
Via di Vigna Murata 605, 00143 Rome, Italy

I. GULER
Yüksel Proje Int. Co., Birlik Mah. 9 Cad. No:41
Çankaya, Ankara, Turkey

V. K. GUSIAKOV
Department of Geophysics, Institute of
Computational Mathematics and Mathematical
Geophysics, Siberian Division, Russian Academy
of Sciences, Pr.Lavrentieva, 6, Novosibirsk
630090, Russia

R. A. HANSEN
Geophysical Institute
University of Alaska Fairbank, USA

K .B. HAUGEN
Department of Agricultural Engineering,
Agricultural University of Norway, 1432 Aas
Norway

A. HAYIR
University of Southern California, Civil
Engineering Department, KAP 216A, Los
Angeles, CA 90089-2531, USA

R. E. HEFFRON
Han-Padron Associates
100 Oceangate, Suite 650
Long Beach, CA 90802, USA

F. IMAMURA
Disaster Control Research Center, Faculty of
Technology, Tohoku University, Sendai 980-
8579, Japan

V. M. KAISTRENKO
Institute of Marine Geology and Geophysics,
Yuzhno-Sakhalinsk, Russia

U. KANOGLU
Department of Engineering,
Middle East Technical University,
06531 Ankara, Turkey

V. KARASTATHIS
Tsunami modeler, National Observatory of
Athens 11810 Athens , Greece

C. KHARIF
Institut de Recherche sur les Phenomenes Hors
Equilibre, BP 146 49 rue Frederic Joliot-Curie
13384 Marseille cedex 13, France

M. A. KLYACHKO
Center on Earthquake and Natural Disaster
Reduction, St. Petersburg, Russia

ROBERT KOENIG
Contributing Correspondent,
SCIENCE Magazine, Published by the American
Association for the Advancement of Science, with
assistance of Stanford University's HighWire
Press, USA

S. V. KOLESOV
Department of Marine Physics, Physical Faculty,
Moscow State, University, Vorobjevy Gory,
Moscow 119899 Russia

Z. KOWALIK
Institute of Marine Science
University of Alaska Fairbanks, USA

E. A. KULIKOV
Tsunami Center, P.P.Shirshov Institute of
Oceanology, 36, Nakhimovsky prospekt, 117851,
Moscow, Russia

U. KURAN
General Directorate of Disaster Affairs,
Earthquake Research Department, Lodumlu,
Ankara, Turkey

W. LETTİS
William Lettis &Association, Wullnutcreek, CA,
USA

B.W. LEVIN
Shirshov Institute of Oceanology, Russian
Academy of Sciences, Nakhimovsky prospekt 36,
Moscow 117851 Russia

P. L. F. LIU
School of Civil and Environmental Engineering,
Cornell University, Ithaca, NY 14853, USA

P. LYNETT
School of Civil and Environmental Engineering ,
Cornell University, Ithaca, NY 14853, USA

A. MARAMAI
Istituto Nazionale di Geofisica e Vulcanologia
Via di Vigna Murata 605, 00143 Rome, Italy

M. MARCOS
Grupo de Oceanografa Interdisciplinar/
Oceanografa Fsica,IMEDEA(CSIC-UIB)
C/Miquel Marques, 21, Esporles, 07190, Balears
Spain)

S.R. MASSEL
Institute of Oceanology of the Polish Academy
of Sciences, Powstańców Warszawy 55, 81-712
Sopot, Poland

M. MATSUYAMA
Central Research Institute of Electric Power
Industry, 1646 Abiko, Abiko-shi, Chiba-ken, 270-
1194 Japan

R. J. McCARTHY
California Seismic Safety Commission
1755 Creekside Oaks Drive, Suite 100
Sacramento, California 95833, USA

R. MEDINA
Grupo de Ingeniería Oceanográfica y de Costas,
Universidad de Cantabria, E.T.S. Ingenieros de
Caminos, Spain

A. L. MERLANI
Dept. di Fisica, University di Roma, "La
Spienza" Italy

T. MILOH
Faculty of Engineering, Tel Aviv University,
Ramat Aviv 69978, Israel

K. MINOURA
Institute of Geology and Paleontology, Faculty
of Science, Tohoku University, Sendai 980-8578,
Japan

H.O. MOFJELD
NOAA/Pacific Marine Environmental
Laboratory, Seattle, Washington

S. MONSERRAT
Departamento de Física, Universitat de les Illes
Balears Ctra Valldemossa km. 7,5 Edificio,
Mateo Orfila, Palma de Mallorca, 07071 Spain

T.S. MURTY
W.F. Baird & Associates Coastal Engineers Ltd.
1145 Hunt Club Road, Suite 500, Ottawa,
Ontario K1V 0Y3, Canada

T. NAKAMURA
Dating and Material Research Center, Nagoya
University, Nagoya 464-0814, Japan

J.C. NEWMAN
Joint Institute for the Study of the Atmosphere
and Ocean, University of Washington, USA

M.A. NOSOV
Department of Marine Physics, Physical Faculty,
Moscow State, University, Vorobjevy Gory,
Moscow 119899 Russia

T. OHMACHI
Department of Built Environment
Interdisciplinary Graduate School of Scinece
and Eng. Tokyo Inst. of Techn. 4259 Nagatsuta-
cho, Midori-ku, Yokohama 226-8502, JAPAN

E. OKAL
Prof. Dr. , Department of Geological Sciences,
Northwestern University,
1847 Sheridan Road Evanston, IL 60208-2150

E. OTAY
Bosphorus University, Civil Engineering
Department, 80815, Bebek, Istanbul, Turkey

I. OZBAY
Ocean Engineering Research Center, Civil
Engineering Department, Middle East Technical
University, 06531 Ankara, Turkey

G. A. PAPADOPOULOS
Institute of Geodynamics, National Observatory
of Athens, P.O. Box 20048, Athens, Greece

M. PARLAKTUNA
Department of Petroleum and Natural Gas
Engineering, Middle East Technical University,
06531 Ankara-Turkey

E.N. PELINOVSKY
Institute of Applied Physics of the Russian
Academy of Sciences
46 Ulyanov Street, 603600 Nizhniy Novgorod,
Russia

T. K. PINEGINA
Institute of Geology and Geochemistry,
Petropavlovsky-Kamchatsky, Russia

G. PLAFKER
Plafker Geohazards Consultants, Menlo Park,
CA, USA

A.B. RABINOVICH
International Tsunami Research, Inc., Sidney,
BC, Canada

P.P. Shirshov Institute of Oceanology
Moscow, 117851 Russia

B. K. RANGUELOV
Geophysical Institute, BAS.
Acad. G.Boncev str. bl.3, Sofia 1113, Bulgaria.

I. RIABOV
Applied Mathematics Department,, Nizhny
Novgorod State Technical University, 24 Minin
Street, 603000, Nizhny Novgorod, Russia

E. SALUSTI
Dept. di Fisica, University di Roma, "La
Spienza" Italy

E. V. SASSOROVA
Shirshov Institute of Oceanology, Russian
Academy of Sciences, Nakhimovsky prospekt 36,
Moscow 117851 Russia

I. T. SELEZOV
Department of Wave Processes
Institute of Hydromechanics, Nat. Acad. Sci. 8/4
Sheliabov Str., Kiev 03680, Ukraine

N. SHUTO
Faculty of Policy Studies, Iwate Prefectural
University, Iwate 020-0193, Japan

E. N. SULEIMANI
Geophysical Institute
University of Alaska Fairbanks, 903 Koyukuk
Drive, P.O.BOX 757320,   Fairbanks,
AK 99775-7320 USA

C.E. SYNOLAKIS
University of Southern California, Civil
Engineering Department, KAP 213, Los Angeles,
CA 90089-2531, USA

F. ŞANAL
KISKA Komandit Co., Mithatpaşa Cad. Lale Apt.
13/1 Yenişehir Ankara, Turkey

T. TAKAHASHI
Department of Civil and Environmental
Engineering, Faculty of Engineering and
Resource Science Akita University, Japan

T. TALIPOVA
Institute of Applied Physics of the Russian
Academy of Sciences, 46 Ulyanov Street, 603600
Nizhniy Novgorod, Russia

D. TAPPIN
British Geological Survey, Kingsley Dunham
Centre Keyworth, Nottingham, NG 12 5GG, UK

A. TETEREV
Dr. Belarus State University Faculty of Applied
Mathematics & Computer Science Belarus State
University F.Skorina avenue,4, Minsk 200050
Belarus

R. E. THOMSON
Institute of Ocean Sciences, Sidney, BC, Canada

S. TINTI
Università di Bologna, Dipartimento di Fisica,
Settore di Geofisica,
Viale Carlo Berti Pichat,8, 40127, Bologna, Italy

V. V. TITOV
NOAA/PMEL/OERD, 7600 Sand Point Way NE
Seattle, WA 98115-6349 USA

M. I. TODOROVSKA
University of Southern California, Civil
Engineering Department, KAP 216A, Los
Angeles, CA 90089-2531, USA

M. D. TRIFUNAC
University of Southern California, Civil
Engineering Department, KAP 216A, Los
Angeles, CA 90089-2531, USA

M. TULIN
Director, Ocean Engineering Laboratory
Univ. of California at Santa Barbara
Santa Barbara, Ca. 93106

P. A. TYVAND
Department of Agricultural Engineering
Agricultural University of Norway, 1432 Aas,
Norway

C. VIDAL
Grupo de Ingeniería Oceanográfica y de Costas,
Universidad de Cantabria, E.T.S. Ingenieros de
Caminos, Spain

A. T. WILLIAMS
Applied Science Department, University of
Glamorgan, Pontypridd, Wales, UK.

A. C. YALÇINER
Ocean Engineering Research Center, Civil
Engineering Department, Middle East Technical
University, 06531 Ankara, Turkey

C. YALTIRAK
Istanbul Technical University, Mining Faculty,
Department of Geology, 80626, Ayazaga,
Istanbul, Turkey.

H. YEH
Dept. of Civil, Constr. & Env. Engng.
202 Apperson Hall, Oregon State University,
Corvallis, OR 97331-2302, USA

N. ZAHIBO
University of Antilles Guyane, Departemente de
Physique, Laboratoire de Physique,
Atmospherique et Tropicale, Point a Pitre,
Guadeloupe

Y. YÜKSEL
Yıldız Tech. Univ.ö Cıvıl Engıneering
Departmentö 80750 Yildiz/Istanbul Turkey

R. ZOBANAKIS
5, P. Nirvana Str., Neo Psychico, Athens 154 51
Greece

Part 1

General Aspects of Tsunami Researches

TSUNAMIS OF SEISMIC ORIGIN
-Science, Disasters and Mitigation-

N. SHUTO
Faculty of Policy Studies, Iwate Prefectural University
Iwate 020-0193, Japan

Abstract
Present knowledge of tsunamis are reviewed and discussed. The tsunami generation mechanism is not yet fully understood. Theories are examined in relation to observed tsunami phenomena. Conditions to be satisfied in numerical simulation are summarized. In addition to loss of human lives, several disasters are tabulated. Defense works including coastal structures, city planning and prevention systems are briefly introduced. Three subjects in urgent needs are the deep-sea measurement of tsunamis, transfer of tsunami knowledge and strengthening of coastal cities against tsunamis.

1. Introduction

Since 1970's, tsunami science and technology have been developing remarkably, assisted by the progress of seismology and computer science. In 1990's, more than ten disastrous tsunamis occurred in the Pacific. International cooperation is successfully established to survey tsunami heights and to understand the tsunamis.

In the present paper, the knowledge commonly used in these works is reviewed and the problems to be solved in the near future are discussed.

2. Science of Tsunamis

2.1 GENERATION OF TSUNAMIS

Since Thucydides, a Greek historian, recognized in 426 BC that a tsunami on Euboea Island was the result of an earthquake, many tsunamis were generated and recorded in the world. Causes are submarine earthquakes, landslides, and volcanic action.

The most of causes are submarine earthquakes. Not the ground shaking but the vertical sea-bottom deformation generates a tsunami. The greater an earthquake is, the larger the vertical displacement is and consequently the greater tsunami is generated. This rule is applicable to the ordinary tsunamigenic earthquake with a serious exception, tsunami-earthquake. Since the Nicaraguan tsunami in September 1992, more than ten tsunamis gave damages in the Pacific. Three of them were tsunami-earthquakes.

In the 1970's, the method to determine the tsunami initial profile generated by submarine earthquake was developed. In 1971, Mansinha and Smylie [1] proposed a method to estimate sea bottom displacement caused by an earthquake if fault parameters

A. C. Yalçıner, E. Pelinovsky, E. Okal, C. E. Synolakis (eds.),
Submarine Landslides and Tsunamis 1-8.
©2003 Kluwer Academic Publishers.

were given. Later [2] gave a more complete set of analytic formulas to compute the surface displacement due to a fault placed in an elastic homogeneous half-space. In 1974, Kanamori and Ciper [3] opened the way to calculate fault parameters from seismic records with the aid of the advancing high-speed computer, by using the 1960 Chilean earthquake as an example. Combined these two, the final displacement caused by an earthquake, i.e. the initial tsunami profile, can be estimated. This method has been most popularly used since then. Sometimes, however, the tsunami initial profile thus determined cannot satisfy the tsunami energy distribution measured along the shoreline. This difference may be due to heterogeneity of fault movement, existence of sub-faults, dynamic movement of fault, or so on.

There is only one example of the measured tsunami profile. Plafker [4] determined the vertical displacement caused by the 1964 Great Alaska earthquake, by displacement on islands and by comparison with a pre-earthquake topography. Along the direction of the short axis of the deformation area, the tsunami initial wave is about 450 km long with one trough and one crest. The wave height is 6 m. This long wave is considered to correlate with the major energy of the earthquake and is, therefore, estimated from the fault parameters. Near its crest, a sharp rise was found. This rise about 6 m high and 30 km wide at its base, although its contribution to tsunami height is apparently important, can not be estimated from seismic information. This rise may be caused by a sub-fault.

These short wave components are not important in case of a far-field tsunami. Entrapped and scattered by islands and seamounts on the route of propagation, their systematic wavy shape will be lost. Only long period components can arrive at distant shores, not disturbed so much by topography along the path.

In case of a near-field tsunami, short period components that are not estimated from seismic information are very important and are considered to be the major reason of the discrepancy between the measured tsunami heights on the shore and the computed. It is often noticed that the initial tsunami profile based upon fault parameters should be made nearly double in order to explain the measured tsunami traces.

The mechanism of tsunami earthquake is not yet clearly understood. Fukao [5] explained two tsunami earthquakes at the Kurile Trench by sub-faults in the thick sedimentary wedge at the leading edge of the continental lithosphere. Tanioka et al. [6] consider that tsunami earthquakes at the Japan Trench are the result of the horst and graben structures of sea bottom, which gives the scattered contact zones in the subducted sediments along the interplate boundary.

The generation mechanism by other causes is not yet clearly understood, although many researchers have been studying.

2.2 PROPAGATION AND RUN-UP

A major tsunami in the ocean is several tens kilometres long and several meters high. It is long and small compared to the ocean depth of several kilometres. The linear long- wave theory is successfully applicable if the travel distance is not long.

For a long travel, e.g. over the Pacific Ocean, other consideration is required.

Firstly, the equations should be described with the spherical co-ordinate system, because the earth is a sphere. Secondly, the Coriolis term should not be missed. Thirdly, the frequency dispersion term of the first order approximation should be included, if the Kajiura parameter relating to the dimension of the source, the travel distance and the water depth is smaller than 4 [7]. Under these conditions, the linear Boussinesq equation is often used.

Approaching the shore, the wave height increases and the water depth decreases. Then, non-linear characteristics of water motion require the shallow-water theory including the bottom friction term, in which the amplitude dispersion term becomes important. The boundary between the linear long-wave theory and the shallow-water theory is approximately the water depth of 200 m.

Approaching further the shore, a tsunami shows many faces in relation to water depth, bottom slope, wave height and wave length. If a tsunami is like a rapid tide without breaking front, the shallow-water theory is successfully applied. If a tsunami with breaking front comes nearly normally to the shore, the shallow-water theory is also applicable if the tsunami runup height is the major concern. If a tsunami with breaking front runs along the shore as an edge bore that follows sometimes the ordinary refraction law depending upon the water depth but sometimes propagates neglecting the law, there is no theory applicable at present. A hydraulic experiment revealed that a slight change in the side boundary condition resulted in a big change in waveform [8]. For a tsunami with an undular-bore front, higher order dispersion terms are required. The Boussinesq equation, Peregrine equation or Goto equation should be used in accordance with the strength of non-linearity. But accuracy and limit of these equations are not yet well determined.

Tsunamis finally run up and down the shore. Since equations in the Eulerian description cannot express the front condition, approximate moving boundary conditions are introduced. For ordinary topography on land where the slope is gentle, the shallow-water theory in which the vertical acceleration is neglected is applicable. In the rare cases of very steep slope, the vertical acceleration of water flow becomes non-negligible and other equations than the shallow-water equations are required.

Tsunamis run up on land and leave sediments as a proof of their existence. Many paleotsunamis were excavated. No work succeeded until now to explain the magnitude and movement of paleotsunamis from sediment samples.

2.3 DESIGN OF NUMERICAL SIMULATION

Difference equations are not the original differential equations but are the approximate equations of the latter. Numerical errors are inevitable. When difference equations are solved, several conditions should be satisfied to ensure the accuracy of the results and the stability of computation. In addition to the CFL condition that is required for the wave equation, several other conditions should be satisfied.

In case of the leapfrog scheme used in FDM computation, more than 20 spatial grids are required within one local wavelength. More than 50 spatial grids are necessary at the front of runup waves, if an approximate moving boundary condition is used [9]. Grid lengths should express well tsunami diffraction and refraction due to local topography. There is no criterion to design spatial length in relation to topography, although several attempts are made.

In case of other schemes, e.g. FEM, numerical inaccuracy due to truncation error should be examined well.

There are two crucial factors in the due evaluation of the computed results. The first one is the initial condition and the second the sea bottom topography. An initial profile constructed from seismic data always requires, particularly in case of a near-field tsunami, some adjustment as described in 2.1. The sea bottom topography that governs refraction then determines the path and convergence of tsunami, is not always precise data. The most popular sea bottom topography is the chart, the main purpose of which is to ensure the route of navigation. The area far from the route is sparsely measured and imprecisely determined by interpolation. This method often misses sea topography important in refraction.

TABLE 1. Kinds, types and causes of tsunami disaster

Human Lives
    Drowned. Buried by sands. Injured hit by debris etc. Disease caused by swallowing alien substances during drifting.
Houses
    Washed away. Destroyed. Flooded. Burnt.
Coastal Structures
    Toe erosion, displacement and overturning of sea walls, sea dikes, breakwaters and quay walls. Scattering and subsidence of concrete blocks.
Traffics

	Railway	Erosion of embankments.
		Displacement of rails and bridges. Rails buried by sands.
	Highway	Displacement and falling down of bridges.
		Overturning of bridge abutment by scouring.
		Erosion of embankment.
		Closure of traffic by debris on roads.
	Harbour	Change in water depth (erosion and deposition)
		Closure of port area due to transported debris and cars.
		Closure of port entrance by fishing gears washed-away.
		Collision of ships in harbour.

Lifelines

	Water supplies	Destruction of hydrants by collision of debris.
	Electricity	Overturning and washed-away of electric poles.
		Power plants flooded.
	Telephone	Damage to telephone lines and poles.
		Cut-off of underground telephone line at the junction to the aerial lines.
		Submergence of telephone receivers.

Fishery
    Damage to fishing boats.
    Destruction and loss of rafts, fishes and shells in aquaculture.
    Loss of fishing nets and other fishing gears.
Commerce
    Depreciation of goods by submergence.
Agriculture
    Physiological damage to crops due to submergence.
    Damage to farms buried by sands.
    Closure of irrigation channels filled by sands and debris.
Forest
    Physical damage (breaking and overturning of trees). Soil erosion.
    Physiological damage by seawater and sands.
Oil spill
    Environmental pollution.
    Spread of fires.
Fire (causes)
    Kitchen fire. Heating. Engine room of fishing boats. Submerged batteries of fishing boats.
    Collision to gasoline tanks. Electricity short circuit caused by seawater.

The validation of the computed results, i.e., tsunami height, is done by comparing them with the tide records and tsunami trace data. Some type of tide gauge has hydraulic filtering effect that reduces short-period components [10]. Tsunami trace data are often biased because many surveyors are attracted to measure only high values.

There is no attempt to validate the computed velocity, wave profile, wave forces and so on, except one [11].

3. Disasters

3.1 KINDS OF DISASTER

Table 1 summarizes disasters caused by tsunamis, collected from documents in the past. Detailed analysis is carried out in relation to tsunami intensity [12, 13].

Loss of human lives depends much upon the action of people on and near the shore when an earthquake occurs. Once caught by a water flow even as thin as 70cm, a person may be swallowed to drown, because the current velocity is quite strong.

Wooden houses are weak. On an average, a wooden house is completely destroyed if the tsunami height above ground exceeds 2 meters. Reinforced concrete buildings are strong. According to records, every reinforced-concrete buildings were not damaged except for windows and gates and were resistant enough to protect weak wooden houses behind them. Destructive force is not only the tsunami force but also impact of lumbers, fishing boats and houses transported by tsunamis. Impact force of one lumber is formulated by hydraulic experiments [14].

Damage to fishing boats moored and/or placed near the shoreline begins with the tsunami height of 2 m. If a ship or a boat meet a tsunami on the sea deeper than 200 meters, they are safe because of small tsunami height and gentle tsunami front slope. Many fishermen want to bring their boats to the deep sea when a tsunami warning is issued. This action may lead them to the very dangerous situation that boats get aground by tsunami ebb near shore and get turned by the next flood before they arrive at the safe place.

Violent currents induced by tsunamis are the cause of erosion and deposition. Not only local scouring at the toe of coastal structures but also topographical change of large scale often occurs. No one has ever succeeded to numerically simulate these topographical changes, partially due to the inaccuracy of computed currents and partially due to the lack of knowledge about the movement of sediment under tsunami effect.

3.2 TSUNAMI, OIL AND FIRE

Even in olden days, tsunami-related fires occurred. At night on January 27, 1700, a tsunami suddenly hit the village of Miyako, Sanriku Coast, Japan, with no precedent earthquake. A fire started from overturned houses and about 40 houses were burnt. The tsunami was generated at the Cascade subduction zone off the western coast of USA.

In the future, the most hazardous effect will be given to the coastal industrial areas by the combination of tsunami, oil and fire. If an earthquake or a tsunami damages oil tanks, and if the oil spread by the tsunami catches fire, or if the burning oil is transported by the tsunami, the result is devastating. There were five examples: three in Alaska, USA, one in California, USA, and one in Niigata, Japan. All of them occurred in 1964. Spread of oil can be numerically simulated, if equations for movement of oil are simultaneously solved with equations for tsunami [15]. An empirical formula is proposed to estimate the size of the burnt area in terms of the volume of stored oil [16].

4. Mitigation of Tsunami Hazards

4.1 COASTAL STRUCTURES

Tsunami countermeasures consist of three parts from hard wares to software: structures, city planning and systems.

Sea walls and coastal dikes are the typical defense structures. They are effective if the tsunami height is lower than their crown height, or a slightly higher (may be, 50 cm or 1 m). If the tsunami height is higher by more than 5 m, they do not work at all [17].

Tsunami breakwaters are constructed at the mouth of a bay where the water is deep, in order to limit the water discharge into the bay. Since they are expensive, only a few are constructed.

A tsunami gate at the river mouth stops the tsunami invasion into the river, in place of heightening long river embankments.

Coastal forests are also one of defense structures. A well-designed coastal forest of the Japanese red-pine trees works well to stop boats and other floated debris if the tsunami height above ground is smaller then 3 meters [12].

4.2 CITY PLANNING

The major items in the city planning are movement of residences to "tsunami-free" high ground and establishment of the tsunami-resistant building zone near the shoreline.

Since olden days, the movement of residence to high ground is one of the most effective methods in tsunami defense work. The highest tsunami runup measured and/or computed for the biggest tsunami in the past is usually used to determine the tsunami-flooded area. Recently in Japan, tsunamis that may be generated by the largest earthquake expected from seismo-tectonics are also taken into consideration. The tsunami-free high ground is outside of the area thus determined.

In many documents and after-tsunami survey reports, it is recorded that reinforced-concrete buildings are strong enough to resist tsunami force. Only one exception is a lighthouse that was hit by the 1946 Aleutian tsunami 20 m high above ground. If the tsunami height above ground is less than 5 m, all the buildings can withstand [12]. This fact leads to the idea of the tsunami-resistant building zone that cannot perfectly stop the tsunami water intrusion but is expected to stop more dangerous floated materials.

In the coastal industrial area, fishing harbor and leisure boats anchorage, storage tanks of inflammable materials should be carefully located and protected against tsunami effects.

4.3 TSUNAMI PREVENTION SYSTEMS

In addition to structures and city planning, software should be taken into consideration to complete the tsunami defense work. The tsunami prevention system consists of forecasting, warning, evacuation, public education, drills, inheritance of disaster culture, and the relief operation after disaster.

The last way to save human lives is an early evacuation based on the forecasting and warning. An earthquake that makes you unable to stand by yourself on a beach is a natural warning of a tsunami. Leave the beach as soon as possible and climb up to a ground higher than 20 m. This is the rule to save the lives from the danger of the near-field tsunami generated by tsunamigenic earthquake. Another rule is "a loud booming noise could mean a tsunami is coming." This noise may be generated by the breaking front of a tsunami higher than 2.5 meters [13].

Many countries have their own tsunami forecasting and warning systems. Most of them use the empirical relationships between earthquake magnitude and tsunami magnitude for tsunamigenic earthquakes. The forecasting system in French Polynesia uses the mantle magnitude and is effective for a tsunami earthquake, too.

In April 1999, the Japan Meteorological Agency renewed its forecasting system based upon numerical simulation for nearly 100,000 cases. This is also for tsunamigenic earthquakes. Records of broadband seismographs are used to make the correction for a tsunami earthquake.

The public education is the crucial key to save human lives. It is very rare for any person to experience plural large tsunamis in his life. If a person has a simple knowledge that a tsunami will come after an earthquake, and if he behaves wisely to climb up to high ground, he will be saved. This knowledge should be transferred to the future generation and to coastal residents in every tsunami-risk areas. The most difficult is to find an effective way to continue this knowledge for several tens or hundreds years.

5. Concluding Remarks

The most urgent subject in research as well as in practical application is the improvement of the method to determine tsunami initial profiles with seismic data alone. In order to solve this problem, we need observation networks in the ocean. Ocean-bottom seismographs near tsunami sources are indispensable to understand the details of fault movements. Deep-ocean tsunami gauges will catch tsunamis at or just after their birth. New technologies such as satellite photometry are desirable to obtain a plan of tsunami profile. With these data, a further development in tsunami research becomes possible.

The second is the continuation of tsunami knowledge to future generation as well as the transfer of it to those who live in tsunami-risk areas but have no clear tsunami history. Every means such as public education, TV media and others should be used. But it is quite difficult is to keep coastal residents' continuous attention. In the Sanriku region, the most tsunami-risky area in the world, special drills once a year have been carried out on the Memorial Day of the past tsunamis with participators constantly reducing in number. To find a way to break this situation is an urgent task.

Thirdly, keeping the fact in mind that disastrous tsunami occurs once per tens or hundreds years, the coastal residents should make their town resistant to tsunamis at every occasion when the town is changed and developed. The knowledge for this reinforcement should be given in terms of building codes and /or manuals.

References

1. Mansinha, L. and D.E. Smylie (1971) The displacement field of inclined faults, *Bulletin of the Seismological Society of America*, **61**, 1433-1440.
2. Okada, Y. (1985) Surface deformation due to shear and tensile faults in a half-space, *Bulletin of the Seismological Society of America*, **75**, 1135-1154.
3. Kanamori, H. and J.J. Ciper (1974) Focal process of the Great Chilean Earthquake, May 22, 1960, *Physics of Earth and Planetary Interiors*, **9**, 128-136.
4. Plafker, G. (1965) Tectonic deformation associated with the 1964 Alaska Earthquake, *Science*, **148**, 1675-1687.
5. Fukao, Y. (1979) Tsunami earthquakes and subduction processes near deep-sea trenches, *J. Geophysical Research*, **84**, 2303-2314

8

6. Tanioka, Y., L. Ruff and K. Satake (1997) What controls the lateral variation of large earthquake occurrence along the Japan Trench, *The Island Arc*, **6**, 261-266.
7. Kajiura, K. (1970) Tsunami source, energy and the directivity of wave radiation, *Bulletin of the Earthquake Research Institute,* University of Tokyo, **48**, 835-869.
8. Uda, T. et al. (1988) Two-dimensional deformation of nonlinear long waves on a beach, *Report No.2627,* Public Works Research Institute, 113p. (In Japanese).
9. Shuto, N. (1991) Numerical simulation of tsunamis-Its present and near future, in E. N. Bernard (ed.), Tsunami Hazard, Kluwer Academic Publishers, Dordrecht, pp.171-191.
10. Satake, K. et al. (1988) Tide gauge response to tsunamis: Measurements at 40 tide gauge stations in Japan, *J. Marine Res.*, **46**, 557-571.
11. Takahashi, T., F. Imamura and N. Shuto (1993) Numerical simulation of topography change due to tsunamis, *Proc. ITS'93*, 243-255.
12. Shuto, N. (1993) Tsunami intensity and disasters, *Advances in Natural and Technological Hazards Research*, 1, 197-216.
13. Shuto, N. (1997) A natural warning of tsunami arrival, *Advances in Natural and Technological Hazards Research*, **9**, 157-173.
14. Matsutomi, H. (1999) A practical formula for estimating impulsive force due to driftwood and variation features of the impulsive force, *J. Hydraulic, Coastal and Environmental Eng.*, Japan Society of Civil Engineering, No.600, II-44, 119-124 (in Japanese).
15. Goto, C. (1985) A simulation method of oil spread due to tsunamis, *Proc. Japan Society of Civil Engineers*, No.357, II-3, 217-223 (in Japanese).
16. Shuto, N. (1987): Spread of oil and fire due to tsunamis, *Proc. ITS'87*, 188-204.
17. Shuto, N. (1995): Tsunamis, disasters and defence works in case of the 1993 Hokkaido-Oki Earthquake Tsunami, *Advances in Natural and Technological Hazards Research*, **4**, 263-276.

NEEDS AND PERSPECTIVES OF TSUNAMI RESEARCH IN EUROPE

S.TINTI
Università di Bologna,
Dipartimento di Fisica, Settore di Geofisica,
Viale Carlo Berti Pichat, 8, 40127, Bologna, Italy

Abstract

Tsunamis are a serious threat for all coastal European countries for they attack countries in the north, such as Norway and Great Britain, as well as countries in the southern belt from Portugal to Greek and Turkey. In the last decade tsunami research in Europe gained strong impulse from mutual collaboration among various groups of several countries chiefly under the sponsorship of the European Union. This led to remarkable progress in physical understanding, in numerical modelling, and in defining and applying standards for data catalogues, and allowed also to better define the objectives still to be achieved and the research to undertake in the next future. This paper will mostly focus on the main issues that have to be faced by future research projects and that demand the establishing of stronger link between experts of physical and social sciences: local tsunamis induced by near-shore earthquakes, tsunamis generated by mass failures and by flank collapses of volcanoes, implementation of instrumental networks and early warning systems, tsunami disaster management, incorporation of the tsunami impact factor in policies of sustainable development of coastal communities.

1. The Past and the Future

Research developments on tsunamis in the last decade have enlarged the recognition that tsunamis can constitute a serious danger for European coasts (see [1] in this volume; see also [2] and [3]). The belt of the southern European countries from Portugal, to the west, to Greece and Turkey, to the east, is mostly affected by tsunamis generated by local earthquakes, and, though less frequently, by volcanic lateral collapses and eruptions. Northern countries may be hit by remotely generated tsunamis, or, as is the case of Norway, by tsunamis caused by local mass failures (ice blocks, rock-falls, landslides, etc). Historical reports date back more than 2000 years for Mediterranean countries and show that tsunamis affected disastrously the European coastal towns and economy several times. Therefore tsunamis, like all other natural events that can be the cause of large disasters, are and must be a topic of interest not only for scientists, but also for society, and studies on tsunamis in Europe are expected to contribute both 1) to the scientific goal of a better understanding of the physical processes, and, equally importantly, also 2) to the objective of defending the European coasts from tsunami attacks, by favouring a correct and wise policy of tsunami hazard evaluation and risk mitigation, which should be suitably integrated in the policy of protection of European society from natural disasters. Both

A. C. Yalçıner, E. Pelinovsky, E. Okal, C. E. Synolakis (eds.),
Submarine Landslides and Tsunamis 9-16.
©2003 Kluwer Academic Publishers.

goals, in order to be achieved, need co-operation on a national and on an international level and a multidisciplinary approach involving scientists of different branches of research, engineers, technicians, experts in civil protection business and in social impact of large events. These concepts have been already discussed and established in the past decade, during which relevant international projects on tsunamis involving several European countries were launched and financed by the European Union, such as GITEC [4], GITEC-TWO [5], INTAS [6], BIGSETS [7]. This financial policy has to continue in the next years, and to be further encouraged to launch joint projects involving collaboration with countries outside Europe (e.g. USA, Japan, northern African countries, and other countries that are members of ITSU) that will be highly beneficial. The topics that are of uttermost interest for Europe and that could form the basis for joint research and activities can be summarised as follows: local tsunamis induced by near-shore earthquakes, tsunamis generated by mass failures and by flank collapses of volcanoes, implementation of instrumental networks and early warning systems, tsunami disaster management and incorporation of the tsunami impact factor in policies of sustainable development of coastal communities.

2. Tsunamigenic Near-Shore Earthquakes

European and national tsunami catalogues ([1], [2], [3]) show that most of the European tsunamis are local tsunamis caused by earthquakes with fault surface located near-shore or in the coastal region. This means that the co-seismic deformation involves zones both on-land and underwater, and that the shore region may experience uplift or subsidence, or both, with a permanent change of the shoreline position and a permanent regression or ingression of the sea (see e.g. the impressive uplift of the Phalasarna harbour in Crete that exceeded 6.5m and was provoked by the tsunamigenic earthquake hitting Crete in 365AD: the uplift caused the disruption of the economic activity of the coastal town and the abandon of the harbour [8]). The changes in surface morphology associated with earthquakes large enough to generate sizeable tsunamis are remarkable and complex, and generally they cannot be entirely explained by simple double-couple point-like or rectangular-fault seismic source models. Non-uniformity of slip or more complicated fault geometries have to be invoked. Quite often a rather complicated picture has to be considered including multiple rupturing taking place in the source region, since beyond the main fault or faults also a number of ancillary faults are involved: these happen to be destabilised by the main series of shocks and, being usually shallow, may induce surface displacements that locally have magnitude as large as or even larger than the ones associated with the main rupture processes. An exemplary case is the complex deformation pattern of the recent tsunamigenic 17 August 1999 Izmit earthquake occurred along the Northern Anatolian Fault in Turkey. The total surface rupture was found to extend about 145-km and to consist of five segments separated by releasing step-overs and was due to a strike slip mechanism with dextral offset up to 5.2 m measured along the Sapanca-Akyazı segment [9], [10]. But it was also accompanied by very shallow normal faulting, with about 2m-offest, observed in the south-eastern Izmit bay region, that could have had effects in producing the tsunami [11], [12]. Another relevant aspect is that near-shore and coastal zones are characterised by large topographic and bathymetric slopes that are prone to gravitational instability. Therefore earthquakes here may trigger submarine or coastal mass failures that can act as additional sources of tsunami. It is worth recalling that the Italian tsunami that was responsible for the largest number of ascertained victims was very likely due to the huge rock-fall that involved the seaward flank of Mount Pacì, at Scilla on the Tyrrhenian side of Calabria. The collapse occurred during a period of intense seismic activity and took place in the night of 6 February 1783 following an aftershock of the catastrophic I=XI (MCS scale) 5 February 1783 earthquake. The tsunami was local, but very lethal, since took more than 1500 human lives according to the abundant coeval documents [2], [13]. The above considerations prove that seismic sources in the coastal areas are expected to be complex sources of tsunamis, and that usually approximating them by simplistic sources would be unsatisfactory, since it would miss important features and leave relevant observations unexplained.

3. Tsunamis Induced by Mass Failures

Tsunamis can be produced by masses entering the sea from altitudes above sea level or by submarine landslides. Recently it has been recognised that this generation mechanism was overlooked by the tsunami research in the past that in countries watered by the Pacific Ocean was probably biased toward tsunamis induced by large earthquakes in the offshore subduction zones and toward the effects of tele-tsunamis. But today, especially after the example of the catastrophic Papua New Guinea 1998 tsunami induced by a huge submarine slump triggered by a coastal earthquake [14], general consensus has been reached on the need to intensify the theoretical developments and to enlarge the basis of observational data for tsunamis induced by masses moving underwater. Mass failures can be caused by seismic load, or more simply by gravity load, but can also be associated with volcanic eruptions. In Europe such events are rare, but they are known to have affected and to have potential to affect most coastal countries (see [1], [4], [5]), and there are even regions, such as Norway, where this mechanism is largely predominant and occasionally disastrous (see the case of the 1934 slide tsunami in the Tafjord fjord [15]), attracting therefore most of the attention of the people involved in tsunami research. Theoretical investigations on waves generated by moving underwater bodies ([16], [17], [18], [19], [20]) help understand the basic process of energy transfer from the body to the wave, but the interaction is certainly very complex and demands for more sophisticated models to simulate the slide, slump or rockfall dynamics (see the Lagrangian approach used by Tinti et al. to model the tsunamigenic volcanic slide of 1988 in the Vulcano island, south Tyrrhenian, Italy [21], and to model the catastrophic 1963 slide displacing most of the water of the Alpine reservoir of the Vajont valley [22], [23]) and to simulate the 2D and/or 3D water wave excitation and propagation ([24] and [25]). This is a field where important progresses are needed, and hopefully expected in the near future.

4. Tsunamigenic Collapses of Volcanic Cones

Tsunamis associated with debris avalanches and with collapses or eruptions of marine or submarine volcanoes are quite rare, but may be catastrophically destructive, since the volumes involved are very huge (up to hundreds of cubic kilometers) and the excited waves can be giant and transoceanic. Several such volcanic islands or underwater cones are of interest for European countries. The most impressive mass failures are expected to take place in the Canary islands (Spain) in the Atlantic Ocean, where at least nine giant landslides have been identified by geologists, that are similar to those observed in the

Hawaiian archipelago ([26] and [27]), and where signs of instability have been recently observed in the volcano Cumbre Vieja in the highly populated island of La Palma [28]: mass collapse could concern the west flank of the volcano with a volume in the order of 500 km^3. Evidence of repeated collapses has been collected also for Stromboli volcano in the Aeolian Islands in the Tyrrhenian Sea, Italy. The last one produced the Sciara del Fuoco scar in north-western flank of the volcano with 1-1.8 km^3 volume involved [29]. Interestingly, a very strong structural similarity can be noticed between Stromboli and the Nishi-yama volcano in the Oshima-Ohshima Island (Japan Sea), as far as regards the scar morphology, the volcanic cone shape, and the underwater topography. The last lateral collapse of this is volcano occurred in historical times during a series of eruptions in 1741 and produced a rather well documented tsunami that attacked the west Japan and was observed in the entire Japan Sea ([30], [31], [32]). The volcanic island of Santorini in the south Aegean sea is known to have been the source of tsunamigenic explosions: the last one was the submarine explosion of the Mount Columbo in 1650 AD producing a tsunami observed in the Aegean sea and in northern Crete ([1], [33], [34]), while the largest one is the eruption culminating in the collapse of the Thera volcano in the Minoan times: this tsunami was speculated to have destroyed all coastal towns of Crete and of the Aegean islands and, consequently, to have caused the irreversible decline of Minoan civilisation. Deposits of paleotsunamis of volcanic origin have been recently identified in various places of the Aegean Sea coasts and attributed to the activity of Thera [35].

5. Instrumental Networks and Early Warning Systems

The most relevant obstacle to the development of knowledge on tsunamis is the endemic lack of instrumental data. This is true world-wide, but it is especially true for Europe where even the traditional gauges to record tides are too few and mostly not organised in centralised data networks. What is needed is investment of resources to place proper instrument on the coasts involved by tsunamis sources as well as at sea. It has to be stressed that not only gauges to record tsunamis are needed, but also instruments to monitor the tsunami sources (active coastal and submarine faults, zones with unstable coastal segments or unstable underwater sediment masses, active volcanoes prone to be tsunamigenic, etc).

The basic elements of a system of warning for tsunamis are the detection of the source and the verification of the tsunami occurrence, with both phases that have to be accomplished soon enough to leave time to launch a useful alert and to take proper actions before the tsunami attacks the target to protect. In case of tsunamis produced by earthquakes an efficient seismic network based on broad-band instruments and a system of coastal and/or offshore buy-based gauges to measure sea level can provide the data needed to recognise that a tsunami has been generated and to assess the tsunami size and propagation fronts through suitable processing (see e.g. the DART system devised and implemented by PMEL for northern Pacific [36], and the TREMORS system conceived by LDG-CEA (Paris) originally for Tahiti and installed in several countries, [37] and [5]). Teletsunamis occurring very far from the target coasts can be identified expectedly with great efficiency by such systems, since travel times from the source to the target exceed several tens of minutes, and there is time to collect, transmit and process data with reliable algorithms. Most of the damaging tsunamis of seismic origin that have occurred in Europe are local tsunamis. The only examples of transoceanic tsunamis are the case of the 1755 Lisbon earthquake tsunami that was observed even in Norway and in the Caribbean Sea

[1], and very likely the 365 AD Crete tsunami that was reported to be seen in the entire eastern Mediterranean as well as in southern Adriatic and the Ionian Sea. Tsunamis generated by coastal near-shore sources attack very soon. This means that we must accept the challenge of devising special tsunami early-warning systems with reaction time as low as a few minutes (1-3 min) to alert and protect the target that in this case is the local communities. The idea is that any coastal community that is located close to a seismic source with tsunamigenic potential should be provided with a local early warning system, capable to detect the event almost at the inception time. Processing time must be minimised by collecting only essential data (such as the signals from strong motion accelerometers) and by adopting very simple detection algorithms based on expert systems, which are able to identify the onset of an anomaly [38] (in this optics the tsunami is viewed as an anomalous departure from the background noise). The local systems are not an alternative to the traditional systems, but rather they complement each other on a regional level: the former ones are only yes- or no-systems (since their goal is to detect whether or not a tsunami has been generated and is on the point to attack the local community), while the latter has the wider scope to assess the size and the propagation path of the tsunami and to alert the communities that are more remote from the source. It should always be borne in mind however that appropriate educational policies of the coastal population are an indispensable prerequisite to reduce the impact of the waves.

6. Scenarios of Disasters

A very useful tool often adopted to plan prevention actions and to reduce natural disaster impact is that of considering scenarios of future large events: for example, a large earthquake triggering slumps and tsunami, a large volcanic eruption producing killing pyroclastic flows and tsunami, etc. These events are usually quite complex, and need very sophisticated modeling and data bases to simulate the physical evolution of the processes and their impact on land and on social communities. Since the results concerning a given scenario are to be used by a very large community of people, the maximum effort is needed to provide the users with tools that are simple and practical, but most importantly, to clarify the assumptions that are at the basis of the elaboration of the scenario, and the reliability of the results. This latter issue is crucial, especially for scenarios concerning infrequent events, such as catastrophic tsunamigenic processes, since it is always difficult and often even impossible to elaborate statistical approaches and to use concepts such as probability of occurrences or return times. An example could help understand this point. If we consider the scenario of a large explosion of the volcano Cumbre Vieja in the Canary islands, entailing a giant collapse of the island of La Palma and if we model the giant tsunami attacking locally the archipelago and remotely the coasts of the Atlantic ocean, including the Caribbean sea countries and the eastern coast of the United States, what is relevant is to provide scientific and technical details on the simulation, and moreover to provide some hints concerning how often such an event is expected to occur in a given future period of time. If this information is missing, the scenario may be a very important scientific contribution, but looses much of its practical value, since it gives the society no elements on the basis of which to take decisions and to formulate a correct policy of future developments. How to reduce or to eliminate this drawback is not an easy task and is certainly one of our major commitments for the future.

7. Conclusions

The topics that have been briefly addressed in the previous sections constitute a serious commitment for researchers involved in the tsunami field and, more generally, in the natural hazards sciences. They are a problem not only for Europe, but for all other countries exposed to tsunami attacks. The solution can be approached successfully only if a network, formal and informal, of collaboration will be established among various research groups world-wide, and if experience and data can be freely exchanged. This is an essential point that is especially relevant for rare events: the whole earth should be seen as a natural laboratory and the occurrence of a tsunami should be seen as a source of experimental data accessible to all researchers. Beyond raw data, collaboration should involve methods, models and elaboration of new concepts. What is even more important, scientists should try to close the gap between earth and physical science and technique on one side, and social sciences on the other side. Only if people pertaining to these worlds will make any efforts to speak a common language with common words and common meanings, and to have common targets, progress on protection of coastal communities from disastrous events will be significant and development plans will be compatible with natural processes and resources.

Acknowledgements

This contribution was carried out on funds from the Gruppo Nazionale di Difesa dai Terremoti (GNDT) of the Istituto Nazionale di Geofisica e Vulcanologia (INGV) and from the Ministero dell'Istruzione, dell'Università e della Ricerca (MIUR), formerly named Ministero dell'Università e della Ricerca Scientifica e Tecnologica (MURST).

References

1. Maramai, A., Graziani, L. and Tinti, S. (2002) Updating and revision of the European tsunami catalogue, this volume.
2. Tinti, S. and Maramai, A. (1996) Catalogue of tsunamis generated in Italy and in Côte d'Azur, France: a step towards a unified catalogue of tsunamis in Europe, *Annali di Geofisica,* 39, 1253-1299 (Errata Corrige, *Annali di Geofisica,* 40, 781).
3. Tinti, S., Baptista, M.A., Harbitz, C.B. and Maramai, A. (1999) The unified European catalogue of tsunamis: a GITEC experience. *Proc.International Conference on Tsunamis, Paris, 26-28 May 1998,* 84-99.
4. GITEC - Genesis and Impact of Tsunamis on the European Coasts (1995) *Final Scientific Report,* European Union Project EV5V-CT92-0175, University of Bologna.
5. GITEC-TWO - Genesis and Impact of Tsunamis on the European Coasts: Tsunami Warning and Observations (1999) *Final Scientific Report,* European Union Project ENV4-CT96-0297, University of Bologna.
6. INTAS-RFBR-95-1000: Tsunami Hazard for the Mediterranean region (1999) *Final Scientific Report,* INTAS Project, University of Bologna.
7. BIGSETS – Big Sources of Earthquakes and Tsunamis in Sw Iberia (1998-1999) European Union Project ENV4-CT97-0547.
8. Pirazzoli, P.A., Ausseil-Badie, J., Giresse, P., Hadjidaki, E. and Arnold, M. (1992) Historical environmental changes at Phalasarna harbour, west Crete, *Geoarchaeology: an International Journal,* 7, 371-392.
9. Barka, A., Akyuz, H.S., Altunel, E., Sunal, G., Cakir, Z., Dikbas, A., Yerli, B., Armijo, R., Meyer, B., de Chabalier, J.B., Rockwell, T., Dolan, J.R., Hartleb, R., Dawson, T., Christofferson, S., Tucker, A., Fumal, T., Langridge, R., Stenner, H., Lettis, W., Bachhuber, J. and Page, W. (2002) The surface rupture and slip distribution of the 17 August 1999 Izmit earthquake (M 7.4), North Anatolian fault, *Bull.Seism.Soc.Am.,* 92, 43-60.

10. Lettis, W., Bachhuber, J., Witter, R., Brankman, C., Randolph, C.E., Barka, A., Page, W.D. and Kaya, A. (2002) Influence of releasing step-overs on surface fault rupture and fault segmentation: Examples from the 17 August 1999 Izmit earthquake on the North Anatolian fault, Turkey, *Bull.Seism.Soc.Am.*, 92, 19-42.

11. Altinok, Y., Tinti, S., Alpar, B., Ersoy, S., Yalçiner, A.C., Bortolucci, E. and Armigliato, A. (2001) The tsunami of August 17, 1999 in the Izmit Bay, Turkey, *Natural Hazards*, 24, 133-146.

12. Piatanesi, A., Tinti, S., Armigliato, A., Bortolucci, E., Altinok, Y. and Yalçiner A.C. (2000) Finite-element numerical modeling of the tsunami induced by the August 17, 1999 Izmit (Turkey) earthquake, *Abstracts of the XXVII General Assembly of ESC, Lisbon 10-15 September 2000*, p.26.

13. Sarconi, M. (1784) *Istoria de' fenomeni del tremuoto avvenuto nelle Calabrie e nel Valdemone nell'anno 1783*, Napoli (in Italian).

14. Synolakis, C.E., Bardet, J.P., Borrero, J.C., Davies, H.L., Okal, E.A., Silver, E.A., Sweet, S. and Tappin, D.R. (2002) The slump origin of the 1998 Papua New Guinea Tsunami, *Proc. Royal Society of London, Series A – Math.Phys.Engng Sciences*, 458, 763-789.

15. Harbitz, C.B., Pedersen, G. and Gjevik, B. (1993), Numerical simulations of large water waves due to landslides, *J.Hydraulic Engng, ASCE*, 119, 1325-1342.

16. Tinti, S. and Bortolucci, E. (2000) Analytical investigation on tsunamis generated by submarine slides, *Annali di Geofisica*, 43, 519-536.

17. Tinti, S., Bortolucci, E. and Chiavettieri, C. (2001) Tsunami excitation by submarine slides in shallow-water approximation, *Pure and Applied Geophysics*, 158, 759-797.

18. Todorovska, M.I., Hayir, A. and Trifunac, M.D. (2002) A note on tsunami amplitudes above submarine slides and slumps, *Soil Dynamics and Earthquake Engineering*, 22, 129-141.

19. Trifunac, M.D. and Todorovska, M.I. (2002) A note on differences in tsunami source parameters for submarine slides and earthquakes, *Soil Dynamics and Earthquake Engineering*, 22, 143-155.

20. Ward, S.N. (2001) Landslide tsunami, *J.Geophys.Res.*, 106, 11,201-11,215.

21. Tinti, S., Bortolucci, E. and Armigliato, A. (1999) Numerical simulation of the landslide-induced tsunami of 1988 on Vulcano Island, Italy, *Bull.Vulcanol.*, 61, 121-137.

22. Bortolucci, E., Tinti, S. and Zaniboni, F. (2001) Lagrangian modelling of the 1963 Vajont catastrophic landslide, *Geophysical Research Abstracts*, 3, Abstracts of the 26th General Assembly of the European Geophysical Society, Nice, France, 25-30 March 2001 (CDROM).

23. Tinti, S., Zaniboni, F., Manucci, A. and Bortolucci, E. (2002) A 2D block model for landslide simulation: an application to the 1963 Vajont case, *Geophysical Research Abstracts*, 4, Abstracts of the 27th General Assembly of the European Geophysical Society, Nice, France, 21-26 April 2002 (CDROM).

24. Grilli, S.T. and Watts, P. (1999) Modeling of waves generated by a moving submerged body. Applications to underwater landslides, *Engnrg. Analysis with Boundary Elements*, 23, 645-656.

25. Heinrich, P., Piatanesi, A. and Hebert, H. (2001) Numerical modelling of tsunami generation and propagation from submarine slumps: the 1998 Papua New Guinea event, *Geophys.J.Int.*, 145, 97-111.

26. Moore, J.G., Normark, W.R. and Holcomb, R.T. (1994) Giant Hawaiian landslides, *Ann. Rev. Earth Planet. Sci.*, 22, 119-144.

27. Keating, B.H. and McGuire, W.J. (2000) Island edifice failures and associated tsunami hazards, *Pure and Applied Geophysics*, 157, 899-955.

28. Day, S.J., Carracedo, J.C., Guillon, H. and Gravestock, P. (1999) Recent structural evolution of the Cumbre Vieja volcano, La Palma, Canary Islands, *J.Volcan. Geotherm. Res.*, 94, 135-167.

29. Kokelaar, P. and Romagnoli, C. (1995) Sector collapse, sedimentation and clast-population evolution at an active island-arc volcano: Stromboli, Italy, *Bull. Volcanol*, 57, 240-262.

30. Satake, K. and Kato Y. (2001) Oshima-Oshima eruption: extent and volume of submarine debris avalanche, *Geophys.Res.Lett.*, 28, 427-430.

31. Tinti, S., Bortolucci, E. and Satake, K., (2000) The 1741 Oshima-Ohshima tsunami, *Abstracts of the XXVII General Assembly of ESC, Lisbon 10-15 September 2000*.

32. Satake, K., Tinti, S. and Bortolucci, E. (2002) Modeling landslide tsunami associated with the 1741 eruption of Oshima-Ohshima volcano in Japan Sea, *Geophysical Research Abstracts*, 4, Abstracts of the 27th General Assembly of the European Geophysical Society, Nice, France, 21-26 April 2002 (CDROM).

33. Galanopoulos, G.A. (1960) Tsunamis observed on the coasts of Greece from antiquity to present time, *Annali di Geofisica*, 13, 369- 386.

34. Dominey-Howes, D.T.M., Papadopoulos, G.A. and Dawson, A.G. (2000) The AD 1650 Mt. Columbo (Thera Island) eruption and tsunami, Aegean Sea, Greece, *Natural Hazards*, 21, 83-96.

35. Minoura, K., Imamura, F., Kuran, U., Nakamura, T., Papadopoulos, G.A., Takahasi, T. and Yalçiner, A.C. (2000) Discovery of Minoan tsunami deposits, *Geology*, 28, 59-62.

36. Bernard, E.N. (2001) Recent developments in tsunami hazard mitigation, in *Tsunami Research at the End of a Critical Decade*, (Ed. Hebenstreit, G.T.), Kluwer Academic Publishers, 7-15.

37. Schindelé, F. (1998) TREMORS : A modern real-time system for tsunami warning, *Proceedings of International Workshop on Tsunami Disaster Mitigation*, Tokyo 19-22 January 1998, 136-139.
38. Piscini, A., Maramai, A. and Tinti, S. (1999) Pilot local Tsunami Warning System in Augusta, eastern Sicily, Italy, *Proc. International Conference on Tsunamis*, Paris, 26-28 May 1998, 137-148.

Part 2

Historical Tsunamis

IDENTIFICATION OF SLIDE-GENERATED TSUNAMIS IN THE HISTORICAL CATALOGUES

V.K. GUSIAKOV

Department of Geophysics, Institute of Computational Mathematics and Mathematical Geophysics, Siberian Division, Russian Academy of Sciences, Pr.Lavrentieva, 6, Novosibirsk 630090, Russia

Abstract

Despite the fact that the parametric tsunami catalogs contain very limited information on a particular event, the preliminary identification of landslide-generated events in the catalogs is possible on the basis of several criteria such as width of an area with the maximum run-up values, a large difference between the tsunami magnitude (on the Iida scale) and the tsunami intensity (on the Soloviev-Imamura scale) and a difference between the observed and the expected tsunami intensity. The latter criterion, introduced in the present study, allows us to divide the Pacific tsunamigenic events into three groups ("red", "green" and "blue"). The geographical distribution of events from the "red" group shows its clear correlation with areas of a high sedimentation rate in the Pacific, thus making possible its interpretation as events where involvement of the slide mechanism into the tsunami generation is essential.

1. Introduction

The world-wide catalog of tsunamis and tsunami-like events covers the period from 1628 B.C. till 2000 A.D. and contains nearly 2250 historical events that occurred in almost all parts of the world ocean, in many marginal seas as well as in lakes and in-land water reservoirs. . The catalog gives many examples of historical events where involvement of subaerial and submarine landslides in the tsunami generation was clearly observed and well documented. Some of these waves were destructive and resulted in a considerable economic damage and numerous losses of lives. One of the largest pre-historic submarine landslides with the estimated volume of mass flow of 1,700 km^3 occurred ca. 7000BC in the Northern Sea at the edge of the continental shelf of Norway (the Storegga slide). The resulted tsunami hit a large part of the Scottish coast with heights up to 6-8 meters. Among the best-known examples of the extreme water splash in the recent history is a well-documented 525-meter run-up in the Lituya Bay (Alaska) caused by a massive landslide occurred after the magnitude 7.8 earthquake of July 10, 1958 in the south-eastern Alaska. Less known cases of the extreme run-up heights in the same bay are the 1936 and 1853 events with the maximum run-up heights of 150 and 120 meters, respectively. One of the most deathful cases occurred on October 9, 1963 in Italy, when a massive rock slide fell into a water reservoir in the Vaiont Valley, destroyed a town and killed 3,000 people. The latest reported event of a similar kind, which resulted in 5 deaths, is the 2-meter waves generated in Lake Coatepeque in El Salvador soon after the magnitude 7.6 earthquake occurred on January 13, 2001 in the Pacific at a distance of 50 km off the coast. It is interesting to note that this earthquake did not generate any observable tsunami in the Pacific basin.

A. C. Yalçıner, E. Pelinovsky, E. Okal, C. E. Synolakis (eds.),
Submarine Landslides and Tsunamis 17-24.
©2003 Kluwer Academic Publishers.

2. Identification criteria

The first indication to the involvement of the slide mechanism into the tsunami generation comes from the absence of any associated seismic activity mentioned in the historical description. The tsunami catalogs contain on average 10-15% such events. The percentage strongly varies with an area, exceeding 50% cases for the waters around England, the North Sea, the Norwegian Sea and the Baltic Sea. Obviously, a considerable part of these events could result from the meteorological and oceanological reasons (storm surges, rogue or freak waves, etc.), however, there are many cases where the catalog compilers specially emphasized that an event occurred at the "clear sky" and in the "calm sea".

The second important feature is the value of the maximum run-up height for the particular event. The results of numerical modeling show that for a typical tsunamigenic earthquake in the magnitude range from 7.0 to 7.5, the coseismic bottom displacement alone can hardly be responsible for the coastal run-ups exceeding 2-3 meters. In fact, the historical data for the last decade give several examples of the shallow-depth earthquakes with magnitudes even above 7.5 when tsunami heights did not exceed several tens of centimeters. Therefore, each case of a seismically-induced tsunami with run-up heights exceeding 4-5 meters can be considered as "suspicious" in terms of the involvement of the slumping mechanism.

Table 1 lists the top ten of Pacific tsunamis with highest run-up values. Five of these ten events have been generated primarily by landslides, in three cases the involvement of slide mechanism was discovered by later studies. Only for two cases (1674 Indonesia and 1737 Kamchatka) the involvement of the slide mechanism has not been proved so far. Both of them are the old historical events with very limited data available.

TABLE 1. List of the Pacific tsunamis with the highest run-up values. M_S – surface wave magnitude, I – tsunami intensity on the Soloviev-Imamura scale, m – tsunami magnitude on the Iida scale, cause of tsunami: T- tectonic, L – landslide, V- volcanic.

Date	M_S	I	m	Hmax, m	Cause	Source Area
10.07.1958		3.0	9.0	525.0	TL	Lituya Bay
27.10.1936		3.0	7.0	150.0	L	Lituya Bay
1853		3.0	7.0	120.0	L	Lituya Bay
06.08.1788	8.0	4.0	6.5	88.0	T	Sanak Is.
24.04.1771	7.4	4.0	6.5	85.4	TL	Ishigaki Is.
17.02.1674	8.0	4.0	6.0	80.0	T	Indonesia
28.03.1964	8.4	5.0	6.0	67.1	TL	Alaska
17.10.1737	8.3	4.0	6.0	63.0	T	Kamchatka
10.09.1899	8.6	3.5	6.0	60.0	TL	Yakutat Bay
01.05.1792		2.5	5.5	55.0	VL	Japan

Another feature that can help to distinguish the tectonically-induced tsunamis from the slump-generated tsunamis is the width of a zone with maximum run-up value. For a typical tectonic tsunamis with the magnitude range 7.0 – 7.5 this area is about the size of the earthquake source, i.e. 100 -150 km. At the same time we know that for typical landslide tsunamis (1998 Papua New Guinea, 1992 Indonesia, 1993 Okushiri) the width of an area with the maximum run-up values is much narrower, quite often not exceeding several tens of kilometers.

The formal criteria for distinguishing the landslide events in the historical catalogs can be a large difference between the tsunami intensity I (on the Soloviev-Imamura scale) and the tsunami magnitude m (on the Iida scale). It should be recalled that the tsunami intensity I is calculated on the basis of the average run-up heights at the nearest coast, while the tsunami magnitude m is calculated as logarithm (by basis 2) from the maximum run-up height. For a typical tectonic tsunami this difference is within 1.0–2.0, while for the slide-generated event it may exceed 3.0 – 4.0 (see the data in Table 1).

A similar feature of involving the slumping mechanism at the tsunami generation stage is the difference of the measured tsunami intensity from the expected intensity calculated on the basis of the moment-magnitude of an event. In the paper [1], the theoretical dependence of the tsunami intensity I on the moment magnitude of an earthquake Mw has been obtained ($I = 3.55Mw - 27.1$). This formula was obtained on the basis of the coseismic bottom displacement model of the tsunami generation. If we know the actually observed intensity I_{obs}, we can estimate the difference ΔI between the observed and the expected tsunami intensity ($\Delta I = I_{obs} - I_{exp,}$).

In the recent study [2], we have introduced a formal classification of the Pacific tsunamigenic earthquakes on the basis of their ΔI parameter. Based on the ΔI value, we divided all tsunamigenic earthquakes with known I and M_w into the three groups: "red" ($\Delta I > 1$), "green" ($-1 < \Delta I < 1$), and "blue" ($\Delta I < -1$). From 293 tsunamigenic events that occurred in the Pacific from 1900 to 1998 and which have both I and M_w values, 90 events fall within the "red" group, 153 are within the "green" group and 50 events are within the "blue" group (Fig.1).

Typical examples of events from the "red" group are: the Ozernoy tsunami of November 22, 1969 in Kamchatka ($M_S=7.7$, $I=3.0$); the Okushiri tsunami of July 12, 1993 in the Japan Sea ($M_S=7.7$, $I=3.0$); the Flores tsunami of December 12, 1992 in Indonesia ($M_S=7.5$, $I=2.7$); the Papua New Guinea tsunami of July 17, 1998 ($M_S=7.1$, $I=3.2$); Typical events from the "green" group are: the Urup tsunami of October 13, 1963 ($M_S=8.1$, $I=2.5$); the Shikotan tsunamis of August 11, 1969 ($M_S=7.8$, $I=2.0$); the Japan tsunami of October 4, 1994 ($M_S=8.1$, $I=2.6$); and the Irian Java tsunami of February 17, 1996 ($M_S=8.1$, $I=1.8$). The "blue" group includes tsunamis from several strongest submarine events of the last decade, such as: the Guam earthquake of August 8, 1993 ($M_S=8.2$, $I=-1.0$); the Tonga earthquake of April 7, 1995 ($M_S=8.1$, $I=-1.5$); and the Balleny Islands earthquake of March 25, 1998 ($M_w=7.9$, $I=-3.0$). Despite the large magnitude value (greater than 7.8) all of these earthquake events generated very unsubstantial tsunamis.

The analysis of average magnitudes, source mechanisms, source and water depths, does not indicate to any significant differences between the three selected groups of the Pacific tsunamigenic earthquakes. With a few exceptions, all of these earthquakes are shallow events with typical subduction zone mechanisms; reverse dip-slip or low-angle thrust. However, analysis of their geographical distribution immediately indicates a clear correlation with the climatic and circum-continental zonation in the oceanic sedimentation as described in Lisitsyn ([3], [4]). Fig.2 shows the geographical distribution of the Pacific tsunamigenic earthquakes categorized into the above three groups overlapped on the map of basic zones of oceanic lithogenesis as taken from Lisitsyn's original monograph "Sedimentation in oceans" [4].

3. Tsunami generation and sedimentation zones in the ocean

The rate of sedimentation and thickness of accumulated bottom sediments in the world ocean are controlled by many factors, the primary being the aggregate quantity of sediments delivered by rivers, patterns of global oceanic circulation and surface wave motions. The resulting sedimentation, extremely variable in general, follows, however, several firmly established law-governed features with two basic of them known as geographical and circum-continental zonation. According to Lisitsyn [4] subsequent treatise, the bulk of oceanic sediments, about 76% of the total, originate in the equatorial humide zone. The next most important sources of ocean sediments are northern and southern humide zones which supply approximately 12%. Located between them, northern and southern aride zones are estimated to contribute only 6% of the ocean sediments. Moreover, within each separate sedimentation zone, the distribution of sediments can vary to the great extent. This effect is known as the circum-continental zonation - increasing sedimentation rate near coastal areas with maximum values near the mouths of main rivers, in the shallow-water marginal seas and near the basement of the continental slope where, ultimately, the main bulk of the terrestrial sediments accumulates.

The map in Fig.2 shows that all the tsunamigenic earthquakes occurring in the marginal seas, as the Yellow, the Japan, or the Bering seas, belong to the "red" group. Also, a similar "red" group of tsunamigenic earthquakes is predominant in the western part of the equatorial humide zone (Indonesia, Philippines, New Guinea). For example, of seven tsunamigenic earthquakes occurred in this century within the Java trench, six belong to the "red" group. At the same time, almost all the tsunamigenic earthquakes occurred within the remote, from continents, subduction zones such as near Guam, the Tonga-Kermadec region or New Zealand belong to the "blue" group. Considerably more tsunamigenic earthquakes from the typical subduction zones, like Kurile-Kamchatka and the Aleutian islands belong to the "green" group. The tsunamigenic potential of these earthquake events corresponds very well to the tectonic displacement model of tsunami generation. The distribution of events from the "red" group along the coast of the Central and South America is also interesting. With one exception, all of the earthquakes are concentrated within a narrow band between 12°N and 10°S, which falls within the equatorial humide zone. The only exception is the November 21, 1927 event (M_S=7.1, I=2.0) which occurred near 45°S and 73°W, within the southern humide zone. For the whole Central and South America coast, there is not a single "red" tsunamigenic event that occurred within northern and southern aride zones.

Another typical example of an event from the "red" group is the recent tsunami that occurred on July 17, 1998 near the north coast of Papua New Guinea. This destructive tsunami was generated by a moderate submarine earthquake with 7.1 magnitude which is below the threshold for issuing a regional tsunami warning. Yet, the earthquake generated a very destructive tsunami with maximum waves of almost 15 meters which killed 2000 people. In Lisitsyn ([3], p.150) there is the following remarkable paragraph related to this area: "On the background of relatively low concentrations of suspended material in the south-western part of the Pacific, we can observe the clear tongue of high concentrations that are traced to the north of the New Guinea Island. Its existence is connected with the presence in this area of strong underwater currents that are moving in the western and north-western directions from the Coral Sea. In addition to a high level of biological productivity of this area, there is a strong flux of suspended sediments that are observed in this area far away from the New Guinea coast". Sediments as well as the steep continental slope on the north coast of Papua New Guinea could be additional enhancing factors in the tsunami generation. A narrow (about 50 km) area of high run-up values on the coast is additional evidence in support of the contribution of submarine slumping to tsunami generation. For typical tectonic tsunamis with a magnitude around 7.0 the width of the affected area is about 100 km.

4. Tsunami generation and bottom topography

In addition to a large amount of sediments accumulating on the oceanic floor, another important factor for the tsunami generation is the underwater topography. Despite the conviction of many marine geologists that the slope steepness is not the key indicator of failure potential, a steep gradient of sloping can, at least, contribute to a higher probability of a rapid movement of sedimentary deposits disturbed by a large earthquake. In the above-mentioned study [2] we investigated the topographic profiles of the source areas of six typical earthquakes from the "red" and "blue" groups. The profiles clearly show that the "red" events occur within the steeper slopes as compared to the events from the "blue" group. In regions where the oceanic floor is relatively flat or has little slope relief, even when substantial sedimentary layers fill completely the entire basin, earthquakes cannot cause the submarine slumping. An example is the shallow Yellow Sea basin which is being filled up with sediments of the Yellow (Huang He) and the Yangtze rivers. In more than 300 years, there have been no destructive tsunamis in this area, in spite of the high regional high seismicity, documented well by historical catalogs. In particular, two of the earliest events shown in the Pacific tsunami catalog in 47 B.C. and 173 A.D., occurred in the Bo Hai Bay in the western region of the Yellow Sea. It is also interesting to note that Soloviev and Go [5] list a total of nine destructive tsunamis between 1076 and 1636 in the coastal region of Korea. Of these nine events five tsunamis are not associated with earthquakes but are listed as being of the "meteorological" origin. Now we can think these tsunamis as being generated by the submarine slumps triggered by small earthquakes or even without seismic disturbance at all. Continuing sedimentary loading from China's two largest rivers and unstable slope conditions resulting in loss of equilibrium could have been responsible for these "meteorological" tsunamis.

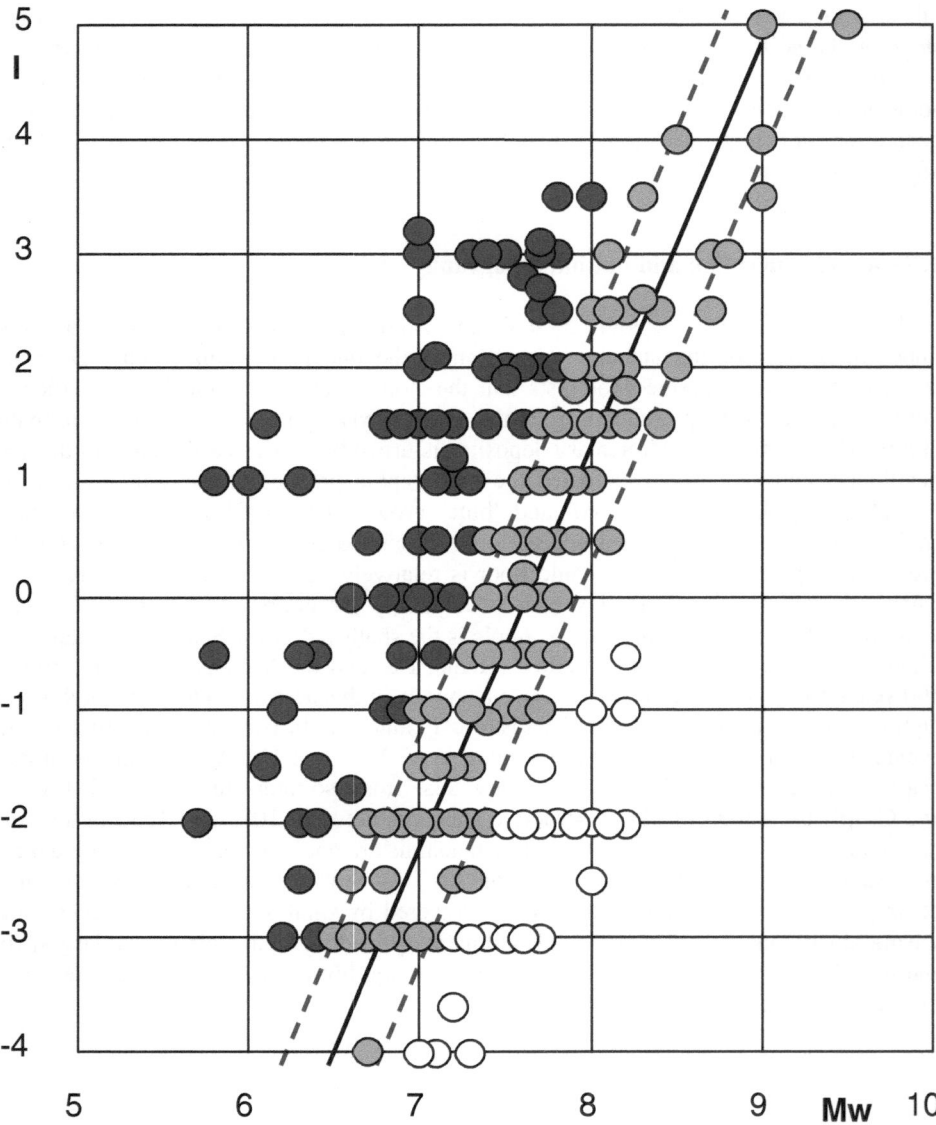

*Figure 1. Tsunami intensity **I** (on the Soloviev-Imamura scale) versus magnitude **M**$_w$ for tsunamigenic earthquakes occurred in the Pacific from 1900 to 1998. Based on the value of the parameter **ΔI** (that is the difference between the observed and the expected tsunami intensity) all events were divided into the three groups - "red" (**ΔI**>1, black circles), "green" (-1≤**ΔI**≤1, grey circles) and "blue" (**ΔI**<-1, white circles). The black line represents the expected tsunami intensity calculated by the formula (1). The dotted lines show the boundaries between the above three groups of events.*

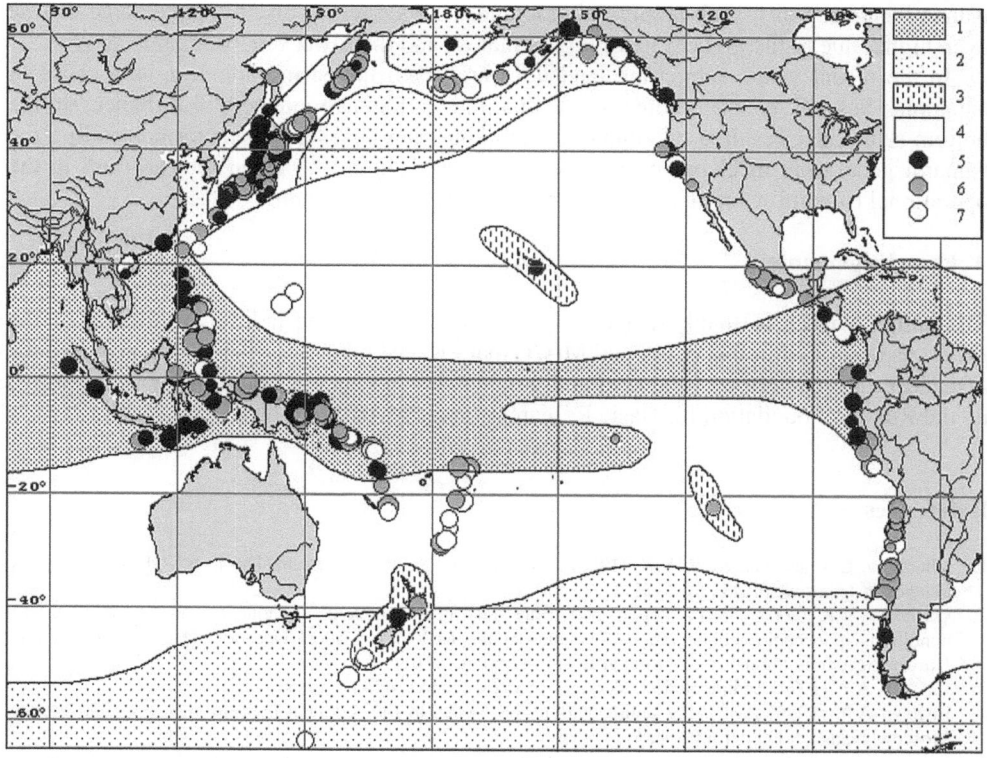

Figure 2. Basic zones of oceanic lithogenesis selected in [4] (equatorial humide zone (1), northern and southern humide zone (2), zones of dominating effusive sedimentation (3), northern and southern aride zones (4)) and the position of "red" (5), "green" (6) and "blue" (7) tsunamigenic earthquakes occurred in the Pacific from 1900 to 1998.

5. Conclusion

The slumping mechanism has been always taken into consideration as a possible source for tsunami generation. Tsunami catalog compilers always marked the tsunamigenic events where the involvement of the landslide components was essential or even dominating. However, such a mechanism was considered exceptional rather than ordinary for the tectonically generated tsunamis. According to the available historical tsunami catalogs ([5], [6], and [7]) only 7% of the total number of tsunamigenic events were generated by a slump mechanism. Possibly, Melekestsev [8] was the first to get an idea about the dominant role of the slumping component in the generation mechanism of destructive tsunamis. The results obtained in this study show that conditions of the oceanic sedimentation are of extreme importance in understanding the tsunami generation mechanism, and that slumping could contribute significantly up to 35% of the historical tsunamis in the Pacific. In other tsunamigenic regions, such as the Atlantic and the Mediterranean, this percentage can be even higher due to the higher rate of sedimentation and the lower rate of seismic activity.

In the recent years, submarine landslides are given increasing attention as a cause of tsunamis. Several recent studies show that the earthquake-induced disturbance of the bottom sediments, resulted in submarine slumping, can be a leading factor controlling the tsunami generation mechanism, and these processes should be taken into account in the operational tsunami warning as well as in the coastal tsunamizoning.

Acknowledgements

The author wishes to thank Academician A.P.Lisitsyn for the helpful discussion of the materials described in this paper and Mrs.Tamara Kalashnikova for her highly professional assistance in preparation of the data and graphics for this paper. This study was supported by the Russian Foundation for Basic Research (Grant 01-07-90199).

References

1. Chubarov, L.B. and Gusiakov, V.K. (1985). Tsunamis and earthquake mechanism in the island-arc regions, *Science of Tsunami Hazard,* 3, 3-21.
2. Gusiakov, V.K. (2001) "Red", "Green" and "Blue" Pacific tsunamigenic earthquakes and their relation with conditions of oceanic sedimentation. Tsunamis at the End of a Critical Decade. G.Hebenstreit (Editor). Kluwer Academic Publishers, Dordrecht-Boston-London, 2001.
3. Lisitsyn, A.P. (1974) Sedimentation in oceans, Moscow, Nauka, 425 pp. (in Russian).
4. Lisitsyn A.P. (1988) Avalanche sedimentation and breaks in the sediment deposition in seas and oceans, Moscow, Nauka, 309 pp. (in Russian).
5. Soloviev, S.L. and Go, Ch.N. (1974). A catalogue of tsunamis on the western shore of the Pacific Ocean. Nauka Publishing House, Moscow, USSR, 310 pp., Can. Transl. Fish. Aquat. Sci 5077, 1984.
6. Soloviev, S.L. and Go, Ch.N. (1975). A catalogue of tsunamis on the eastern shore of the Pacific Ocean. Nauka Publishing House, Moscow, USSR, 204 pp., Can. Transl. Fish. Aquat. Sci 5077, 1984.
7. Lander, J.F. (1996). Tsunamis affecting Alaska, 1737 - 1996, Boulder, Colorado, National Geophysical Data Center, 195pp.
8. Melekestsev, I.V. (1995). On the possible source of the November 23, 1969 Ozernoy tsunami in Kamchatka, *Vulkanologiya i Seismologiya*, No.3, 105-108 (in Russian).

UPDATING AND REVISION OF THE EUROPEAN TSUNAMI CATALOGUE

A.MARAMAI [1], L. GRAZIANI [1], S.TINTI [2]
[1] Istituto Nazionale di Geofisica e Vulcanologia
Via di Vigna Murata 605, 00143 Rome, Italy
[2] Università di Bologna, Dip.to di Fisica, Settore di Geofisica,
Viale Carlo Berti Pichat, 8 40127, Bologna, Italy

Abstract

The first catalogue of the European tsunamis was built in the frame of the EU Projects called GITEC and GITEC-TWO. The catalogue was implemented as a FoxPro 2.5 database, and it can be fully used on PC with Windows 3.1 or with the first versions of Windows 95. In the present work, we describe the new version of the database that has been totally rebuilt within the Visual FoxPro 6.0 DBMS environment in order to make it suitable for the operating systems currently on the market (i.e. Windows 98, Windows 2000, Windows NT, etc.). The general structure of the previous data base version is preserved. The catalogue is accessible through a main screen containing a) functional buttons to perform the basic inquiry and browser actions, and b) parametric and textual data concerning a specific event that can be selected by the user. The data base has been enriched by including new tsunami entries, as the result of revision of historical sources, and by adding a new category of data, namely graphical data, such as digitised tide-gauge records, photos, relevant maps, etc., that form a special section accessible from the main screen. Moreover, the auxiliary data base of the references has been updated by introducing all the contributions, such as papers and scientific studies that have been published in the last years.

1. The European Tsunami Catalogue

Tsunami catalogues are indispensable means to assess the tsunami potential of a given region and to evaluate vulnerability and risk exposure of coastal areas and environments.

The first European tsunami catalogue was the fruit of the work of various research groups co-operating in the frame of the projects finance by the European Union named GITEC and GITEC-TWO and was delivered in 1998. It was not built by simply gathering and putting together the available national catalogues, but it was the result of efforts aimed at establishing a standard format and structure for the data base and applying uniform criteria to evaluate the informational sources and to parameterise the data. The catalogue was implemented as a digital data base through the DBMS FoxPro 2.5, running under Windows 3.1 and the first version of Windows 95, but not usable correctly in PCs with the operating systems distributed today. The problem of migrating the catalogue to the current

A. C. Yalçıner, E. Pelinovsky, E. Okal, C. E. Synolakis (eds.),
Submarine Landslides and Tsunamis 25-32.

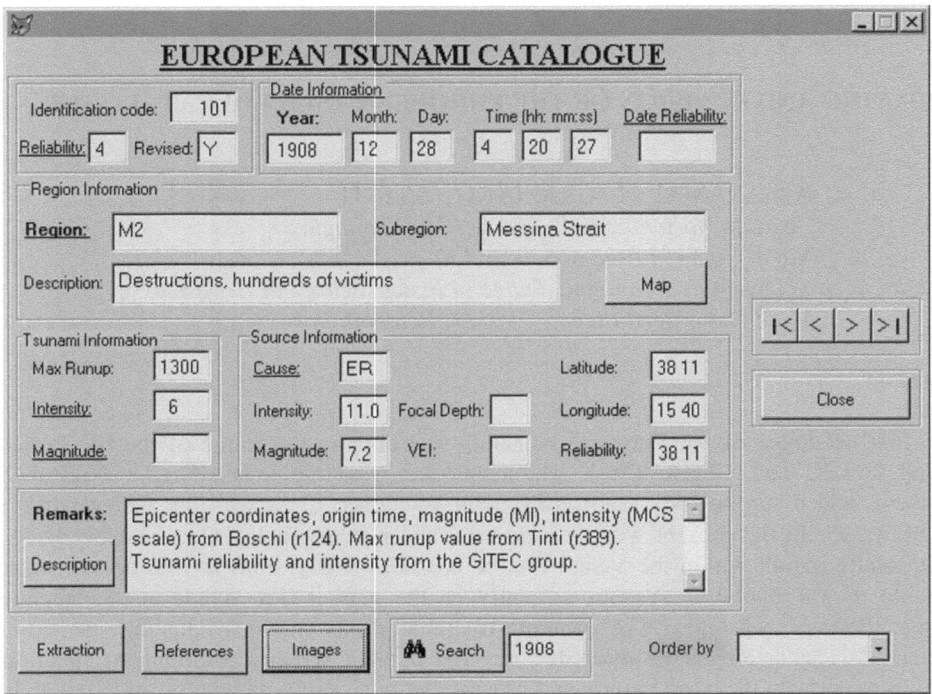

Figure 1. The Catalogue Interface Screen (CIS) is the main user interface of the data base: this example displays the data of the Messina Strait, December 1908 tsunami.

PC systems (Windows 98, Windows 2000, Windows NT and so on) has been solved by implementing a new electronic version of the catalogue under Visual FoxPro 6.0. Since this latter is a DBMS substantially different from FoxPro 2.5, the new catalogue could not be obtained by automatically converting the old data base, but all the basic data entry had to be done again.

The general architecture of the catalogue is the same as the old version, to favour the user approach. Each event is presented through the main Catalogue Interface Screen, CIS, (displayed here below in Figure 1), showing all the event data available in the database either directly or indirectly through the activation of further linked windows.

Examples of event data that are immediately displayed on the screen are: identification code, date, reliability, source region and subregion, tsunami information (runup, intensity and magnitude), and source information (cause, coordinates, etc). A short one-line verbal text provides a description of the tsunami. But further details can be learned through the button "Description": by pressing it, a new temporary window, containing a brief note on the cause, the account of the tsunami and the list of the references, opens on the CIS, and the user can later close it by clicking the mouse on any points of the screen.

In addition to the "Description" button, the CIS includes other functional buttons that enable the user to perform operations on the event data base or to have access to other ancillary data bases of the catalogue, namely the data bases of the references and of the images. This latter database is a new addition of the present version of the catalogue, consisting in a set of graphical data such as maps, photos and mareograms of the most relevant tsunamis that can be selected and examined by the user. In Fig.2 an example of one of the tide-gauge figures inserted in the images section is given.

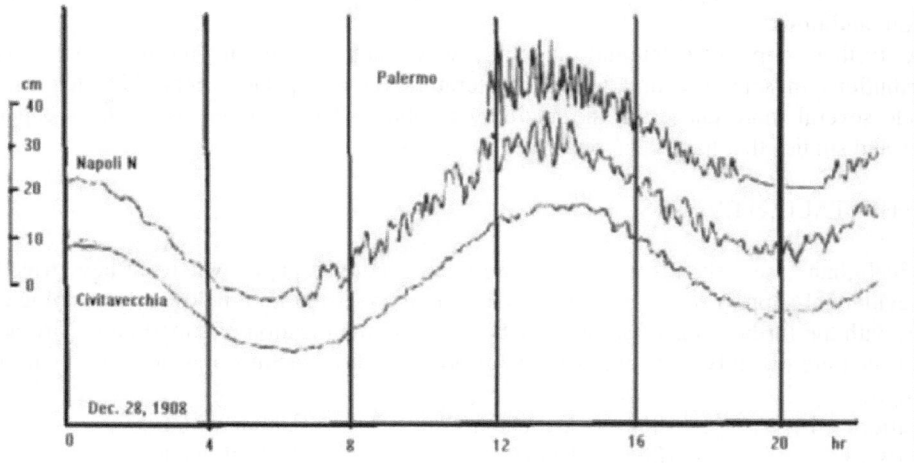

Figure 2. Picture included in the images database of the catalogue. It shows the mareograms with the signal of the Messina Strait 1908 tsunami, recorded by the tide-gauge stations of Palermo, Napoli and Civitavecchia, Italy.

2. Updating: state of the art

If we compare a conventional catalogue published in the form of a printed book, booklet or paper, to a digital data base, we see that the main difference existing between them is that the former is static and cannot be changed, whilst the second is subject by its nature to a continual process of modification and updating, that in a broad sense is part of the daily maintenance activity. The expression "state of the art" we include in the title of this section underlines intentionally the concept that a snapshot of the catalogue as it is "today" will be provided in this section. Up to now, updating of the catalogue resulted in addition of new events, in the definite decision that some cases cannot be included in the catalogue and in inclusion of additional entries in the database of the references.

We started from the re-examination of the Italian events and, in particular from 34 cases of the XIX and XX century that were reported in the Caputo and Faita [1] catalogue of the Italian tsunamis, but that were not inserted in the Italian section of the GITEC catalogue [2], basically because they were supported by extremely insufficient sets of data. With the chief purpose of acquiring new data, a very minute, careful and exhausting search has been undertaken for each event, by consulting the most important national libraries and archives, examining with special attention the newspapers collections (about 50 different newspapers). Despite our efforts, we were not able to find any additional data for 32 cases. On considering the number and importance of the data archives where no mention exists of them, we definitely conclude that these cases are not tsunamis and cannot be entries in the European tsunami catalogue. On the other hand, we discovered valuable data on 2 cases that in our view have to be considered tsunami events: their occurrence dates are July 3, 1809 (Ligurian Sea) and January 15, 1940 (Northern Sicily).

Case history studies and collaborative relationships with European tsunami specialists have been advantageous to extend the catalogue with inclusion of new well supported events, one that occurred in Italy in 1988 and six that took place in Norway between 1867 and 1998. The following two subsections will be used to provide more details on the new tsunami additions.

A further step of catalogue updating was made by going through the recent contributions in scientific and technical literature on European events. This led us to include several more entries in the ancillary database of the references, such as papers, books and studies that have been published in the last 3 years.

2.1. THE ITALIAN EVENTS

The 34 Italian cases taken from Caputo and Faita catalogue [1] of which we have tried to find evidential support through an extensive search are listed here below in chronological order, with the further indication of the subregion and the location of supposed occurrence. Underlined are the only 2 events that we consider worth of inclusion as new entries in the catalogue.

January 4	1802	Dalmatian Coasts	Dalmazia Istria
July 31	1804	Campania	Gulf of Naples
August 26	1806	Latium	Ardea
<u>July 3</u>	<u>1809</u>	<u>Liguria Côte d'Azur</u>	<u>La Spezia</u>
July 27	1809	Campania	Gulf of Naples
June 28	1812	Liguria Côte d'Azur	Marseille Harbour
April 7	1813	Central Adriatic	Ancona
Nov. 12-15	1816	Liguria Côte d'Azur	Genoa
January 8	1819	Liguria Côte d'Azur	Genoa
July 23	1820	Liguria Côte d'Azur	Genoa
March 20	1822	Northern Sicily	Marsala
April 10	1822	Eastern Sicily	Catania
	1824	Campania	Gulf of Naples
March 18	1826	Central Adriatic	Pesaro Senigallia
July 20	1828	Liguria Côte d'Azur	Genoa
May 26	1831	Liguria Côte d'Azur	Ventimiglia
July 2	1831	Sicily Channel	Sciacca
August 10	1838	Dalmatian Coasts	Dalmatia
October 11	1843	Dalmatian Coasts	Ragusa
June 18	1845	Messina Strait	Messina
December 29	1854	Liguria Côte d'Azur	Genoa
November 24	1862	Liguria Côte d'Azur	Genoa
Oct. –Nov.	1870	Mediterranean Sea	
January 17	1871	Western Liguria	
December 23	1876	Liguria Côte d'Azur	Nice
April 25	1880	Western Liguria	
Nov. 15-16	1892	Campania	Ponza
December 27	1894	Northern Sicily	Filicudi
October 16	1896	Liguria Côte d'Azur	Sanremo
May 13-14	1903	Northern Sicily	Palermo
January 27	1939	Northern Sicily	Filicudi
<u>January 15</u>	<u>1940</u>	<u>Northern Sicily</u>	<u>Palermo</u>
March 16	1941	Northern Sicily	Palermo
June 22	1978	Central Adriatic	Marche

The 1809 event occurred in the La Spezia gulf, Ligurian Sea. We cannot attribute a certain cause for the tsunami generation, but on the basis of the sea bottom morphology, we can suggest the occurrence of a submarine landslide in the gulf. On the grounds of the gathered information we assigned it reliability 2 (meaning "questionable tsunami" in the reliability catalogue scale, i.e. to say that doubts on the tsunami occurrence can be legitimately raised) and tsunami intensity 2, referred to the Ambraseys-Sieberg scale [3], that means "light tsunami" with waves generally noticed on very flat shores.

A short piece of the original description quoted in the "Gazette Nationale" [4] is reported here: "...the inhabitants of La Spezia and those living in the whole gulf observed an extraordinary tide on July 4. ...At about 8 a.m. the sea, that till that time was absolutely calm, suddenly rose about 1 m above its usual limit. This extraordinary tide lasted for about 15-20 min. rising and falling. No apparent cause was observed...The tide was so strong and quick that the sea water flew up to the city of La Spezia through a small canal that crosses the city itself. Some merchants that were settled in the embankments ran away. The seafloor in the gulf was submerged and immediately afterwards large parts of the shallow beach were left dried and some big fish were drag by the water and trapped in the dried beach...The first flux of the sea water was followed by 4 or 5 others that gradually diminished their strength. ...We can suppose that the effects of this extraordinary tide were due to some seismic shock or submarine or in land." Observe that the newspaper reporter makes the hypothesis of an event with seismic origin, on which we cannot agree, being rather in favour of a generation mechanism related to a submarine mass failure.

The January 15, 1940 event, in Palermo, northern Sicily, was due to a strong shock (VIII MCS), with epicentre located in land about 20 km away from the coast, which caused one victim and severe damage. Many localities were involved. According to "L'Ora" [5] and the "Ufficio Centrale di Meteorologia e Geodinamica" [6], in Palermo some people noted a strong agitation of the sea just a few seconds before the shock and immediately after some sudden sea waves were seen in the gulf. This event has been attributed tsunami reliability 2 and tsunami intensity 2.

One more Italian event, that occurred in 1988 at Vulcano (Aeolian Islands) and that was never inserted in previous catalogues, has been studied and finally introduced in the data base. A specific study on this case has been published by Tinti et al. [7]. On April 20, 1988 a large landslide occurred on the north-eastern flank of volcano La Fossa on the island of Vulcano, during a period of increasing volcanic and seismic activity. The landslide entered the sea generating a tsunami. Instrumental data do not exist since no tide gauges were present in the area.

Observational data are from eyewitnesses, who upon specific interviews gave detailed reports. A fisherman, who at the time of the landslide was in his boat just inside the generation area, reports that at about 5:30 am local time he heard a big noise and, in looking at the coast, he had the perception that the mountain was running towards him. A positive, approximately 1-2 m high wave was excited that raised the boat without damaging it. Then he felt the boat to go down, and other smaller oscillations followed the first wave. Another boat in the same bay but farther from the source was impacted by the tsunami with no damage. In the harbour of Porto di Levante many people observed sudden waves entering the harbour, similar to those produced in the sea by a storm, but the weather was fine with no wind. Wave amplitude was estimated to be approximately 0.5 m. Information about observed waves with 0.5m amplitude in some places of Lipari Island is further available. The reliability attributed to this event is 4, which means "definite tsunami" with 100 percent certainty on its occurrence.

2. 2. THE NORWEGIAN EVENTS

In the Norwegian section of the European tsunami catalogue 6 new events have been added, chiefly thanks to the revision carried out by Dr. C. Harbitz from Norwegian Geotechnical Institute, who made his results available to us.

On May 7, 1867 after a strong earthquake, an unusual withdrawal of the sea, followed by oscillations for about half an hour, was observed at several locations between Haugesund and Lindesnes, south-west Norway. The wave came from SSW, and caused runup heights of 2-6 feet. Boats were turned around. Eyewitnesses also report roaring and shaking of the ground [8]. This event has been attributed reliability 4 and tsunami intensity 3 ("rather strong, generally noticed, with some shipwreck").

The second studied tsunami occurred on August 31, 1940 in northern Norway, and has a two phase occurrence. According to eyewitnesses a landslide fell into a fjord, causing a railway embankment to disappear in the sea, and determining a tsunami. About 5 minutes later the sediments in the inner part of the fjord were involved in a tremendous subaqueous slide and disappeared together with a landing pier and a small harbour, causing a second impressive wave that was observed at the failure point of the railway embankment a couple of minutes later [9]. The reliability attributed to this event is 4 and the tsunami intensity is 2.

The January 9, 1952 tsunami was engendered by a gravitative landslide in Mid-Norway. A 1-2 m high wave swept past a dregder within the area of sliding in Follafjorden, Nord-Trøndelag [9]. From the gathered information, this event has reliability 4 and tsunami intensity 2.

On October 6, 1979 a rockslide of 5000 cubic meters dropped vertically into the fjord Bindalsfjorden at Hildringen near Terråk, Nordland (northern Norway). The upper part was released from 110 m a.s.l. The runup height at the island Øksninga (1.5 km to the north) was more than 2 m at high tide, and caused damage to boat houses and harbours [10, 11]. The detailed description suggests event reliability 4 and tsunami intensity 3.

The August 18, 1983 western Norway tsunami was generated by a 150,000 cubic meters of rocks and by about the same amount of scree that fell from Kleppura, in the mountain Middagshaugen, into the fjord Årdalsfjorden. A surface elevation of about 2.5-3.5 m was measured in the fjord, and runup heights of 5-7 m in the harbour area of Årdalstangen, 1-1.5 km to the north-east. Small sailing vessels and light structures suffered damage, while parts of quay structures were lifted or damaged [12, 10]. On the basis of the above data, the event deserves reliability 4 and tsunami intensity 4.

The last event examined occurred on March, 19, 199. A gravitative rock-slide with mass of 20-30,000 cubic meters in Sørefjorden, Hyllestad, Sogn og Fjordane (western Norway) produced runup heights exceeding 6 m on the opposite side of the fjord. The sources referred extensive damage on quays and boat-houses, and no victims were registered [13].

3. Conclusions

The 1998 FoxPro 2.5 European catalogue of tsunamis has been replaced by a new data base implemented in the environment of Visual FoxPro version 6.0, in order to provide a catalogue suitable for the operating systems distributed today. The new version has been enriched with the insertion of new events and with the creation of a specific data section containing digitized tide-gauge records, photos and maps for the most interesting tsunamis.

The Italian and Norwegian sections of the European tsunami catalogue have been revised and updated.

As regards the Italian part, 34 cases from 1800 to 1978 that were not included in the 1998 catalogue due to scarcity of information have been re-examined; 2 of them have been inserted as new entries in the data base, whilst the others have been definitely eliminated. Further, one more Italian event that took place in 1988 at the Vulcano Island has been added to the tsunami catalogue.

As regards the Norwegian section, the data base has been enriched with 6 new tsunami events, from 1867 to 1998.

Moreover, the European tsunami catalogue updating also involved the references data base, with inclusion of papers, books and studies published in the last years.

At present two more events are under investigation, both causing tsunami effects along the Italian coast but probably originated in the southern Mediterranean area.

The next steps will be in the first place the revision of all events in the data base with reliability 0 and 1, and then the extension of the updating to all European events that were not revised in implementing the GITEC version of the catalogue.

Acknowledgements

The authors are indebted to Dr Carl Harbitz from the Norwegian Geotechnical Institute for information on Norwegian tsunamis. This research was carried out on funds from the Gruppo Nazionale di Difesa dai Terremoti (GNDT) of the Istituto Nazionale di Geofisica e Vulcanologia (INGV) and from the Ministero dell'Università e della Ricerca Scientifica e Tecnologica (MURST).

References

1. Caputo, M. and Faita, G. (1984) Primo catalogo dei maremoti delle coste italiane, *Atti Accademia Nazionale dei Lincei, Memorie Classe Scienze Fisiche, Matematiche, Naturali, Roma, serie VIII*, 17, 213-356.
2. Tinti, S. and Maramai, A. (1996) Catalogue of tsunamis generated in Italy and in Côte d'Azur, France: a step towards a unified catalogue of tsunamis in Europe, *Annali di Geofisica*, vol.XXXIX, n.6, 1253-1299.
3. Ambraseys, N.N. (1962) Data for investigation on seismic sea waves in the Eastern Mediterranean, *Bull Seism.Soc.Am.*, 52, 895-913.
4. *Gazette Nationale*, (1809) July 18, Paris.
5. *L'Ora*, (1940) January 16, Palermo.
6. *Ufficio Centrale di Meteorologia e Geodinamica*, (1940) Notizie sismiche pervenute all'UCMG per l'anno 1940, Roma.
7. Tinti, S., Bortolucci, E. and Armigliato, A. (1999) Numerical simulation of the landslide-induced tsunami of 1988 on Vulcano Island, Italy, *Bull.Vulcanol.*,61, 121-137.

32

8. Mhon, H (1867) Meddelt i Videnskabsselskapet i Christiania den 24 mai 1867, *Meddelelse angaaende en usædvanling Bevægelse af havet paa Norges Vestkyst.*
9. Bjerrum, l (1971) Subaqueous Slope Failures in Norwegian Fjords, *Norwegian Geotechnical Institute,* 88, 2-8.
10. Larsen, J.O. (1999) Tension cracks and landslides in steep hard rock mountains in the Norwegian fjord districts, *Dept. of Geotech. Engn., NTNU, Trondheim, Norway.*
11. Larsen, J.O. (1979) Hildringsflauget i Bindal kommune-befaring etter skred, *Norwegian Geotechnical Institute*, Report n° 79923-1, April 23, pp. 11.
12. Lied, K., Larsen, J.O., Schieldrop, B., Gjevik, B. and Pedersen, G. (1984) Årdalstangen, Årdal kommune - Vurdering av fare for flodbølge ved skred i Kleppura, *Norwegian Geotechnical Institute*, Report n° 83483-1, Oct. 12.
13. Harbitz, C. and Domaas, U. (1999) Hyllestad kommune-Vurdering av skredfare og bølgehøyder i Åfjorden, *Norwegian Geotechnical Institute*, report n° 981014-1.

TSUNAMI OF ŞARKOY-MÜREFTE 1912 EARTHQUAKE: WESTERN MARMARA, TURKEY

Y. ALTINOK[1] , B. ALPAR[2] and C. YALTIRAK[3]
[1]University of İstanbul, Engineering Faculty, Department of Geophysics, 34850 Avcılar, İstanbul, Turkey.
[2]University of İstanbul, Institute of Marine Sciences and Management, Müskile Sok., 34470 Vefa, İstanbul, Turkey.
[3]İstanbul Technical University, Mining Faculty, Department of Geology, 80626, Ayazağa, İstanbul, Turkey.

Abstract

One of the largest earthquakes in the Balkans is the 09.08.1912 (01:29:00 UTH) Şarköy-Mürefte Earthquake which occurred on the active Ganos fault zone (Ms=7.4). The eastern termination of the associated faulting is in the Marmara Sea. On the basis of archiving and library studies, geological field surveys and offshore investigations, the 1912 earthquake produced a tsunami. The mechanism of this co-seismic tsunami may be assigned to an underwater failure along the southern slopes of the Tekirdağ Trough rather than a seabed dislocation.

1. Introduction

The Ganos Fault zone is composed from southward bending fault segments and forms the main structural frame of the Western Marmara region [1] (Figure 1). It is bounded by the Tekirdağ Trough to the east, where counter-clockwise rotation of the system causes compressional forces to the south while extensional forces are dominant to the north. Further west, the Ganos fault makes land at Gaziköy (Ganos), at the south foot of the Işıklar (Ganos) mountain and crosses the land area via Guzelköy (Miljo), Mürselli (Miseli, Musala), Yayaköy (Fila), Yorguç, Gölcük (Joldzik) and Evreşe.

A. C. Yalçıner, E. Pelinovsky, E. Okal, C. E. Synolakis (eds.),
Submarine Landslides and Tsunamis 33-42.
©2003 Kluwer Academic Publishers.

34

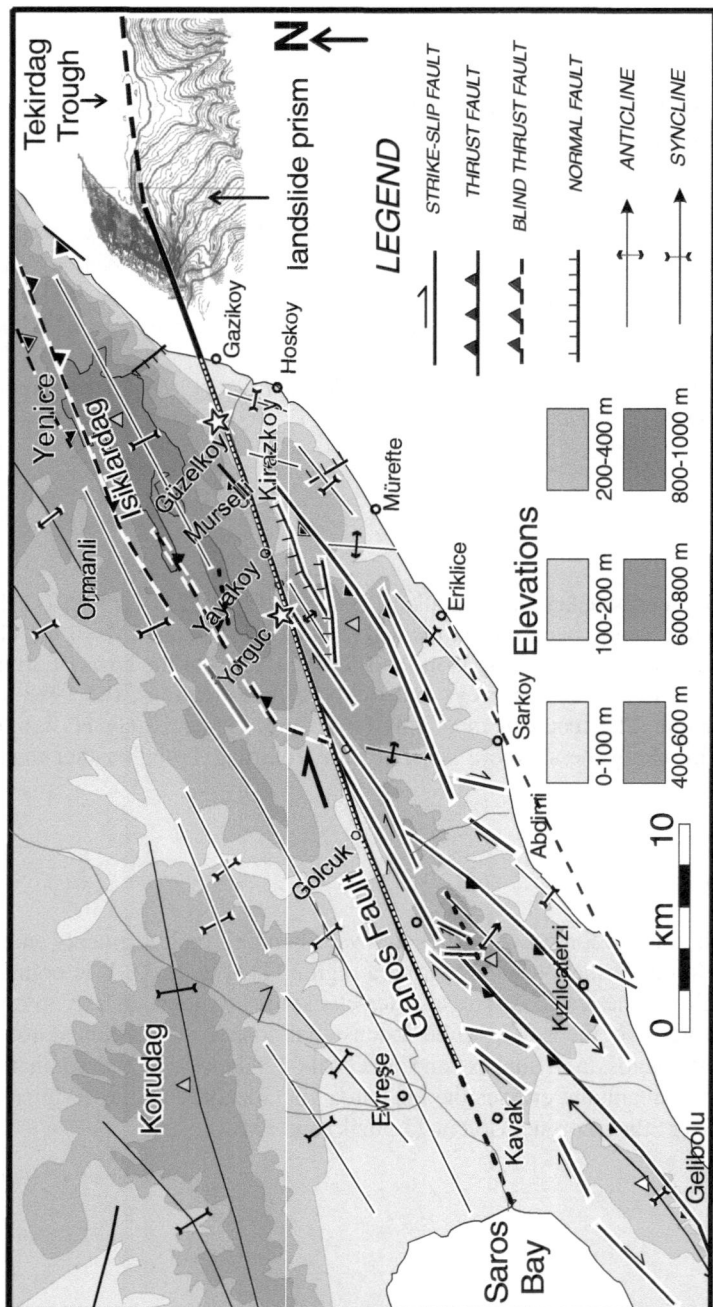

Figure 1. Morphotectonic map of the region along the Ganos fault (modified from [1]). The surface rupture of 1912 earthquake is defined. White dots on the Ganos Fault represent surface rupture of 1912 earthquake on land and stars represent places where maximum displacement (3.5-5.5 m) observed. Detailed bathymetry deducted from multibeam data shows landslide prisms.

Photo1. Hoşköy (Hora) village was completely destroyed in Şarköy-Mürefte 1912 earthquake. Masonry structures in the vicinity of surface ruptures were totally destroyed, timber-frame houses were severely damaged [C.E.S. Palmer and N. Parodi, L'illustration Journal Universal, 24 August, 1912, Paris].

The fault reaches into the Saros Bay at the mouth of the Kavak creek. Its western extent in the gulf remains controversial, and has been postulated as either along the northern edge of the Gelibolu Peninsula [1, 2, 3] or along the northern continental slope [4]. The Ganos Fault zone continues its development as a negative flower structure under the control of the master fault [5].

The latest earthquake on this active structural element is the Şarköy-Mürefte Earthquake of 09.08.1912, one of the largest in the Balkans. The shock was felt within a circle with 400 km radius. It totally destroyed more than 300 villages, 12,600 houses and caused enormous losses in Şarköy (Peristasis), Mürefte (Myriophyto), Hoşköy (Hora), Gelibolu (Gallipolis), Tekirdağ (Rodosto), Çanakkale (Dardanelles), resulting in 2800 deaths and about 7000 injuries (Photo 1). A maximum intensity of X (MSK) was assigned to NE trending area between Tekirdağ and Gökçeada (Imbros) on the latest and probably the most reliable isoseismal map (Figure 2). Later, the surface-wave magnitude of the main shock was determined as Ms=7.4± (0.3) from mechanical instruments mainly from European stations [6].

Because the earthquake coincided with political uneasiness in the Balkans, there is not much detailed information about this earthquake nor any specific document for a tsunami event, except for an affirmative one-sentence statement given by Ambraseys and Finkel [6]. In this paper, on the basis of our new field observations and new marine geophysical data, the tsunami of 1912 Şarköy-Mürefte Earthquake will be reviewed. In addition, Ottoman archives and libraries were investigated for unknown documents, sources and clues. Some locally distributed minorities were interrogated for unknown or forgotten places encountered in these documents.

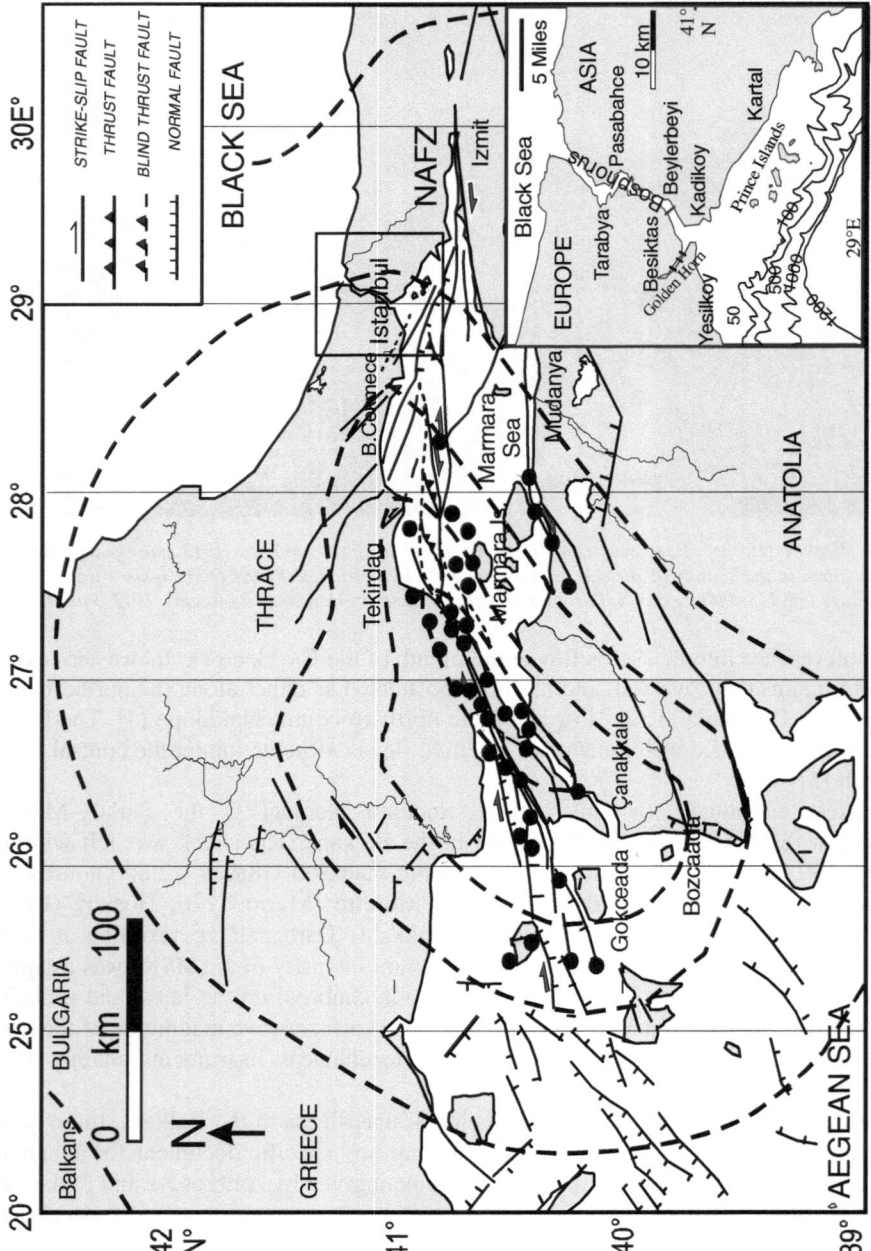

Figure 2. Isoseismal map of 1912 earthquake [6] and active faults in the Marmara Sea [4]. Possible epicentres of the historical earthquakes of the Western Marmara Sea region (from 427 B.C. to present) were interpreted from [11, 9, 27, 8, 10] and superimposed. Inset shows Bosphorus region where abnormal waves observed.

2. Tsunamis in the Region

During the last 41 centuries, more than 300 earthquakes have taken place in the Marmara Sea region. Eight of them are large earthquakes, occurring at intervals of 150 to 420 years [6]. Considering historical records going back to 120 AD, many tsunamis are known in the Marmara Sea [7]. The most acknowledged of them similar to the 1912 tsunami are given below in some detail.

24/25/26.09.477/480: With the earthquake which affected Çanakkale, İstanbul, İzmit, Gelibolu and Bozcaada, a tsunami had occurred in İstanbul and damaged coastal areas. Wild sea waves rushed right in, engulfed a part of what had formerly been land, and destroyed several houses [8, 9].

18.10.1343: The earthquake affected most of the Thrace region and İstanbul. To an extent that could be traced on the basis of available sources, tsunami waves reached Beylerbeyi (Stauros) in the Bosphorus. Inundation distance was reported to be 2-2.2 km (10-12 stadion) [10]. Anchored sea vessels were destroyed in the bays and coastal area. Fishes, some dragged farm animals and even people were found dead on land after the sea receded.

14.10.1344 (06.11.1344 Saturday): The earthquake affected almost all of the coastal area of Thrace and its inner parts. Teikhos castle in the Marmara Island collapsed completely. Gaziköy and Hoşköy areas were severely affected. More than 300 people perished in the ruins of their houses [10]. The earthquake was associated with a tsunami [11, 12, 13]. The sea inundated 2000 m [14, 15].

3. Surface Ruptures and Ground Displacements of 1912 Earthquake

Although the 1912 Şarköy-Mürefte Earthquake occurred in one of the most densely populated part of Turkey, it is one of the less known events due to political conflicts. There are some reports written after the event by [16, 17, 18, 19, 20]. The epicentral zone was reported to be about 60 km long and 10 km wide, extending NE-SW from Gaziköy (Ganos) to Gelibolu. In his unpublished report, Prof. Rothe connected the rupture with the fault passing through the Marmara Sea [21]. On the basis of different field studies, although surface ruptures were well defined on land (Table 1), the total length of the main segment remains controversial. It is generally considered to be between 50 km [6] and 84 km [22] long within which fault breaks of a dextral strike slip nature showed displacements, mainly associated with a large normal component, of up to 3 m.

Our detailed field surveys showed right lateral strike-slip motion, with offsets as much as 4-5 m, which were not exactly clarified in earlier studies. These were best observed in and around Yorguç, Yayaköy, Mürselli and Güzelköy (Figure 1). To the west of Güzelköy, the northern block has been uplifted about 1.5 m with respect to the southern one. The fault dips 80-85° northwest, indicating an oblique component beyond its dextral character. On the other hand, to the south of Gaziköy, the southern block uplifted about 0.5 m with respect to the mean sea-level. Maximum ground displacements of 5.5 m are between eastern part of Mürselli and Güzelköy. They become smaller west of the Kavak village, where the Ganos Fault offsets the Kavak creek right-laterally. Excavation studies by Rockwell et al. [23] in this locality indicated that the southern block was uplifted and also pushed a little northward. On the basis of radiocarbon dates, they concluded that this segment ruptured during the A.D. 484, 824 and 1343 (or 1766) earthquakes. They did not manage to find any rupture related to 1912 event. Therefore, we may assume that the rupture terminates somewhere between Kavak and Evreşe villages. This means that the lateral distance between this western termination of the rupture and the place where we observed the maximum ground displacement (Güzelköy-Yorguç) is about 35 km. We may postulate that the 1912 rupture should be similarly extending from Gaziköy eastward into the sea at least 30-35 km. Recent multibeam bathymetry (Figure 1) and marine geophysical data (Figure 3) favour of this model. These data indicate en-échelon landslide prisms, which should not be misinterpreted as thrusts, on the southern flanks of the Tekirdağ Trough. Consequently, the total length of the surface rupture is possibly more than 70 km. The landslide complex to the southwestern flank of the Tekirdağ Trough (Figures 1 and 3) coincides with the eastern termination of the rupture and may be the cause of 1912 tsunami.

TABLE 1. Some important surface ruptures observed on land between Saros Bay and Marmara Sea. [a] [17 and 19] cited in [20]. [b] [17 and 18] cited in [28]. [c] [16]. [d] Dişbudak (Estelyanos) is between the villages of Ormanlı (Kestanbol) and Yenice. [e] Total length of en echelon faults. [e] [29]. [f] Downthrown block.

Place	Length(m)	Opening (m)	Deep (m)	Vertical Disp. (m) [f]	Lateral Disp. (m)	Remarks
Gaziköy	484	0.8		2		b
Güzelköy	840	0.8	2	S – block		a, b
Dişbudak [d]	400	5	10			c
Ormanlı	300-347	6.0 - 6.2	12	1		a,b
Kirazköy	960	1.2	7.5	0.4		a, b, c
Mürselli	2000	0.5				a, b
Yayaköy					3.5-5.0	e
Gölcük	6000 [e]			1.8		a, b, c
Kavak		1.0	8 – 10	1 - 1.5, S – block		a

4. Effects of 1912 Tsunami and Some Abnormal Features

Documents about the tsunami event of 1912 earthquake are scarce due to critical political developments in the Balkans at that time. Turks and Greek minorities were living together in the disaster area. In addition, regional research encountered difficulties because of the migration of the people of the area due to "exchange of population" policy implemented after the establishment of the Republic of Turkey, as well as of the fact that the control of various districts was turned over to the control of different provinces in new administrative system. Most of the sources reached are in the Ottoman and French languages. The most significant findings about the tsunami event were classified depending on their observation place.

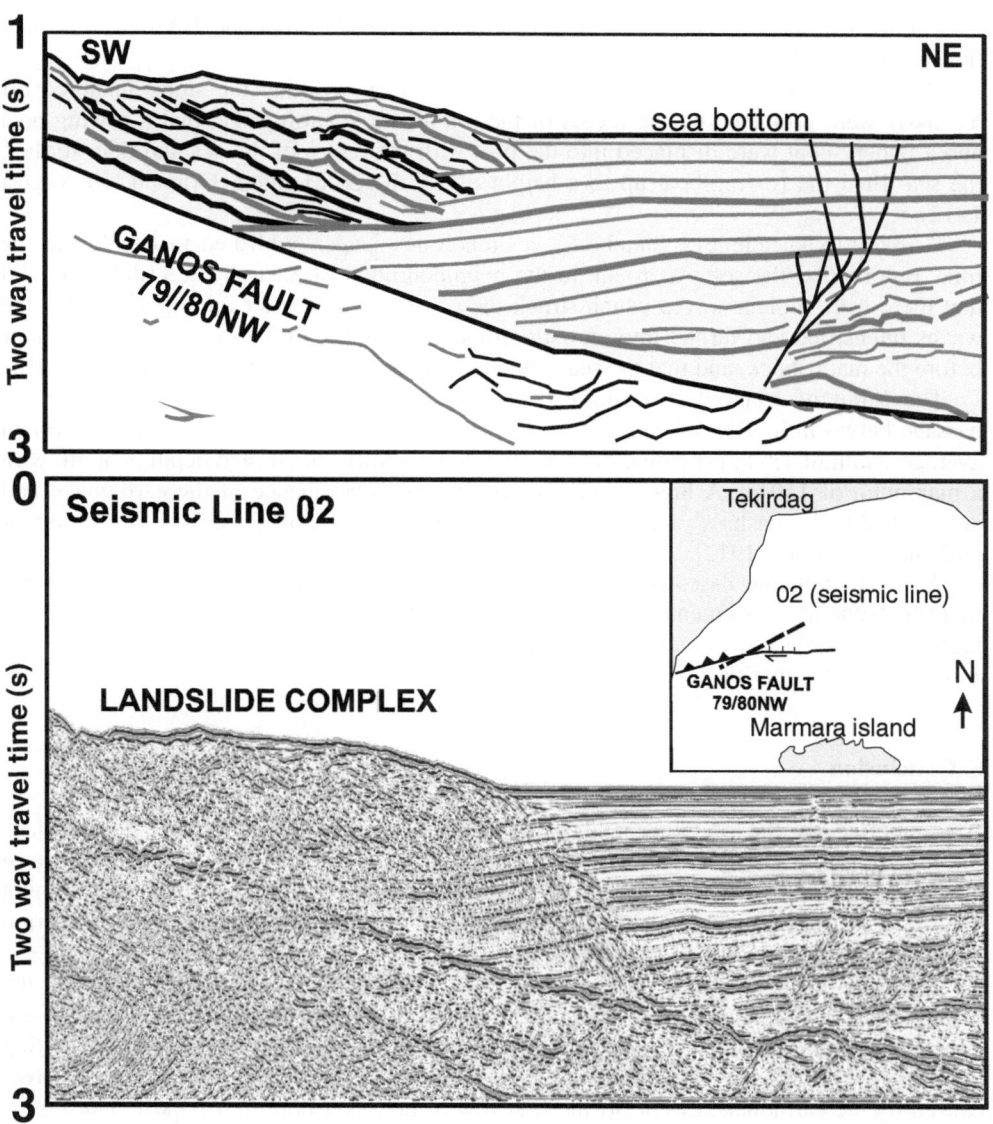

Figure 3. Seismic section showing underwater landslide complex as re-interpreted from [2]. Inset shows position of seismic line and extension of the Ganos Fault into the Marmara Sea.

Abdimi: It was an old seashore village between Şarköy and Kızılcaterzi. During the earthquake hot water sprung up. A sulphur smell came from the sea [Le Moniteur Oriental, 10 August 1912].

Eriklice (Heraclitza): Tar-like materials of pea size jerked out of the sea. These are not original materials [Le Moniteur Oriental, 17 August 1912].

Gaziköy (Ganos): Ships passing off Gaziköy during the course of earthquake reported a wide layer of oil floating on the sea [16].

Tekirdağ (Rodosto): The Sea receded after the earthquake along the Tekirdağ coast and then returned with some force causing no damage [6].

Yeşilköy (San Stefano): Engineer M. Ed. Schneider says that the ships anchored off Yeşilköy were grounded by the recess of the sea after the earthquake, that a rowing-boat and a fishery-boat were displaced into the port with the return of the sea and that, further, the sea lifted the rowing-boat up to a height of 2.7 m (Le Moniteur Oriental, 10 August 1912).

Kadıköy (Khalcedone): "Stamboul", a French newspaper published in İstanbul at the time, reports that the sea was rough in the neighbourhood of Kadıköy at 11:00 p.m., and that there were violent waves which struck the coast one hour after midnight and that the waves intensified and turned into an extraordinary event at 02:00 a.m., one and half hour before the main shock, and that the sea calmed down half an hour after the earthquake [16].

The Strait of İstanbul (Bosphorus): This shallow and narrow strait makes a water passage between the Black Sea and the Marmara Sea. Its average length is 31 km with an average width of 1.6 km (ranging between 0.7 and 3.5 km). Its average depth is 36 m, with a maximum of 110 m. A high water occurred as the consequence of the earthquake and demolished Hidiv Pasha's yacht named "Mahrussa", anchored off Pasabahce [Le Moniteur Ortiental, 10 August 1912].

As it is seen, the cases for Abdimi, Eriklice, Gaziköy and Kadıköy do not clearly indicate tsunami phonemena. However, we also include these descriptions in order to present a complete documentation of the pre-seismic and post-seismic effects of the earthquake to the marine environment.

5. Conclusion

As supported by the previous studies, our detailed field surveys indicate that surface rupture from the Şarköy-Mürefte Earthquake of 9 August 1912 extends in an area from Kavak village to the western end of the Tekirdağ Trough. In addition to 5.5 m maximum dextral strike-slip movement, vertical downthrusting of up to 2-3 meters has also occurred along the south block of the faulting couple.

Observations in Yeşilköy during the 1912 event makes a detailed image of the formation of the tsunami, with a run-up level of 2.7 m. Similar tsunami observations were experienced in the same place as a consequence of the İstanbul Earthquake of 1894. During the 1894 event, the sea receded about 10 minutes before the earthquake and huge waves caused by the earthquake swept off the first row of houses ashore in Yesilkoy. During the 1894 event, the recess of the sea followed by inundation to its original level was also confirmed in the Prince Islands. Similar observations are valid along the coastline from Kartal to B. Çekmece. However, the waves inundated into the Golden Horn [18, 24, 25, Le Monitor Oriental, 10 July 1894].

Pre-seismic wave movements observed at the southern entrance (e.g. Kadıköy) during the 1912 event can be interpreted as a sign of tsunami. In addition, the sea waves during this event on the Bosphorus resemble those recognised during the İstanbul Earthquake of 22.05.1766 in Beşiktaş and the inner parts of the Bosphorus [26]. That earthquake was destructive mostly around eastern Marmara and its associated tsunami also caused considerable damage in the Gulf of Mudanya [9, 27]. The relation of the pre-seismic event at the southern entrance and the post-seismic one in the central part of the Bosphorus deserves further studying.

All these pictorial representations given above imply that a tsunami took place during the course of the 1912 Şarköy-Mürefte Earthquake. The mechanism of the tsunami may be assigned to an underwater failure rather than a seabed dislocation. In addition, en-échelon landslide prisms on the southern flanks of the Tekirdağ Trough (Figure 1) seem to be the main cause of the 1912 tsunami.

These results should be enlightened by tsunami modelling and paleotsunami surveys. Modelling designed to predict the formation of waves which might find their way into the inner parts of the Bosphorus would be desirable.

Acknowledgements

This study was partly supported by Research Fund of İstanbul University (O-1131/09112001, B-991/31052001, B-999/31052001). Authors thank to the Turkish Republic Prime Ministry, General Directorate of the State Archives, İstanbul Library, and İstanbul Ataturk Library. We also thank Dr. Sinan Öngen for his kind helps. Authors also thank reviewers for making helpful suggestions.

References

1. Yaltirak, C. (1996) Tectonic history of the Ganos Fault System. *Bull. Turkish Assoc. Petrol. Geologists* **8**, 137-156.
2. Yaltirak, C. (2002). Tectonic evolution of the Marmara Sea and its surroundings, Marine Geology, 190/1-2 (in press).
3. Armijo, R., Meyer, B., Hubert, A., and Barka, A.A. (1999) Westward propagation of the North Anatolian fault into the northern Aegean: Timing and Kinematics, *Geology* **27**, 267-270.
4. Yaltirak, C., Alpar, B., Sakinc, M., and Yuce, H. (2000) Origin of the Strait of Canakkale (Dardanelles): regional tectonics and the Mediterranean–Marmara incursion. *Marine Geology* **164/3-4**, 139-156 with erratum **167**, 189-190.
5. Yaltirak, C., Alpar, B., and Yuce, H. (1998) Tectonic elements controlling the evolution of the Gulf of Saros (Northeastern Aegean Sea), *Tectonophysics* **300**, 227-248.
6. Ambraseys, N.N., and Finkel, C.F., (1987) The Saros-Marmara earthquake of 9 August 1912, *Earthquake Eng. and Struct. Dyn.* **15**, 189-211.
7. Altinok, Y., and Ersoy, S. (2000) Tsunamis observed on and near the Turkish coast, *Natural Hazards* **21(2-3)**, 185-203.
8. Guidoboni, E., Comastri, A., and Trania, G., (1994) Catalogue of Ancient Earthquakes in the Mediterranean Area up to 10th Century, Instituto Nazionale di Geofisica, Rome.
9. Ambraseys, N.N., and Finkel, C.F., (1991) Long-term seismicity of İstanbul and of the Marmara Sea region, *Terra Nova.* **3**, 527-539.
10. Ozansoy, E. (2001) Bizans kaynaklarina gore 1200-1453 İstanbul depremleri, In: Tarih Boyunca Anadoluda Dogal Afetler ve Deprem semineri (ed: I. Sahin), 22-23 May 2000, Globus Dunya Basimevi, İstanbul, 1-29.
11. Soysal, H., Sipahioglu, S., Kolcak, D., and Altinok, Y. (1981) Turkiye ve Cevresinin Tarihsel Deprem Katalogu (MO 2100-MS 1900) Tubitak, Project Number TBAG 341, İstanbul.
12. Soysal, H. (1985). Tsunami and tsunamis that effect Turkish coasts, *Bulletin of Institute of Marine Sciences and Geography* **2**, 59-66.
13. Soloviev, S.L., Solovieva, O.N., Go, C.N., Kim, K.S. and Shchetnikov, N.A. (2000). Tsunamis in the Mediterranean Sea 2000 B.C. – 2000 A.D., Kluwer Academic Publishers, Netherlands, 237pp.
14. Ambraseys, N.N. (1962) Data for the investigation of the seismic sea-waves in the Eastern Mediterranean, *Bull. Seism. Soc. Am.* **52**, 895-913.
15. Papadopoulos, G.A., and Chalkis, B.J. (1984) Tsunamis observed in Greece and the surrounding area from antiquity to the present times, *Marine Geology* **56**, 309-317.
16. Sadi, (1912) Marmara Havzasinin 26-27 Temmuz hareket-i arzi, 15 Eylul 1328, İstanbul, Resimli Kitap Matbaasi, 45p (Ottoman Language).
17. Macovei, R., (1912) Sur le Tremblement de Terre de la Mer de Marmara le 9 aout 1912, *Bull. Section Sc. Academic Roumaine* **1**, 1-9.

42

18. Mihailovic, J., (1927) Memoire-Sur les Grands Tremblements de Terre de la Mer de Marmara, Belgrade, 215-222.
19. Mihailovic, J., (1933) La seismicite de la Thrace, de la Mer de Marmara et de l'Asie Mineure, Monograph. Et Travaux Sci. Inst. Seism. No. 2B, Belgrade.
20. Tabban, A., and Ates, R. (1976) 9 Agustos 1912 Şarköy-Mürefte depremi calismalari on raporu, T.C. Imar ve Iskan Bakanligi, Deprem Arastirma Enstitusu Baskanligi, Ankara, 20 p.
21. Pinar, N., and Lahn, E. (1952) Turkiye depremleri izahli katalogu. Bayindirlik ve Iskan Bakanligi, Imar Reisligi yayinlari, Seri 6, Sayi 36.
22. Ambraseys, N.N., and Jackson, J.A., (2000) Seismicity of Sea of Marmara (Turkey) since 1500, *Geophys. J. Int.* **141**, F1-F6.
23. Rockwell T., Barka, A., Thorup, K., Dawson, T., and Akyuz, S., (1997) Paleoseismology of the Gaziköy-Saros segment of the North Anatolian fault, Northwestern Turkey: Implications of regional seismic hazard and models of earthquake recurrence, International Symposium on Recent Development on Active Fault Studies, ITU, İstanbul, pp. 34-48.
24. Eginitis, D. (1894) 1310 Zelzelesi Hakkinda Rapor, Cev: Bogos, İstanbul, 21 Agustos, 1310, Basbakanlik Arsivi Genel Mudurlugu, Yildiz Esas Evraki, Kisim 14, Evrak C. Zarf 126, Karton 11, s 1-29 (manuscript).
25. Batur, A., (1999) Muharrem ayinda bir Sali gunu, Yapi ve Kredi Yayinlari, Deprem Ozel Sayisi, *Cogito* **20**, 42-59, İstanbul.
26. Cesmi-zade, M.R. (1766-1768). Cesmizade Tarihi (1766). Ed. B.S. Kutukoglu, İstanbul Fetih Cemiyeti Yayinlari 1993.
27. Ambreseys, N.N., and Finkel, C.F., (1995) Seismicity of Turkey and Adjacent Areas, A Historical Review, 1500-1800. Eren Yayincilik ve Kitapcilik Ltd., 240p.
28. Gundogdu, O. (1986) Turkiye Depremlerinin Kaynak Parametreleri ve Aralarindaki Iliskiler, Ph.D. Thesis, Engineering Faculty of İstanbul University, İstanbul, 120p.
29. Altunel, E., Barka, A.A., and Akyuz, S. (2000) Slip distribution along the 1912 Murefte-Sarkoy Earthquake, North Anatolian Fault, Western Marmara, paper in the book, The 1999 Izmit and Duzce Earthquakes: Preliminary Results, Editors: Barka A., Kozaci O., Akyuz S., Altunel E., İstanbul Technical University Turkey, 341-349.

SPATIAL AND TEMPORAL PERIODICITY IN THE PACIFIC TSUNAMI OCCURRENCE

E.V. SASSOROVA & B.W. LEVIN
Shirshov Institute of Oceanology, Russian Academy of Sciences
Nakhimovsky prospekt 36, Moscow 117851 Russia

Abstract

Recent studies of the latitude distribution of earthquake numbers and energy have shown that the region seismic activity depends on the geographic latitude of the area and it varies with time. The time-dependent variations of seismicity suggest a 6-year-period of activity. Such time interval is well known as a period of Chandler pole motion. To analyze the Pacific tsunami distribution, we use the Historical Tsunami Database compiled by V.Gusiakov looking more closely at the local seismic regions of the Pacific area. By choosing 139 events with magnitude M>6.0 out of the Pacific zone tsunami list for the last 50 years, we subdivided these events into two groups: northern part and southern one. The analysis showed that there is temporal regularity in the tsunami occurrence: the event series occurred in turn for Northern Hemisphere and then for Southern one and so on. The typical periods of the change over from one to another hemisphere were 6 and 18 years approximately. The assessment of probability that a tsunami may occur at a given region over the next 6 years is very important for tsunami long-term forecast. The advance of the study gives new perspective for the tsunami warning system development and for implementation of improved mitigation measures.

1. Introduction

The general problem of predictability and unpredictability of the earthquake and tsunami events has been discussed before [1, 2] and remains an active area of study. Recent research has shown that the Earth's seismicity and probability of the earthquake occurrence depends on the astronomical reasons: the geographical latitude of the event [3, 4], the relative attitude of the Earth and the Moon [5], the position of the Earth on ecliptic [6, 7]. Hence the connection between the Chandler pole motion as a global process and some geophysical phenomena is investigated. It should be mentioned that such effects as the crust deformation induced by polar motion [8], the earthquake energy accumulation [9,10], the shift of the Earth's center of mass [11-12], the temporal changes in the seismic activity and the atmospheric processes, all correlate with the Chandler wobble periods [13].

The main goal of the study is the searching of the patterns for the Pacific tsunami event occurrence, the identification of special periods typical for the temporal distribution, and the assessment of the possibility for improvements in the understanding of tsunami generation process that may be useful in the solution of the tsunami prediction problem in future.

A. C. Yalçıner, E. Pelinovsky, E. Okal, C. E. Synolakis (eds.),
Submarine Landslides and Tsunamis 43-50.
©2003 Kluwer Academic Publishers.

2. Formulation of the problem

At present, sequences of tsunami events in defined areas of the Pacific have not yet been identified. For analyzing of a spatio-temporal distribution of the Pacific tsunami events, we used the tabulated seismic measurement data in time interval 1950-2000 given in the History Tsunami Data Base (HTDB) compiled by V.K.Gusiakov [14]. First and foremost we should make a separation of the Pacific whole area into several regions (lineaments), each with similar tectonic characteristics and response to the stress and deformation of the Earth's crust that lead to an earthquakes. Hence for determining the boundary of each lineament, we took into account the geological, tectonic, geophysical characteristics of plates, blocks, and other parts of the Pacific region that should be included into the certain lineament. Moreover, some structural peculiarities of the planet have been considered.

Basing on expert opinions, we divided the whole of the Pacific area into four lineaments (Fig.1). The first was the north-western lineament (I) including the coast of Alaska, Aleutian arc, Kamchatka, Kuril Islands, Japan and Taiwan with the Pacific area up to 150° W. The north-eastern lineament (II) contained the coast of the Northern America until its intersection with the latitude of 8°N and the adjacent part of the Pacific. The attitude of the south-western lineament (III) defined its content: the Philippines Islands, Indonesia, Australia and New Zealand with the adjacent ocean area until to the longitude 150° W. The south-eastern lineament (IV) consists of the coast of the Southern America with the corresponding part of the ocean.

Since the large-scale lineaments of the Pacific seismic zone have been defined, the main goal of the study has been reduced to the finding of a temporal sequence of the tsunami generation events or the groups of events for each lineament, determination of the statistical validity of results, and calculation of the typical periods for this tsunami temporal distribution.

3. Analysis of the observation data

To analyze the process of tsunami occurrence on the Pacific we selected the tectonic events that were registered in the time interval 1950-2000 according to the HTDB [7] with magnitude M>6.0 and the tsunami wave intensity (I) more or equal to -1. The total number of events that was equal to 139 was distributed amongst the lineaments in the following manner: I - 61, II - 15, III - 44, IV - 19. A fragment of the chart with an attitude of the lineaments and the tsunami source displacement is presented in Figure 1.

It was shown that the group of the events as well as a single event occurs either at one lineament or at other by turn almost without intersection. The cases of the tsunami occurrence at the different lineaments simultaneously during one year run close to 10%.

A study of the temporal behavior of the tsunami event sequence showed that the seismic activity of the neighboring lineaments is turned on one after another. In all cases considered, the general rule was inferred where a tsunami event group in the Northern

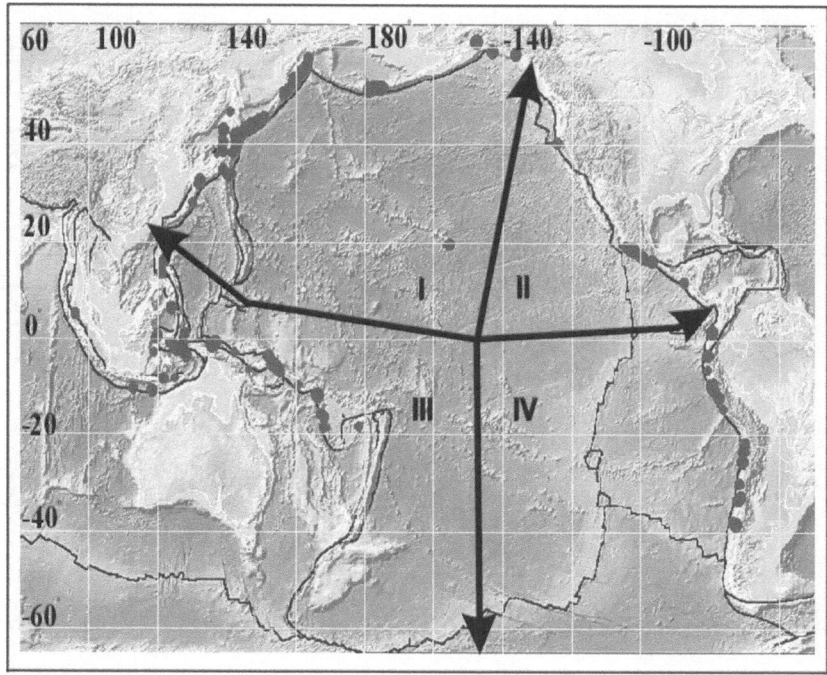

Figure 1. Four lineaments in the Pacific area (dark gray circles - the earthquake epicenters).

Hemisphere changed to the occurrence of an event series into Southern one and vice versa. The temporal behavior of tsunami events in the Northern and the Southern Hemisphere is presented in Figure 2. As noted previously, one can see a clearly defined periodicity.

On the Figure 2 horizontal axis is time axis and it is situated between two rectangles (upper for the Northern Hemisphere and lower for the Southern Hemisphere). Tsunami is presented as switching function for two Hemispheres. Every event corresponds to one vertical line started from the time axis. If tsunami occurs in Northern Hemisphere so correspondent vertical line directed upwards and it directed downwards for the tsunami which occurs in the Southern Hemisphere.

Thus the time sequence of the tsunami events is now described as time series with the significance +1 and −1. The nonparametric run test was used and the distribution-free statistics was calculated, showing that with the probability 99.8% sequence shown is not a purely random and it includes nonrandom component.

On the figure 2 the whole catalogue is subdivided into tree parts (time intervals: 1950 - 1967, 1968 - 1985, 1986 - 2000). Here, it is shown also the envelope for the observed data. The envelope was computed for switching rectangle function with the smoothing procedure.

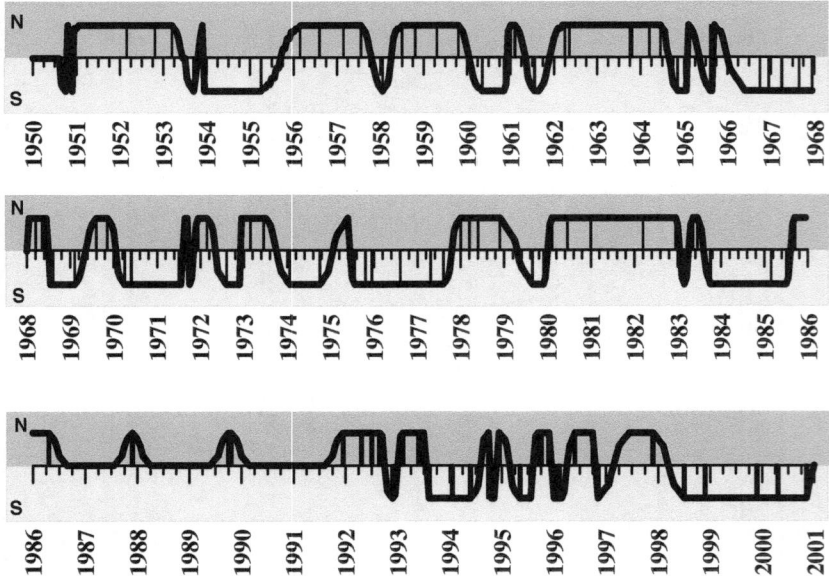

Figure 2. The temporal behavior of tsunami event generation in the Northern and the Southern Hemisphere

The gain-frequency spectrum for the envelope is presented in Figure 3 and it reflected temporal characteristics of the tsunami occurrence process. The main maximum of the spectrum corresponds to the period of 6 years and one of the maximums is close to period of 412-437 days. Both of the mentioned periods are well known as periods of the Chandler pole motion [15]. The spectrums computed for the envelope as a rectangle switched function and for the envelope line computed with the smoothing procedure show inessential difference for the frequencies from 0.01 up to. 2.5 cycles per year.

Bold dotted line indicates the upper confidence boundary U (1- α) of the level (1- α), equal to 0.95 (or 95%). One can see that the spectral peaks for the frequencies more than 0.5 cycle/year (period equal to 2 years) are below U (1- α), while the peaks for the frequencies less than 0.5 cycle/year are clearly in excess of U (1- α).

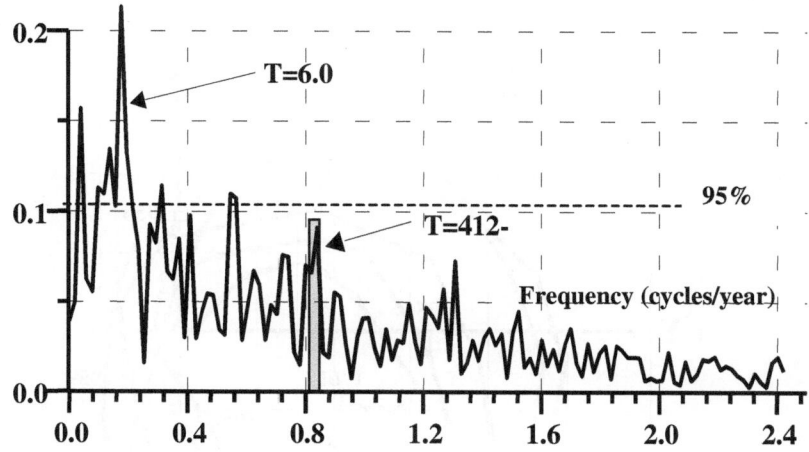

Figure 3. The gain-frequency spectrum of the Pacific tsunami occurrence (for the events 1950 - 2000 y).

According to the Avsjuk's hypothesis [11, 16] the mass center of the Earth shifts because of Earth's inner core displacement in the ecliptic plane. The change of the Earth's mass center position is caused by the effect of the Sun and the Moon and it leads to the movement of the rotation axis of the Earth. The schematic model of this process is presented on the figure 4.

In the astronomical observations this process is treated as the Chandler pole wobble. The typical samples of pole trajectories for the time interval 1955 - 1961 are given in Figure 5. The temporal variations of pole trajectory radius (R - modulus of radius vector) from 1898 to 2000 are presented in Figure 6.

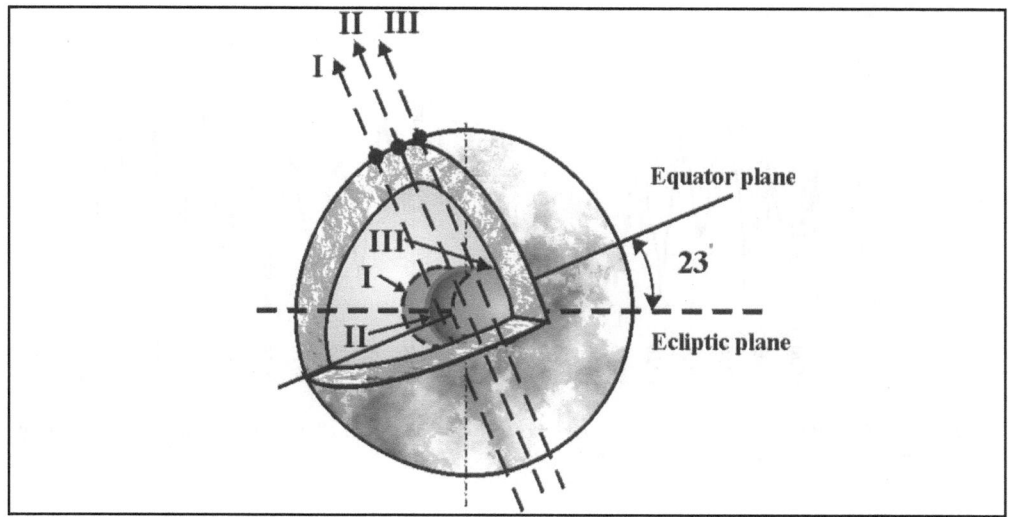

Figure 4. The inner core mover and the shift of the Earth rotation axis and the shift of Earth's pole.

48

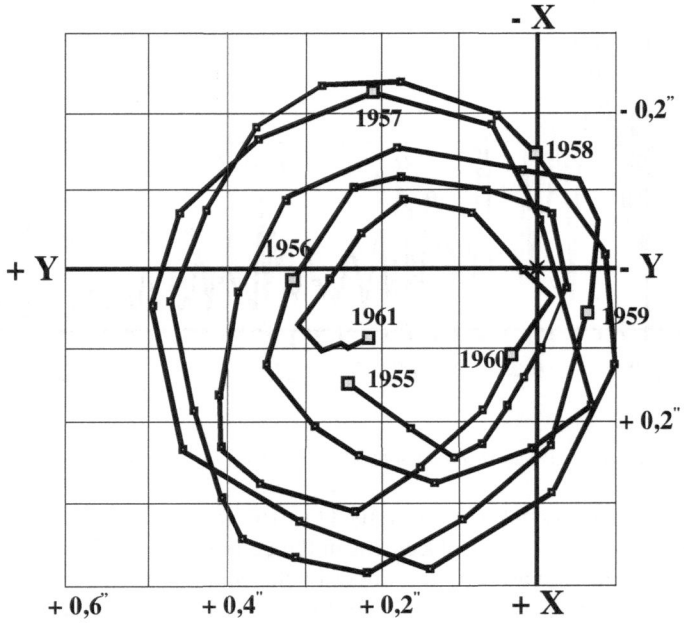

Figure 5. The Chandler pole motion from 1955 up to 1961 years.

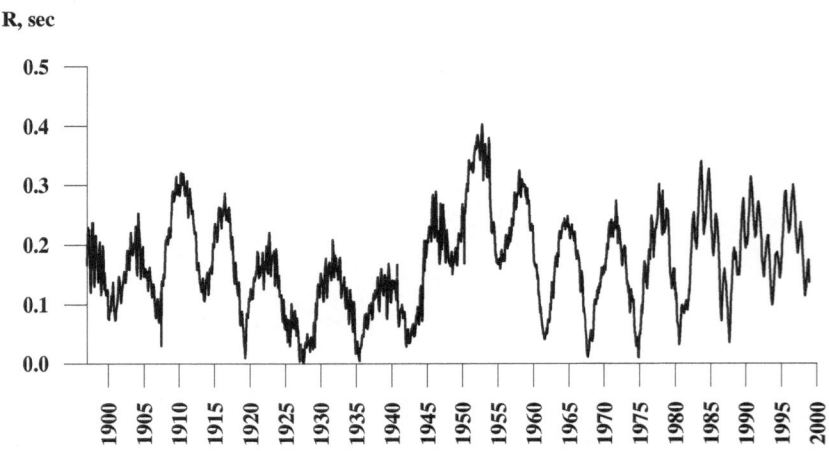

Figure 6. The pole trajectory radius (R - modulus of radius vector) from 1898 to 2000 years. The variations of the radius vector are presented as angular value.

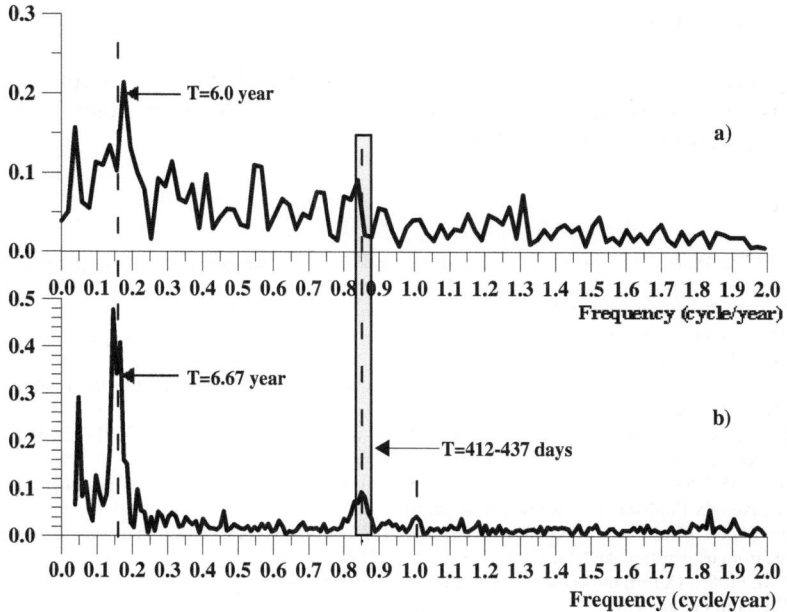

*Figure 7. **a)** - Spectrum of the Pacific tsunami occurrence, and **b)** - the spectrum of Chandler pole motion - R(x,y).*

The inner core motion and its regular intersection of the plate of equator can lead to an alternation of a supplementary stress into different hemispheres of the Earth. The envelope on the Figure 2 may reflect an accumulation and a switching of a supplementary stress between the Northern and Southern Hemispheres. These processes are very complicated.

Two spectrums (tsunami occurrence in two Pacific Hemispheres and R- variations of the pole motion) are presented at the same time on the figures 7a and 7b correspondingly. By comparing the two spectrums, we deserve some coincidence between main periods of the Pacific tsunami variation and the Chandler pole motion.

4. Discussion and conclusions

By grouping into certain lineaments have found an unknown effect of alternating tsunami activity between the Northern and Southern Hemispheres of the Earth. This alternative process is periodic in time and the typical periods are comparable with the well-known periods of the Chandler pole motion (6-7 years and 412-437 days).

The coincidence between the typical periods of tsunami activity in the Pacific and the Chandler pole motion periods obtained in our investigation supports our hypothesis that certain earthquake and tsunamis can be triggered by perturbation of the Earth's rotation axis. We believe that the similar analysis may help to understand the physics of earthquakes and suggest new methods for improvements of tsunami warning systems.

Acknowledgements

We are grateful to Y.N. Avsjuk, V.K. Gusiakov, and V.P. Pavlov for useful discussions and comments. The work was partially supported by the Russian Foundation for Basic Research, Grants 99-05-64218 and 00-15-98583.

References

1. Geller, R.J., Jackson, D.D., Kagan, Y.Y. and Mulargia, F. (1997) Earthquakes cannot be predicted. *Science* **275**, 1616-1617.
2. Wyss, M. (1997) Cannot earthquake be predicted?, *Science* **278**, 487-488.
3. Levin, B.W. (2001) Can the Earth's inner core be a conductor of the seismic activity? *Zemlja i Vselennaja* **3,** (in print); (in Russian).
4. Levin, B.W. and Chirkov, Ey.B. (2000) The latitude distribution of seismicity and the rotation of the Earth, *Volc. Seis.*, **21**, 731-739.
5. Volodichev, N.N., Podorolskii, A.N., Levin, B.W., and Podorolskii, V.A.. (2001) Correlation of major earthquake sequences with new moon and full moon phases, *Volc. Seis.*, **1**, 60-67.
6. Fedorov, V.M. (2000) *Gravitation factors and astronomical chronology of geospherical processes*, Moscow State University Publishers, Moscow (in Russian)
7. Gor`kavyi, N.N., Minin, V.A., Tajdakova, T.A., and Fridman, A.M. (1989) Are there any astronomical reasons for the strongest earthquake? *Astronomical circular*, **1540**, 35-36.
8. Wahr, J.M. (1985) Deformation induced by polar motion, *J. Geophys. Res.*, **90**, B11, 9363-9368.
9. Chao, B.F. and Gross, R.S. (1995) Changes in the Earth`s rotational energy induced by earthquakes, *Geophys. J. Int.* **122**, 776-783.
10. Chao, B.F., Gross, R.S. and Dong, D.N. (1995) Changes in global gravitational energy induced by earthquakes, *Geophys. J. Int.* **122**, 784-789. 6.
11. Avsjuk, Y.N. (1973) On the Earth's inner core motion, *Doklady Academii Nauk SSSR* **212**, 5, 1103-1105. (in Russian).
12. Avsjuk, Y.N. and Levin, B.W. (1999) To M.V. Lomonosov question on the displacement of the Earth's mass center, *Herald of RFBR (Russian Foundation for Basic Research)* **2(16),** 4-11. (in Russian)
13. Levin, B.W., Savelieva, N.I., Semiletov, I.P., and Chirkov, Ey.B. (2000) The Chandler pole wobble and its role in geophysical and meteorological processes, in V.A. Akulichev and I.P.Semiletov (eds.), *Hydrometeorological and biogeochemical research in the Arctic*, Pacific Oceanological Institute of RAS, Vladivostok, pp.158-165 (in Russian).
14. Gusiakov, V.K. (2000) Historical Tsunami Data Base (http//tsun.sscc.ru/htdbpac).
15. International Earth Rotation Service (IERS) Annual Report of 1997 (1998), Central Bureau of IERS - Observatoire de Paris, France.
16. Avsjuk, Y.N. (1996) *Tidal forces and natural processes,* Russian Academic Sc. Press., Moscow. (in Russian).

Part 3

Submarine Landslides and Tsunami Generation

SUBMARINE LANDSLIDE GENERATED WAVES MODELED USING DEPTH-INTEGRATED EQUATIONS

PATRICK LYNETT, PHILIP L.-F. LIU
School of Civil and Environmental Engineering
Cornell University, Ithaca, NY 14853, USA

Abstract

A mathematical model is derived to describe the generation and propagation of water waves by a submarine landslide. The model consists of a depth-integrated continuity equation and a momentum equation, in which the ground movement is a forcing function. These equations include full nonlinear, but weakly dispersive effects. The model is also capable of describing wave propagation from relatively deep water to shallow water. A numerical algorithm is developed for the general fully nonlinear model. As a case study, tsunamis generated by a prehistoric massive submarine slump off the northern coast of Puerto Rico are modeled. The evolution of the created waves and the large runup due to them is discussed.

1. Introduction

In this paper, we shall present a new model describing the generation and propagation of tsunamis by a submarine landslide. In this general model only the assumption of weak frequency dispersion is employed, i.e., the ratio of water depth to wavelength is small or $O(\mu^2) << 1$. However, by choosing a proper representative velocity in the governing equations the applicability of these model equations may possibly be extended to reasonably deep water (or a short wave). Moreover, the full nonlinear effect is included in the model, i.e., the ratio of wave amplitude to water depth is of order one or $\varepsilon = O(1)$. Therefore, this new model is more general than that developed by [1], in which the Boussinesq approximation, i.e., $O(\mu^2) = O(\varepsilon) << 1$ was used. The model is applicable for both the impulsive slide movement and creeping slide movement. In the latter case the time duration for the slide is much longer than the characteristic wave period.

This paper is organized in the following manner. Governing equations for flow motions generated by a ground movement are summarized in the next section. A numerical algorithm is then presented to solve the general mathematical model. As a case study, a large prehistoric slide off the northern coast of Puerto Rico, whose attributes have been well documented [2], is examined.

A. C. Yalçıner, E. Pelinovsky, E. Okal, C. E. Synolakis (eds.),
Submarine Landslides and Tsunamis 51-58.

2. Approximate Two-Dimensional Governing Equations

The three-dimensional boundary-value problem will be approximated and projected onto a two-dimensional horizontal plane. In this section, the nonlinearity is assumed to be of $O(1)$. However, the frequency dispersion is assumed to be weak, i.e.

$$O(\mu^2) << 1 \tag{1}$$

Using μ^2 as the small parameter, a perturbation analysis is performed on the primitive governing equations. The resulting approximate continuity equation is

$$
\frac{1}{\varepsilon}h_t + \zeta_t + \nabla \cdot (Hu_\alpha)
$$

$$
-\mu^2 \nabla \cdot \left\{ H \left[\left(\frac{1}{6}(\varepsilon^2 \zeta^2 - \varepsilon \zeta h + h^2) - \frac{1}{2}z_\alpha^2 \right) \nabla(\nabla \cdot u_\alpha) \right.\right.
$$

$$
\left.\left. + \left(\frac{1}{2}(\varepsilon \zeta - h) - z_\alpha \right) \nabla \left(\nabla \cdot (hu_\alpha) + \frac{h_t}{\varepsilon} \right) \right] \right\} = O(\mu^4) \tag{2}
$$

in which $H = h + \varepsilon \zeta$. Equation (2) is one of three governing equations for ζ and u_α. The other two equations come from the horizontal momentum equation, and are given in vector form as

$$
u_{\alpha t} + \varepsilon u_\alpha \cdot \nabla u_\alpha + \nabla \zeta + \mu^2 \left\{ \frac{1}{2}z_\alpha^2 \nabla(\nabla \cdot u_{\alpha t}) + z_\alpha \nabla \left[\nabla \cdot (hu_\alpha)_t + \frac{h_{tt}}{\varepsilon} \right] \right\}
$$

$$
+\mu^2 z_{\alpha t} \left\{ z_\alpha \nabla(\nabla \cdot u_\alpha) + \nabla \left[\nabla \cdot (hu_\alpha) + \frac{h_t}{\varepsilon} \right] \right\}
$$

$$
+\varepsilon \mu^2 \left\{ \left[\nabla \cdot (hu_\alpha) + \frac{h_t}{\varepsilon} \right] \nabla \left[\nabla \cdot (hu_\alpha) + \frac{h_t}{\varepsilon} \right] \right.
$$

$$
-\nabla \left[\zeta \left(\nabla \cdot (hu_\alpha)_t + \frac{h_{tt}}{\varepsilon} \right) \right] + (u_\alpha \cdot \nabla z_\alpha) \nabla \left[\nabla \cdot (hu_\alpha) + \frac{h_t}{\varepsilon} \right]
$$

$$
+z_\alpha \nabla \left[u_\alpha \cdot \nabla \left(\nabla \cdot (hu_\alpha) + \frac{h_t}{\varepsilon} \right) \right]
$$

$$
+z_\alpha (u_\alpha \cdot \nabla z_\alpha) \nabla(\nabla \cdot u_\alpha) + \frac{z_\alpha^2}{2} \nabla [u_\alpha \cdot \nabla(\nabla \cdot u_\alpha)] \right\}
$$

$$
+\varepsilon^2 \mu^2 \nabla \left\{ -\frac{\zeta^2}{2} \nabla \cdot u_{\alpha t} - \zeta u_\alpha \cdot \nabla \left[\nabla \cdot (hu_\alpha) + \frac{h_t}{\varepsilon} \right] \right.
$$

$$+\zeta\left[\nabla\cdot(hu_\alpha)+\frac{h_t}{\varepsilon}\right]\nabla\cdot u_\alpha\right\}$$

$$+\varepsilon^3\mu^2\nabla\left\{\frac{\zeta^2}{2}\left[(\nabla\cdot u_\alpha)^2-u_\alpha\cdot\nabla(\nabla\cdot u_\alpha)\right]\right\}=O(\mu^4)$$

(3)

Equations (2) and (3) are the coupled governing equations, written in terms of u_α and ζ, for fully nonlinear, weakly dispersive waves generated by a seafloor movement. We reiterate here that u_α is evaluated at $z=z_\alpha(x,y,t)$, which is a function of time. The choice of z_α is made based on the linear dispersion characteristics of the governing equations [3]. Assuming a fixed seafloor, in order to extend the applicability of the governing equations to relatively deep water (or a short wave), z_α is recommended to be evaluated as $z_\alpha=-0.531h$. In the following analysis, the same relationship is employed.

3. Numerical Model

In this section, a finite difference algorithm is described for the general model equations. This model has the robustness of enabling slide-generated surface waves, although initially linear or weakly nonlinear in nature, to propagate into shallow water, where fully nonlinear effects may become important. An alternative is to use different sets of governing equations, switching from a linear set to a nonlinear set as nonlinear effects become important. Although this approach may lead to significant computational benefits, one must empirically determine the switch point, which could differ for different physical setups. This numerical tuning is time consuming in itself, and could cancel the computational benefit of running the equation-switching model.

The structure of the current numerical model is very similar to [5] and [6], with the added effects due to changing water depth in time. A high-order predictor-corrector scheme is utilized, employing a third order in time explicit Adams-Bashforth predictor step, and a fourth order in time Adams-Moulton implicit corrector step [7]. The implicit corrector step must be iterated until a convergence criterion is satisfied. All spatial derivatives are differenced to fourth order accuracy, yielding a model that is numerically accurate to $(\Delta x)^4,(\Delta y)^4$ in space and $(\Delta t)^4$ in time. The governing equations are dimensionalized for the numerical model, and all variables described in this and following sections will be in the dimensional form. Runup and rundown of the waves generated by the submarine disturbance will also be examined. The moving boundary scheme employed here is the technique developed by [8]. To simulate the effects of wave breaking, the eddy viscosity model [9], [10] is used here. Readers are directed to [10] for a thorough description and validation of the breaking model, and the coefficients and thresholds given therein are used for all the simulations presented in this paper.

4. Modeling a Submarine Slump

As a case study to apply this model, a prehistoric, massive submarine slump off the northern coast of Puerto Rico is investigated. A measured depth profile along the centerline of the failure region is shown in Figure 1. According to [2], the slump was approximately 57 km wide, occurring on a steep slope (roughly 1/10) with a length of about 40 km; the top of the failure slope is at a depth of 3000 meters, the bottom at 7000 meters. The catastrophic failure is estimated to involve over 900 km^3 of soil. With this information and the evidence of a circular slip, the maximum decrease in water depth along the slope is estimated at 700 m. Assuming solid body motion of the mass and using the estimated soil density given by Grindlay, the duration of the movement is calculated to be on the order of 10 minutes.

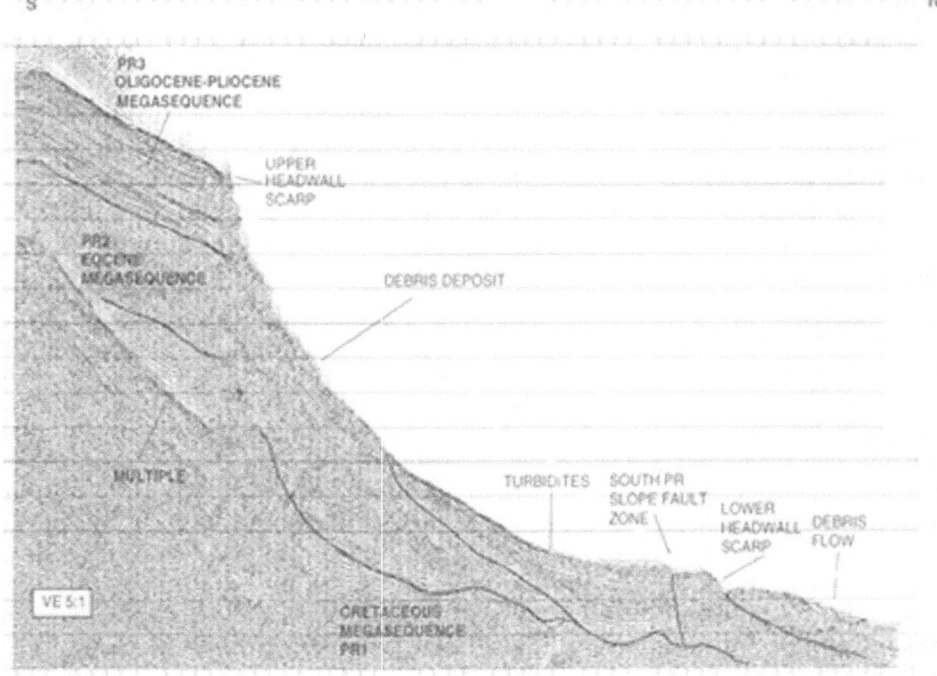

Figure 1. Seafloor profile along the centerline of the failure region (taken from Grindlay).

To implement a landslide in the model, the evolution of the bottom movement must be completely known beforehand. There are different slide mechanisms, which of course will determine the free surface response. In this analysis, a rotational slip is examined. Figure 2 shows the numerical representation of this type of submarine slide. Along the steep slope where the slide is to occur, a circular slip line develops, usually due to the shaking of an earthquake. The soil above the slip rotates downward, and at the bottom of the slope, translates away from the steep slope. This type of slide is most likely the type that occurred off the coast of Puerto Rico, where there is a large circular cutout of a steep slope (Grindlay 1998).

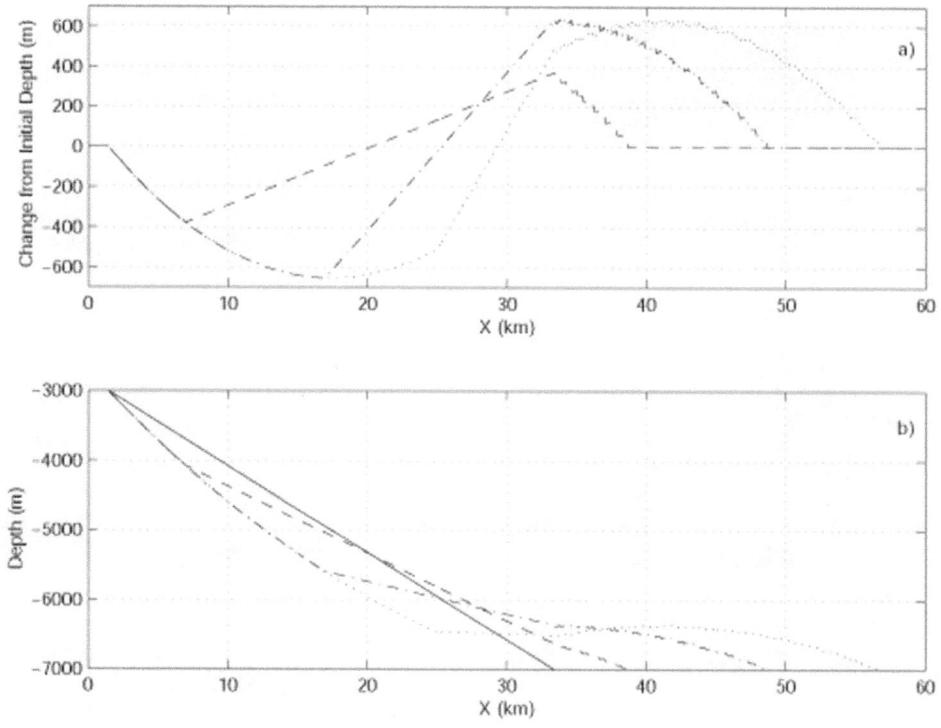

Figure 2. Numerical representation of a large rotational submarine slide. The top subplot shows the cumulative change in water depth at successive times; the lower subplot shows the actual bottom profile, where the solid (—) line is the initial water depth.

In order to numerically model a two horizontal dimension slide, the centerline of the slide is identical to Figure 2, and the surrounding region is described with a Gaussian distribution of the centerline profile. When modeling a two-dimensional slide, the distribution is determined such that the numerical soil volume of the slide is equal to the estimated actual volume of 900 km^3.

A number of snapshots of the free surface from the numerical results are shown in Figure 3. After 3 minutes, a large depression wave has been created along the top of the failure slope, measuring about 35 m in wave amplitude. Also, an elevation wave of height 18 m has formed above the deep-water region, where the depth is decreasing. Roughly 8 minutes after the initiation of the slide, the leading depression wave has reached the north coast of Puerto Rico, where it has shoaled to a depression of 45 m. This wave is actually an N-wave, with a trailing elevation wave roughly 8 m. After the depression wave reflects off the island, the trailing positive elevation wave generates extremely large runup heights along the coast. The greatest free surface elevations, nearly 70 m, are reached about 15 minutes after the submarine slide motion initiates. The positive elevation wave continues to flood the coast more than 25 minutes after the slide start. In fact, at this time, the tsunami is just beginning to impact the populous eastern half of the north coast of the island.

56

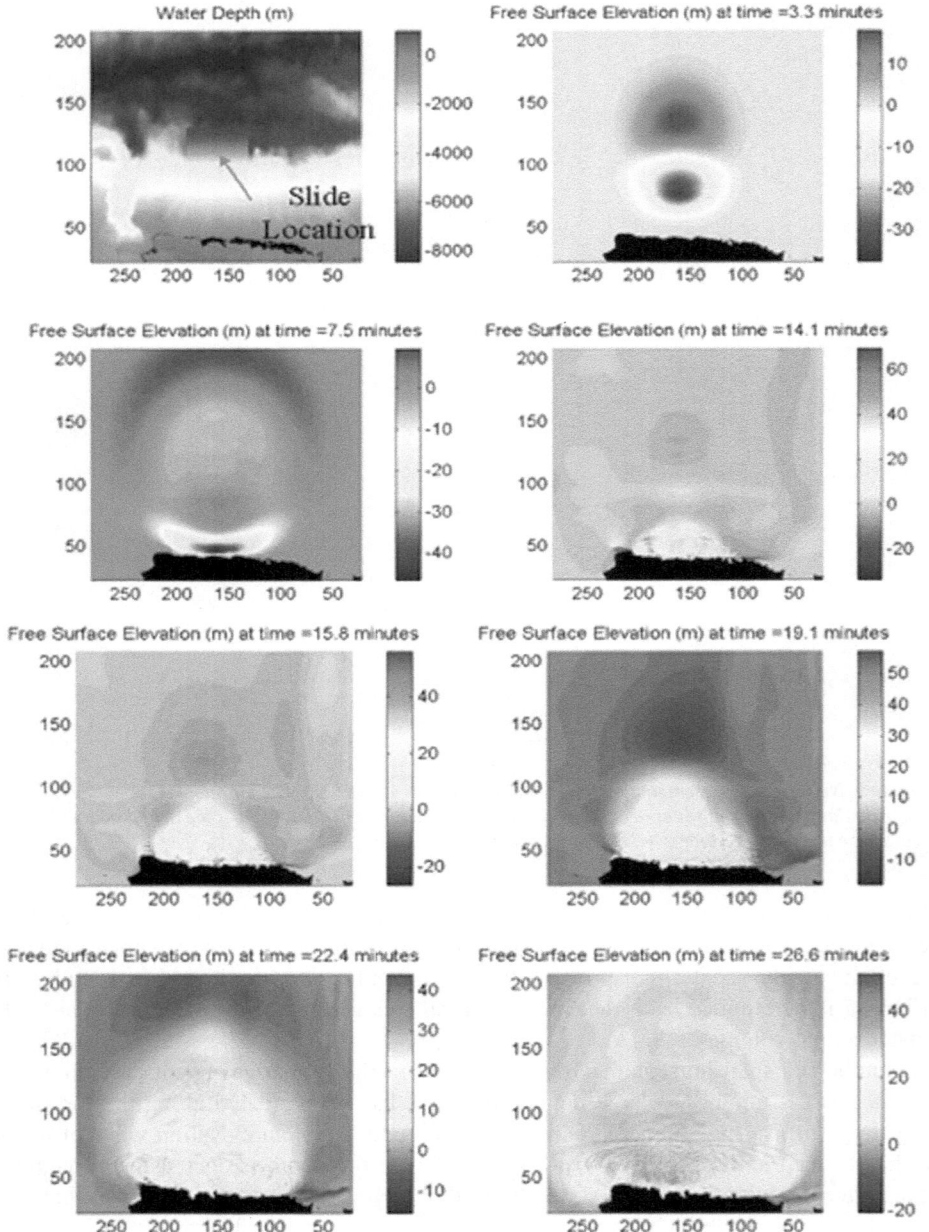

Figure 3. Plan-view snapshots of the waves generated by a submarine slump. The subplot in the upper left shows the water depth profile. The island of Puerto Rico is located on the bottom of each subplot.

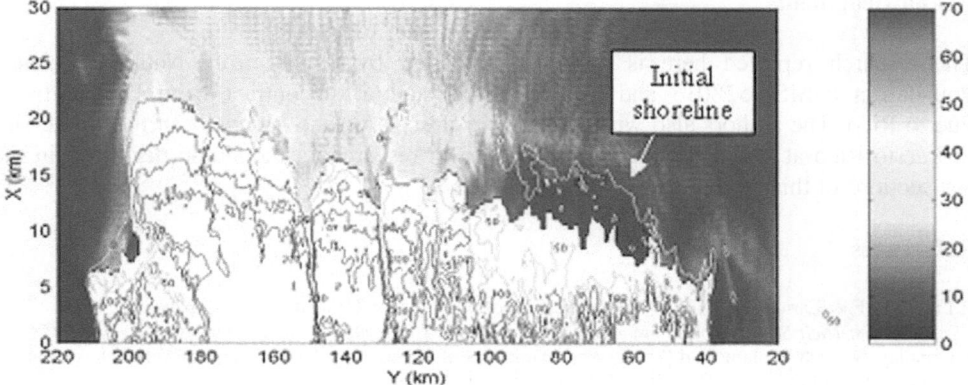

Figure 4. The maximum free surface elevation recorded near the coast of Puerto Rico.

Figure 4 shows a closeup of the maximum recorded free surface elevation very near the coast of Puerto Rico. The initial shoreline is noted on the plot, and the land that remains dry for the entire duration of the tsunami event is shown by the solid white coloring. This plot shows maximum elevations near 70 m. The largest free surface elevations are localized near the western half of the island (from around $Y=110$ km to $Y=190$ km). The effects of the tsunami are focused on the northern coast of Puerto Rico. The maximum free surface elevations on the western side of the island (near $Y=200$ km) are relatively small, only reaching single digits values. The finger-like intrusions of runup (at $Y=150$ km and $Y=130$ km) are actually the tsunami traveling up river channels. Inundation distance in this vicinity is on the order of 5 km. Also note that along the eastern half of the island, the maximum elevation is not that great (5-10 m), but the inundation distance is also large. This is due to the fact that this area is a gradually sloping coastal plain, with land elevations only a few meters above sea level.

5. Conclusions

A model for the creation of fully nonlinear long waves by seafloor movement, and their propagation away from the source region, is presented. The general fully nonlinear model can be truncated, so as to only include weakly nonlinear effects, or model a non-dispersive wave system. Rarely will fully nonlinear effects be important above the landslide region, but the model has the advantage of allowing the slide-generated waves to become fully nonlinear in nature, without requiring a transition between governing equations. A high-order finite difference model is developed to numerically simulate wave creation by seafloor movement. The model is applied along the north coast of Puerto Rico, recreating a large, ancient submarine slump.

58

Acknowledgment

The research reported here is partially supported by Grants from National Science Foundation (CMS-9528013 and CTS-9808542) and a subcontract from University of Puerto Rico. The authors also wish to thank Professor Aurelio Mercado, of the University of Puerto Rico at Mayaguez, for his assistance in researching the slump discussed in the last sections of this work.

References

1. Liu, P. L.-F. & Earickson, J. 1983 "A Numerical Model for Tsunami Generation and Propagation", in *Tsunamis: Their Science and Engineering* (eds. J. Iida and T. Iwasaki), Terra Science Pub. Co., 227-240.
2. Grindlay, N. 1998 "Volume and Density Approximations of Material Involved in a Debris Avalanche on the South Slope of the Puerto Rico Tench", Puerto Rico Civil Defense Report.
3. Nwogu, O., 1993 "Alternative Form of Boussinesq Equations for Nearshore Wave Propagation", J. Wtrwy, Port, Coast and Ocean Engrg., ASCE, 119(6), 618-638.
4. Chen, Y., & Liu, P. L.-F., 1995. "Modified Boussinesq Equations and Associated Parabolic Model for Water Wave Propagation." J. Fluid Mech., , 351–381.
5. Wei, G. & Kirby, J. T. 1995. "A Time-Dependent Numerical Code for Extended Boussinesq Equations." Journal of Waterway, Port, Coastal and Ocean Engng., , 251-261.
6. Wei, G., Kirby, J.T., Grilli, S.T., & Subramanya, R., 1995. "A Fully Nonlinear Boussinesq Model for Surface Waves. Part 1. Highly Nonlinear Unsteady Waves," J. Fluid Mech. , 71–92.
7. Press, W.H., Flannery, B.P., & Teukolsky, S.A. 1989. "Numerical Recipes," Cambridge University Press, 569-572.
8. Lynett, P., Wu, T-W., & Liu, P. L.-F., 2002. "Modeling Wave Runup with Depth-Integrated Equations," Coast. Engng, 46(2), 89-107.
9. Zelt, J. A. 1991. "The runup of nonbreaking and breaking solitary waves." Coast. Engrg., 15, 205-246
10. Kennedy, A. B., Chen, Q., Kirby, J. T., and Dalrymple, R. A. 2000. "Boussinesq modeling of wave transformation, breaking, and runup. Part I: 1D." Journal of Waterway, Port, Coastal and Ocean Engng., 126(1), 39-47.

NEAR-FIELD AMPLITUDES OF TSUNAMI FROM SUBMARINE SLUMPS AND SLIDES

M.I. TODOROVSKA, A. HAYIR and M.D. TRIFUNAC
University of Southern California, Civil Engineering Department, KAP 216A, Los Angeles, CA 90089-2531, USA

Abstract

Tsunami generated by submarine slumps and slides are investigated in the near-field, using simple source models that consider the effects of source finiteness and directivity. Five simple two-dimensional kinematic models of submarine slumps and slides are described mathematically as combinations of spreading constant or sloping uplift functions. Tsunami waveforms for these models are computed using linearized shallow-water theory for constant water depth and transform method of solution (Lapace in time and Fourier in space). Results for tsunami waveforms and tsunami peak amplitudes are illustrated for selected model parameters, in the near-field, for a time window of the order of the source duration.

1. Introduction

The common mechanisms for triggering failure of submarine slopes are over-steepening due to rapid deposition of sediments, generation of gas created by decomposition of organic matter, storm waves, and earthquakes, which are the major cause of landslides on continental slopes. The failed material is driven by the gravity forces. If the moving sediment resembles a viscous fluid, the process is called *mass flow* (bottom of Fig. 1a). Translational or rotational movement of essentially rigid segments with many discrete slope planes within the moving mass constitutes a *slide* (bottom of Figs 1b and 1c). *Slumps* are slides in which the blocks of failed material rotate along curved slip surfaces. The end product of disintegrating slides may become *debris flow*, when the sediment is heterogeneous. *Turbidity currents* transport diluted suspension of sediment grains supported by fluid turbulence.

Submarine landslides in fjords (glacially eroded valleys fed by sediment-laden rivers that drain glaciers) can generate giant waves damaging the coastal communities (e.g. Valdeaz and Seward, [1]). Undersea landslides can occur near major sedimentary deposition. For example, Yukon River transports 60 million tons per year. Glacially fed rivers deposit 450 million tons of sediments per year into the Gulf of Alaska. Mississippi river contributes 2 to 7×10^8 tons of sediment per year, building the delta seaward by 50 to 100 m per year. This results in sediment accumulation up to 1 m per year.

A. C. Yalçıner, E. Pelinovsky, E. Okal, C. E. Synolakis (eds.),
Submarine Landslides and Tsunamis 59-68.
©2003 Kluwer Academic Publishers.

Landslides originate at all depths (less than 2,500 m), with most occurrences initiated between 800 and 1000 m water depth. Many landslides also originate at the base of a slope (2,000 to 2,200 m water depth). Likewise, landslides can terminate over the entire depth range. Measured lengths of landslides range from 0.3 to 380 km (the mode of distribution is at 2 to 4 km), and widths from 0.2 to 50 km (the mode is at 1 to 2 km). The thickness of the sedimentary section near its origin varies from 10 to 650 m. Landslide areas reach up to 2×10^4 km^2, but most have an area of about 10 km^2. Most landslides (56 percent) have occurred at slopes of 4° or less. Large landslides (more than 10^2 km^2) tend to occur on gentle slopes (3° to 4°) [2]. Most frequent of landslide types are *debris slides* (35 percent), *flow bowls* (20 percent), *slab slides* (17 percent) and *block slides* (11 percent). *Debris flow* and *carpet slides* each have the frequency of occurrence of about 8 percent.

In this paper, we investigate waveforms of tsunami in the near-field generated by submarine slumps and slides. We use five two-dimensional, kinematic source models that consider the effects of source finiteness and directivity.

1.1. SIMPLIFIED 2D MODELS OF SUBMARINE SLIDES AND SLUMPS

Figs 1a through 1d show vertical cross-sections (through $y=0$) of the mathematical models of submarine slides and slumps we consider in this study, as those evolve for time $t \leq t^*$, where t^* is the instance when the motion stops. Following [3] we consider three models, of which Models 1 and 3 have two variants, A and B, i.e. total of five models: 1.A, 1.B, 2, 3.A and 3.B (see Fig. 1). All models are characterized by sliding or slumping in one direction, without loss of generality coinciding with the x-axis, and tsunami propagating in the x-y plane. All slides are assumed to have constant width, W. The spreading can be unilateral or bi-lateral, i.e. in the positive and negative x-direction. The vertical displacement, ζ, is negative (downwards) in zones of depletion, and positive (upwards) in zones of accumulation (Fig. 1).

Models 1.A (Fig. 1a,b), 3.A and 3.B (Fig. 1d) represent mass movement triggered at $x =0$ and spreading *unilaterally* in the positive x-direction (down hill) with velocities c_L and c_R, respectively for the zones of depletion and of accumulation. Models 1.B and 2 (Fig. 1a, b and c) represent mass movement starting at any point (including the foot of the slide) and spreading *bilaterally*. For these models, the zone of accumulation spreads with velocity c_R in the positive x-direction, and the zone of depletion spreads with velocity c_L in the negative x-direction and with velocity c_C in the positive x-direction. For all examples illustrated in this paper, the balance of mass is assumed to be constant, i.e. the volume of the "accumulation" zone is equal to the volume of "depletion" zone, except for Model 2 if $c_L \neq c_R$, and for Model 3.B which does not have a depletion zone. In Models 1.A and 1.B, the zones of accumulation and depletion have uniform amplitudes, ζ_0 (accumulation) and ζ_1 (depletion), equal to the average amplitude over the area of these zones. The volumes of the uplifted and removed material are A_1W and A_2W where A_1 and A_2 are areas of the vertical cross-sections of these zones, as shown in Fig. 1. For Model 1.A, the final $L_{accum}=t^* (c_R- c_L)$ and $L_{depl}=t^* c_L$, their ratio is $L_{depl} / L_{accum}= c_L /(c_R- c_L)$, $A_1= \zeta_0 L_{accum}= \zeta_0 t^* (c_R- c_L)$ and $A_2= \zeta_0 L_{depl}= \zeta_1 t^* c_L$. Conservation of mass then implies $\zeta_1/ \zeta_0= (c_R- c_L) / c_L$. Characteristic length of this model is $L_R = c_R t^*$. For Model 1.B,

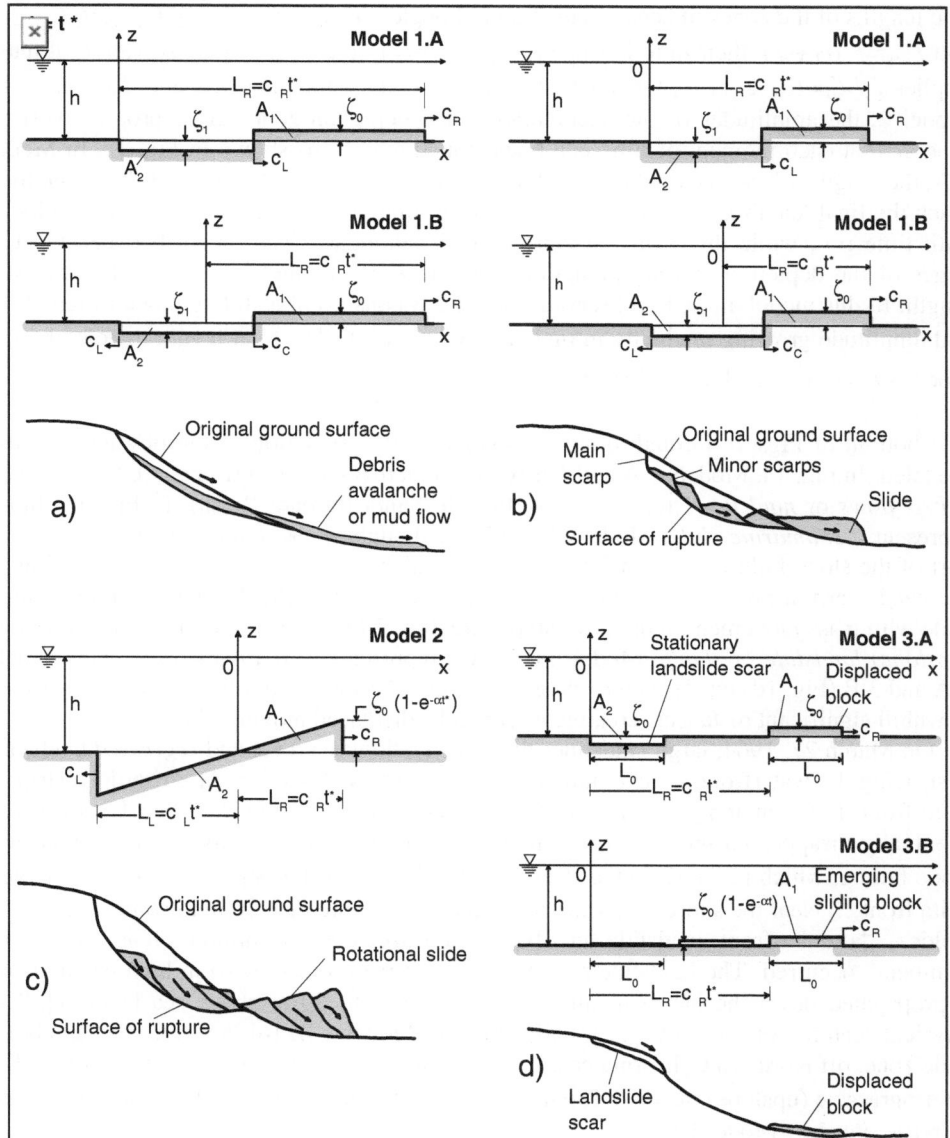

Figure 1. (**a**) *Models I .A (top) and I.B (center), and a schematic cross-section of debris avalanches, flows or mud flows (bottom) those models represent. Model I A represents sliding down hill, while Model IB can represent spreading of the source area up hill and down hill, at rates specified by c_L, c_C and c_R. It is assumed that $A_1 = A_2$. Distance $L_R = c_R$ t^* is characteristic length.* (**b**) *Models I .A (top) and I.B (center) and a schematic cross-section of a submarine slide (bottom) that those models represent.* (**c**) *Model 2 (top) and a schematic representation of a rotational slide (bottom). In general, $A_1 \neq A_2$ unless $c_R = c_L$. Distance $L_R = c_R$ t^* is characteristic length.* (**d**) *Models 3.A (top) and 3.B (center), and a schematic representation of a landslide (bottom) represented by Model 3.A. The landslide may travel significant distance downhill, thus creating a "scar" and a moving displaced block. In Model 3.A, $A_1 = A_2$. Model 3B represents emergence of "large" blocks (of increasing thickness), sliding downhill and with $0 \leq A_1 \leq \varsigma_0 L_0$. Distance $LR = c_R$ t^* is characteristic length.*

The lengths of the zones of accumulation and of depletion are respectively $L_{accum}=t^*(c_R-c_C)$ and $L_{depl}=t^*(c_L+c_C)$, their ratio is $L_{depl}/L_{accum}=(c_L+c_C)/(c_R-c_C)$, and conservation of mass implies $\zeta_1/\zeta_0=(c_R-c_C)/(c_L+c_C)$. Characteristic length of this model is also $L_R = c_R\, t^*$. In Model 2, the amplitudes of the accumulation and depletion zones grow progressively in time and, at each moment of time is a linear function of x, as shown in Fig. 1c. In Model 3.A, the lengths of the accumulation and depletion zones grow until time $t=L_0/c_R$ when they reach the final length L_0. For time $t > L_0/c_R$, the accumulation zone slides as a "rigid block" until time $t=t^*$, while the depletion zone remains stationary. The length between the left edges of the depletion and accumulation zones is $L_R=c_R\, t^*$, and is used as characteristic length. In this model, mass is conserved. Model 3.B consists of a sliding block of length L_0 and amplitude growing gradually in time as $\varsigma=\varsigma_0(1-e^{-\alpha})$. The total length traveled by the block is $L_R=c_R\, t^*$, which is used as characteristic length.

The bottom of Figs 1a through d shows schematic representation of the physical process modeled. In Fig. 1a, Models 1.A and 1.B represent *debris avalanches* (e.g. see Fig. 3 in [4], *debris flows* or *mud flows* (e.g. the mud flow in Santa Barbara Basin). In Fig. 1b, those represent a *submarine slide*. Model 1B could be thought of as approximating the eastern part of the slope failure in Santa Barbara Basin, which was enlarged by the degradation of the head scarp, a process referred to as *retrogressive failure* [5]. Model 1.A represents a slide with mass movement in the downslope direction. Model 1.B represents a *retrogressive (upslope) landslide* or *slump*. Model 2 (Fig. 1c) represents a *rotational slide*, and Models 3.A and 3.B (Fig. 1d) represent a *moving "block slide"* (this is a *landslide* which may travel downhill significant distances, creating a scar and a displaced moving block).

On March 27, 1964, large landslides originated below sea level and *regressed* landward destroying the waterfront at the communities of Seward and Valdez. At Seward, "a strip of waterfront 1,200 m long and 15 to 150 m wide started to subside, slice by slice and eventually disappeared into the bay. This was a consequence of landward regression of a slope failure, which initiated on the steep (20° to 35°, to a water depth of 50 m) submerged delta front…. Near the shore, the water depth increased more than 30 m in some places". At Valdez, "a delta-front landslide involving an estimated 75 million cubic meters of sediment" occurred. The landslide retrogressed shoreward creating 10 m high tsunami. As it propagated down the bay tsunami amplitude surged to 50 m above sea level [1]. The physical features of these events we illustrate by Model 1.B. [6] describe the Humboldt slide zone, off Northern California coast, which consists of large slump blocks that failed in a retrogressive (upslope) manner. The motion on 8×9 km^2 Humboldt slide zone also could be represented by Model 1.B.

The "block slide" represented by Model 3.A (Fig. 1d), for example, could be used to represent motion of collapsed blocks at the Blake Escarpment, east of Florida. In this area "deep sea floor has enormous agglomeration of blocks, commonly 10 km across, that appear to have fallen from the face of the cliff" (see Fig. 7 in [7]). Other examples are the block slide at the base of Middle Canyon along the Beringian Margin in Alaska (see Fig. 8 in [8], and the Sur submarine landslide, where intact sections of slope sediment as large as office buildings (≤ 25 m high) moved 5 km down the continental slope, and 20 km across the gentle 0.5° incline of Monterey fan. Smaller, house-sized (≤ 10 m high) blocks were transported as much as 30 km farther. The Nuuanu debris avalanche appears to contain many large blocks. "Below the amphitheater is a tongue-shaped mass studded with giant blocks that are tens of kilometers in maximum dimension and rise 0.5 to 1.8 km above the regional slope… The largest block, Tuscalousa Seamount, 90 km northwest of Oahu, is 30 km long, 17 km wide and has a broad flat summit about 1.8 km above its base" [4]. Model 3.B in Fig. 1d may illustrate the water waves created by a large block, gradually emerging above the avalanche surface, as it moves down slope.

1.2. MATHEMATICAL MODEL OF TSUNAMI GENERATION AND PROPAGATION

The mathematical model we consider is a fluid layer of constant depth, h, bounded by the rigid ocean floor at $z = -h$ and by the free surface at $z = 0$, and excited by a "small" uplift, $\zeta(x, y; t)$, at $z = -h$. The uplift of the free surface is $\eta(x, y; t)$. The motion of the fluid layer is such that the fluid velocity potential $\phi(x, y, z; t)$ satisfies the Laplace differential equation. A linearized shallow water solution (for water depth much smaller than the tsunami wavelength) can be obtained by the Fourier-Laplace transform. Transformation of the equation of motion and boundary conditions, and the assumptions of linearity and shallow water lead to the solution for the transform of η, $\overline{\eta}(\vec{k}; s)$, in terms of the transform of ζ, $\overline{\zeta}(\vec{k}; s)$ [9]. The solution for $\eta(x, y; t)$ is obtained as follows: (1) $\zeta(x, y; t)$ is transformed to obtain $\overline{\zeta}(\vec{k}; s)$, (2) $\overline{\eta}(\vec{k}; s)$ is computed from $\left[(s^2 + \omega^2) \cosh kh\right]^{-1} s^2 \overline{\zeta}(\vec{k}; s)$ after substitution for $\overline{\zeta}(\vec{k}; s)$, and (3) $\eta(x, y; t)$ is computed by performing inverse transform. For the models we consider in this paper, the forward and inverse Laplace transforms are evaluated analytically using tables, and the Fourier transforms are evaluated numerically using Fast Fourier Transform (FFT). Details of the solutions for all five models are presented in [3].

2. Example Results and Discussion

Detailed results showing two-dimensional tsunami waveforms as functions of time and of the model parameters, and peak amplitudes versus selected model parameters are shown in [3]. Here we summarize only the elementary results for several models.

2.1 DEPENDENCE OF WAVE FORMS ON c_R/c_T – AMPLIFICATION CAUSED BY FOCUSING

We illustrate the results on waveforms along $y=0$ at $t=t^*$ computed for different values of the spreading velocity. For Model 1.B, Fig. 2a, and for Model 2 in Fig. 2b. The results show that, in all cases, the largest positive peak of the tsunami amplitude at time t^* occurs when $c_R/c_T \sim 1$, for predominant direction of sliding in the positive x-direction with velocity c_R.

A comparable negative peak occurs when $c_L \neq 0$, i.e. for sliding initiated at the base of the slope and spreading "uphill" with velocity $c_L \sim c_T$ (e.g. see Fig. 2a for $c_R/c_T = 0.5$ and $c_L/c_R = 2$). For the slides that spread "rapidly" ($c_R/c_T \geq 20$), the displacement of the free surface resembles the displacement of the ocean floor at time t^*. Then $\eta/\zeta_0 \approx 1$, which is the common assumption in numerical simulations that ignores the source time dependence. As c_R/c_T decreases towards 1, the largest tsunami waveform has progressively larger peak amplitudes, and higher frequency content. For $c_R/c_T < 1$ the peak amplitudes decrease, but

64

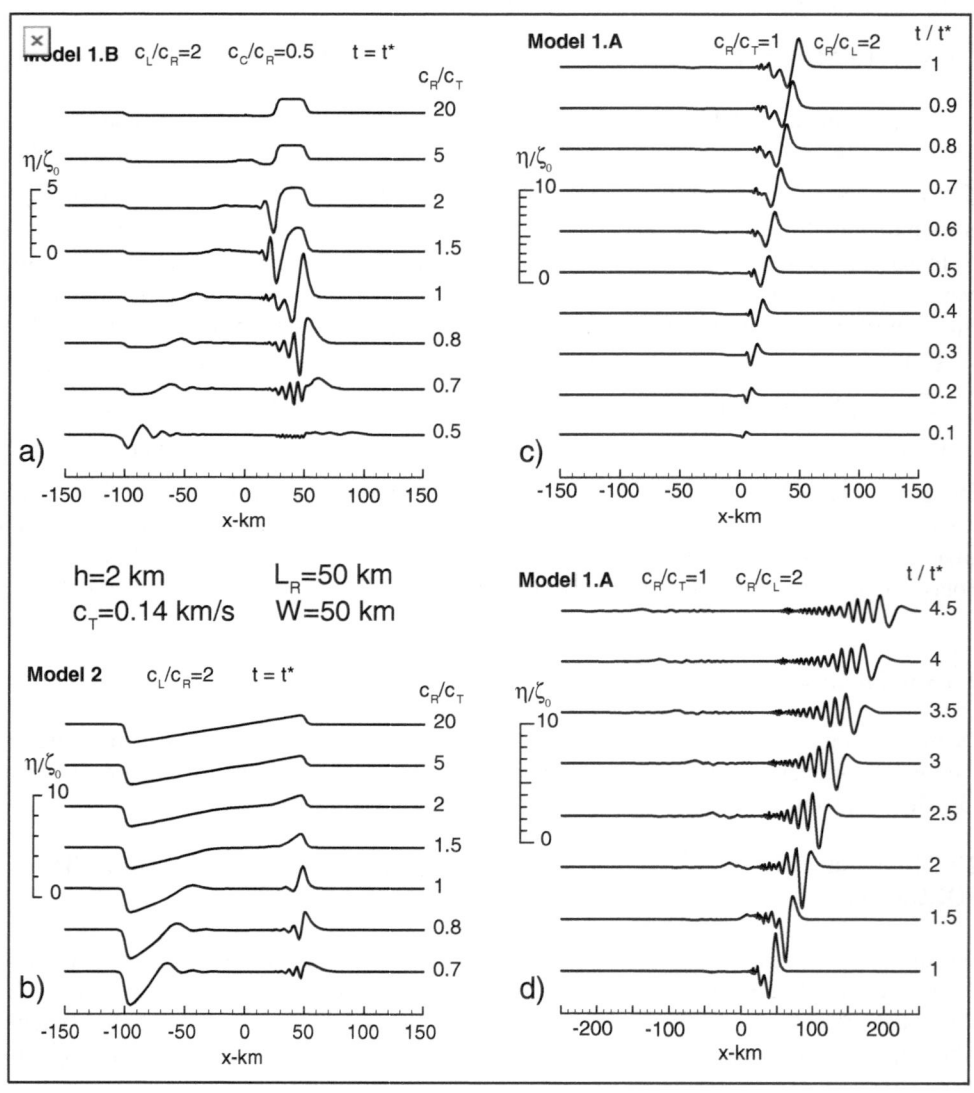

Figure 2. (a) Normalized tsunami waveforms $\eta(x,0;t)/\varsigma_0$ at time $t=t^*=$time when the spreading of the slide stops, for $c_L/c_R = 2$ and $c_0/c_R = 0.5$ and for 8 values of c_R/c_T between 0.5 and 20. The ocean depth is $h=2$ km, and slide has width $W=50$ km and characteristic length $L_R = 50$ km. (b) Normalized tsunami waveforms $\eta(x,0;t)/\varsigma_0$ at time $t = t^* =$ time when the spreading of the slide stops, for $c_L/c_R = 2$ and for 8 values of c_R/c_T between 0.7 and 20. The ocean depth is $h=2$ km, and slide has width $W=50$ km and characteristic length $L_R = 50$ km. (c) Tsunami waveform $\eta(x,0;t)/\varsigma_0$ at selected times $t \leq t^*$ where $t^*=$time when the slide spreading stops, for $c_R/c_T = 1$ and $c_R/c_L = 2$. The ocean depth is $h=2$ km, and the slide has width $W=50$ km and characteristic length $L_R=50$ km. (d) Same as Fig. 2c but for selected times $t \geq t^*$.

the high frequency (short) waves continue to be present. This is a result of constructive interference, and as the process lasts longer, the peak amplitudes increase. When the tsunami is faster than the uplift ($c_R/c_T < 1$) the initial wave "escapes" ahead of the currently uplifted water and amplification does not occur.

2.2 PEAK TSUNAMI AMPLITUDES AS FUNCTION OF L/h

Next, we compare the results with those of [3] and show the peak positive tsunami amplitude $\eta_{R,max}/\zeta_0$ and the peak negative tsunami amplitude $\eta_{L,min}/\zeta_0$ as functions of L/h for all models, computed when $t=t^*$ and for $c_R/c_T = 1$ (L is the characteristic length of the model and h is the depth of the ocean). The positive peak amplitude $\eta_{R,max}/\zeta_0$ is essentially *model independent*. Also, for $c_L/c_T \sim 1$, the negative peak wave amplitudes are approximately equal to the positive peak amplitudes ($\eta_{L,min}/\zeta_0 \approx \eta_{R,max}/\zeta_0$). In Fig. 3 (left) we compare $\eta_{R,max}/\zeta_0$ for all five models with those for the unilaterally spreading uplift of ocean floor, with constant amplitude ζ_0; [9], for $W/L = 0.25$, 0.5 and ≥ 1. For $W/L \geq 0.5$ (and $c_R = c_T$), the peak amplitudes are governed by focusing (piling up of water above the spreading uplift). In the vicinity of the largest positive peak, the waveforms $\eta(x,y; t^*)$ for all models considered here are similar. What is different are the periods of the waves near $\eta_{R,max}$.

2.3 LONG PERIOD LIMIT OF TSUNAMI AMPLITUDES

The long wavelength limit $\lim_{k \to 0} \tilde{\eta}(\vec{k};t) = V$, total volume of the displaced ocean floor, following a slide. In the case of earthquakes, $V \neq 0$, and has a trend parallel to M_0/μ (=source area × average dislocation), but is one to two orders of magnitude smaller [3]. For the models of slides and slumps illustrated here, $V = 0$ for Models 1.A, 1.B, and 3.A, by our definition of these models, $V \neq 0$ but is small for Model 2, and $V \neq 0$ for Model 3.B. The total volume of the moving material during submarine avalanches, slides and slumps can be comparable to and larger than the volume displaced by shallow earthquakes [3,10]. However, the differences in the nature of motion between earthquakes and slides and slumps will result in substantial values of V for earthquakes and in $V \approx 0$ for most slides and slumps. In future, when instrumental recordings of tsunami waveforms $\eta(x,y;t)$ become more ubiquitous, it will be possible to compute $\tilde{\eta}(\vec{k};t)$ for $\vec{k} \to 0$ and to use this result for discrimination of the physical nature of tsunami sources. At present, this is possible only via time consuming mapping of amplitudes and of the geographical extent of tsunami runup.

2.4 EVOLUTION OF TSUNAMI WAVEFORMS IN TIME

Trifunac et al. [3] show many examples of tsunami waveforms $\eta(x,0;t)$ along the axis of symmetry y=0, as they evolve in time from $t=0$ to several times the duration of the source process. Figure 2c illustrates one such case for Model 1.A. Figure 2d shows how the large (short wavelength) peak caused by focusing is dispersed during $t > t^*$.

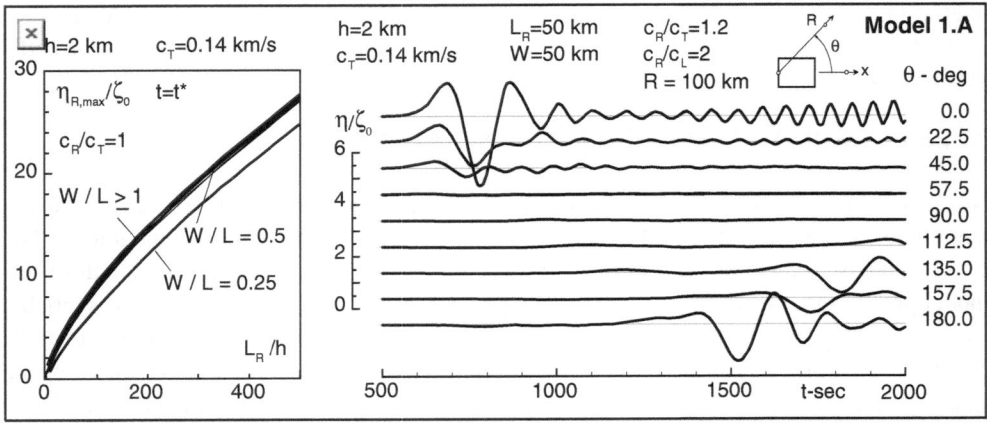

*Figure 3. (**left**) Comparison of normalized peak tsunami amplitudes η_{max}/ς_0 for Models 1.A, 1.B, 3.A and 3.B (light dotted lines) with the results of [9] for a spreading uplift of ocean floor with constant amplitude for $c_R/c_T = 1$, and for W/L=0.25, 0.5 and 1 (heavy dashed lines). It is seen that al $\eta_{R,max}/\varsigma_0$ amplitudes in this paper, except for Model 2, are very near those for W/L=0.5 and 1. (**Right**) Tsunami waveform η at distance R=100 km from the origin for different azimuths θ, measured counter-clockwise from the x-axis (θ =0 is the direction of spreading. The ocean depth is h=2 km, the slide has width W=50 km and characteristic length L_R=50 km, and c_R/c_T = 1.2 and c_R/c_L = 2.*

2.5 ANALOGY BETWEEN WAVE FOCUSING AND RESONANCE OF A SDOF OSCILLATOR

Todorovska and Trifunac [9] showed that when $c_R \sim \sqrt{gh}$, amplification of the peak amplitude η_{max}/ς_0 occurs, which when plotted versus c_T/c_R resembles amplitude versus frequency behavior for a single degree of freedom system. This amplification increases with L/h, where L depends on the model, and is the largest for $W/L \rightarrow \infty$. However, the largest amplitudes are essentially realized for $W/L > 1$. The ratio $h / (WL)$ then may be thought of as representing "damping". For the examples considered in this work, the "resonance peak" occurs for $h/WL \leq 0.001$. No "resonance" peak occurs for $h/WL \geq 0.001$ and thus there is no amplification by focusing. Trifunac et al. [3] show such results for all five models, for $\eta_{R,max}/\varsigma_0$ and for $\eta_{L,max}/\varsigma_0$.

2.6 RADIATION PATTERN

Radiation pattern of tsunami wave amplitudes in the near-field can be evaluated starting with data on the inundation heights. In the past, this has been combined with inverse refraction diagrams to estimate the distribution of wave amplitudes in the source region. In long-period teleseismic seismological studies, radiation pattern has been used to infer the velocity and direction of a spreading dislocation. In the near-field the radiation pattern of short tsunami waves depends on the geometry of the source and on the distribution of the amplitudes of uplift of the ocean floor, and therefore it must be analyzed for each source case separately. In this work we did not analyze the radiation patterns for the five slide and slump models because they are not general and not detailed enough to help visualize radiation patterns from actual submarine slides and slumps. An example of a radiation pattern from a rectangular spreading uplift of the ocean floor can be found in [9]. In Fig. 3 (right), we illustrate variation of wave amplitudes for Model 1A and for variation of azimuths of observation station, within $0 \leq \theta \leq 180°$.

3. Summary and Conclusions

In this note, we illustrated the solutions for five simple kinematic models of submarine slides and slumps, spreading along one axis only. The purpose was to contribute to understanding of the nature of waveforms of tsunami in the near-field. At present, it is difficult to describe the details of movements at the ocean floor during sliding and slumping, because there are no high frequency inverse studies of ground deformations in the source area of past tsunami [11]. Therefore, we address only the basic ideas and illustrate the range of possible tsunami amplitudes using most elementary forward models. Five models were considered, representing submarine slides, slumps, rotational slide and "sliding block" slides. For examples on tsunami wave forms from submarine slides and slumps spreading in two dimensions see [12,13], and for the effects of nonuniform spreading velocity of submarine slumps and slides [14].

The main differences between earthquake and tsunami sources are that (1) the earthquake rupture usually spreads with larger velocities, and (2) the balance of the total uplifted material in case of slides is small or zero while for earthquakes it can be considerable. This affects the long period tsunami amplitude, which is equal to the total uplifted volume.

For given constant water depth, the peak tsunami amplitude depends on the ratios of velocities (c_R/c_T for unilateral spreading, and c_R/c_T and c_L/c_T for bilateral spreading), and on the volumes of accumulated and depleted material. The peak tsunami amplitude also depends on the water depth so that even a small area source can generate a large amplitude tsunami if the water is shallow. The largest positive peak of the tsunami amplitude at time t^* occurs when $c_R/c_T \sim 1$, for predominant direction of sliding in the positive x-direction with velocity c_R. A comparable negative peak occurs when $c_L \neq 0$, i.e. for sliding initiated at the base of the slope and spreading "uphill" with velocity $c_L \sim c_T$. A plot of the peak amplitude η_{max}/ζ_0 versus c_T/c_R has a peak at $c_T/c_R =1$ and resembles amplitude versus frequency behavior for a single degree of freedom oscillator. This amplification increases with L/h, where L depends on the model, and is the largest for $W/L \rightarrow \infty$. The ratio $h/(WL)$ can be thought of as representing damping. For the examples illustrated in this work, the "resonance peak" occurs for $h/(WL) \leq 0.001$ No "resonance" peak occurs for $h/(WL) \geq 0.001$.

For the slides that spread rapidly ($c_R/c_T \geq 20$), the displacement of the free surface resembles the displacement of the ocean floor at time t^*. Then $\eta/\zeta_0 \approx 1$, a common assumption in numerical simulations that ignore the source finiteness. As c_R/c_T decreases towards 1, the largest tsunami waveform has progressively larger peak amplitudes, and higher frequency content. For $c_R/c_T < 1$ the peak amplitudes decrease, but the high frequency (short) waves continue to be present. It is a result of constructive interference, and as the process lasts longer, the peak amplitudes also increase.

68

For all the models in this work it has been assumed that the spreading of submarine slides or slums is in one direction only, here chosen to coincide with positive and/or negative x direction. The methods outlined in this work can be generalized to study also the tsunami generated by slides and slumps spreading in two directions [12] and with variable spreading velocity [14].

References

1. Hampton, M.A., Lemke, R.W. and Coulter, H.W. (1993). Submarine Landslides that had a Significant Impact on Man and His Activities: Seward and Valdez, Alaska, in "Submarine Landslides: Selected Studies in the U.S. Exclusive Economic Zone, *U.S. Geological Survey Bulletin 2002*, U.S. Dept. of the Interior, Denver, Co., 80225, 123-134.
2. Booth, J.S. O'Leary, D.W., Popenoe, P., and Danforth, W.W. (1993). U.S. Atlantic Continental Slope Landslides: Their Distribution, General Attributes, and Implications, in Submarine Landslides: Selected Studies in the U.S. Exclusive Economic Zone, *U.S. Geological Survey Bulletin 2002*, U.S. Dept. of Interior, Denver, Co. 80225, 14-22.
3. Trifunac, M.D., Hayir, A. and Todorovska, M.I. (2001)*. Near-Field Tsunami Wave Forms from Submarine Slumps and Slides, Dept. of Civil Eng. Report No. CE 01-01, Univ. of Southern California, Los Angeles, California.
4. Normark, W.R., Moore, J.G. and Torresan, M.E. (1993). Giant Volcano-Related Landslides and the Development of the Hawaiian Islands, in "Submarine Landslides: Selected Studies in the U.S. Exclusive Economic Zone, *U.S. Geological Survey Bulletin 2002*, U.S. Dept. of the Interior, Denver, Co., 80225, 184-196.
5. Edwards, B.D., Lee, H.J. and Field, M.E. (1993). Seismically Induced Mudflow in Santa Barbara Basin, California, in "Submarine Landslides: Selected Studies in the U.S. Exclusive Economic Zone, *U.S. Geological Survey Bulletin 2002*, U.S. Dept. of the Interior, Denver, Co. 80225, 167-175.
6. Field, M.E. and Barber, J.H. (1993). A Submarine Landslide Associated with Shallow Sea-Floor Gas and Hydrates Off Northern California, in "Submarine Landslides: Selected Studies in the U.S. Exclusive Economic Zone," *U.S. Geological Survey Bulletin 2002*, U.S. Dept. of the Interior, Denver, Co. 80225, 151-157.
7. Dillon, W.P., Risch, J.S., Scanlon, K.M., Valentine, P.C. and Higgett, Q.J. (1993). Ancient Crustal Fractures Control the Location and Size of Collapsed Blocks at the Blake Escarpment, East of Florida, in "Submarine Landslides: Selected Studies in the U.S. Exclusive Economic Zone," *U.S. Geological Survey Bulletin 2002*, U.S. Dept. of the Interior, Denver, Co. 80225, 54-68.
8. Carlson, P.R., Karl, H.A., Edwards, B.D., Gardner, J.V. and Hall, K. (1993). Mass Movement Related to Large Submarine Canyons Along the Beringian Margin, Alaska, in "Submarine Landslides: Selected Studies in the U.S. Exclusive Economic Zone" *U.S. Geological Survey Bulletin 2002*, U.S. Department of Interior, Denver Co. 80225, 104-116.
9. Todorovska, M.I. and Trifunac, M.D. (2001). Generation of Tsunamis by Slowly Spreading Uplift of the Sea Floor, *Soil Dynam. and Earthq. Eng.*, **21**(2), 151-167.
10. Trifunac, M.D. and Todorovska, M.I. (2002). A Note on Differences in Tsunami Source Parameters for submarine Slides and earthquakes *Soil Dynam. and Earthq. Eng.*, 22(2), 143-155.
11. Jordanovski, L. and Todorovska, M.I. (2002). Inverse Studies of the Earthquake Source Mechanism from Near-Field Strong Motion Records, *J. Indian Soc. Earthquake Technology*, (in press).
12. Trifunac, M.D. Hayir, A., and Todorovska, M.I. (2001)*. Tsunami Waveforms from Submarine Slides and Slumps Spreading in Two Dimensions, Dept of Civil Eng. Report No. CE 01-06, Univ. of Southern California, Los Angeles, California.
13. Trifunac, M.D. Hayir, A. and Todorovska, M.I. (2002). Was the Grand Banks Event of 1929 in Slump Spreading ij Two Dimensions? *Soil Dynam. and Earthq. Eng.*, (in press).
14. Trifunac, M.D. Hayir, A. and Todorovska, M.I. (2002). A Note on the Effects of Nonuniform Spreading Velocity of Submarine Slumps and Slides on the Near-Field Tsunami Amplitudes, *Soil Dynam. and Earthq. Eng.*, 22(3), 167-180.

NUMERICAL MODELING OF TSUNAMI GENERATION BY SUBMARINE AND SUBAERIAL LANDSLIDES

ISAAC V. FINE[1, 2, 4], ALEXANDER B. RABINOVICH[2, 3, 4],
RICHARD E. THOMSON[4], and EVGUENI A. KULIKOV[2, 3, 4]
[1] *Heat and Mass Transfer Institute, Minsk, Belarus*
[2] *International Tsunami Research, Inc., Sidney, BC, Canada*
[3] *P.P. Shirshov Institute of Oceanology, Moscow, 117851 Russia*
[4] *Institute of Ocean Sciences, Sidney, BC, Canada*

Abstract

Recent catastrophic tsunamis at Flores Island, Indonesia (1992), Skagway, Alaska (1994), Papua New Guinea (1998), and İzmit, Turkey (1999) have significantly increased scientific interest in landslides and slide-generated tsunamis. Theoretical investigations and laboratory modeling further indicate that purely submarine landslides are ineffective at tsunami generation compared with subaerial slides. In the present study, we undertook several numerical experiments to examine the influence of the subaerial component of slides on surface wave generation and to compare the tsunami generation efficiency of viscous and rigid-body slide models. We found that a rigid-body slide produces much higher tsunami waves than a viscous (liquid) slide. The maximum wave height and energy of generated surface waves were found to depend on various slide parameters and factors, including slide volume, density, position, and slope angle. For a rigid-body slide, the higher the initial slide above sea level, the higher the generated waves. For a viscous slide, there is an optimal slide position (elevation) which produces the largest waves. An increase in slide volume, density, and slope angle always increases the energy of the generated waves. The added volume associated with a subaerial slide entering the water is one of the reasons that subaerial slides are much more effective tsunami generators than submarine slides. The critical parameter determining the generation of surface waves is the Froude number, Fr (the ratio between slide and wave speeds). The most efficient generation occurs near resonance when $Fr = 1.0$. For purely submarine slides with $\rho_2 \leq 2.0$ g·cm^{-3}, the Froude number is always less than unity and resonance coupling of slides and surface waves is physically impossible. For subaerial slides there is always a resonant point (in time and space) where $Fr = 1.0$ for which there is a significant transfer of energy from a slide into surface waves. This resonant effect is the second reason that subaerial slides are much more important for tsunami generation than submarine slides.

.

A. C. Yalçıner, E. Pelinovsky, E. Okal, C. E. Synolakis (eds.),
Submarine Landslides and Tsunamis 69-88.
©2003 Kluwer Academic Publishers.

1. Introduction

Submarine landslides, slumps, rock falls, and avalanches may produce catastrophic tsunami waves in coastal areas of the World Ocean. Although landslide generated tsunamis are much more localized than seismically generated tsunamis, they can produce destructive coastal run-up and cause severe damage to coastal emplacements [1, 13]. Recent catastrophic tsunamis at Flores Island (1992), Papua New Guinea (1998), and İzmit, Turkey (1999) apparently originated with local landslides triggered by earthquakes [2, 3, 4, 5]. These events, as well as the non-seismic catastrophic event in 1994 in Skagway Harbor, Southeast Alaska [6, 7, 8], have significantly increased scientific interest in landslides and slide-generated tsunamis.

Submarine landslides are ineffective at tsunami generation compared with subaerial slides [9]. Subaerial slides displace a considerable volume of water at relatively high speed as they slide into the water from the foreslope. The famous event of July 10, 1958 in Lituya Bay, Southeast Alaska was initiated by a subaerial rockslide at the head of the bay that caused a giant tsunami which impacted the sides of the inlet to a height of 525 m [10, 11]. The destructive Skagway event of November 3, 1994 was associated with a subaerial slide component which generated a series of large amplitude waves, estimated by eyewitnesses to have heights of 5-6 m in the harbor and 9-11 m at the shoreline [11, 6, 8]. Because the efficiency of tsunami generation is inversely proportional to the water depth, subaerial slides are particularly effective wave generators. *Raichlen et al.* [12], using laboratory modeling, examined the 1994 Skagway tsunami and demonstrated that the subaerial component of the slide caused a significant increase in the slide-generated wave amplitudes.

Numerical modeling of tsunamis caused by submarine slides and slumps is a much more complicated problem than simulation of seismically-generated tsunamis. The durations of the slide deformation and propagation are sufficiently long that they affect the characteristics of the surface waves. As a consequence, coupling between the slide body and the surface waves must be considered. Moreover, the landslide shape changes significantly during slide movement, causing the slide to modify the surface waves it has generated. *Jiang and LeBlond* [1,13] appear to have been the first to formulate models that account for all submarine landslides effects, including the coupling of the landslide and associated surface waves. We have corrected minor errors in the governing equations of the three-dimensional viscous-slide shallow-water model proposed by *Jiang and LeBlond* [1] (herein JLB94) and generalized the model to include arbitrary bottom topography (see [14] and [15] for details). Customized versions of this model were used to simulate the 1999 PNG tsunami [4, 16, 17] and the tsunami caused by the slumping of the Nice harbour extension (1979) [18].

The principal advantage of our extended model over the JLB94 model is the inclusion of a subaerial component of the slide. The main problem concerning the numerical modeling of subaerial slides is that 'wet' and 'dry' areas change during the slide/wave motions, so that there are variable boundaries between these areas. Only a few papers [19, 20] deal with this problem. Here, we have effectively bypassed the moving boundary problem and successfully used our model to simulate subaerial tsunamis for realistic bathymetry and coastline geometry, with application to the 1966 and 1994 tsunamis in Skagway Harbor, Alaska [15]. We also made similar modifications to the commonly used rigid-body (frictional) shallow-water slide model [21, 22]. The main purposes of the present study are to examine the influences of subaerial landslides on tsunami wave generation and to compare the tsunami generation efficiency of viscous and rigid-body slide models. We also examine the influence of various slide parameters, such as slope angle, slide position, water depth, and friction, on surface wave generation.

2. Governing Equations and Model Description

Surface wave generation by a moving slide is affected by the water depth, gravity, and fundamental characteristics of the slide [1, 13]. The principal mechanism for energy transfer from the slide motion to the surface waves, water displacement, is readily incorporated using the long-wave (shallow-water) approximation [22]. The main assumptions for the present models (viscous and rigid-body) are the following:

(1) The surface waves and slides satisfy the long-wave (hydrostatic) approximation, implying that the wavelength of the water waves is much greater than the water depth, and that the width and length of the viscous slide is much greater than the slide thickness.

(2) The viscous slide is an incompressible, isotropic, laminar, quasi-steady viscous fluid; the viscous regime is rapidly reached in any failure and in the steady-state regime, the horizontal velocities have a parabolic vertical profile.

(3) The rigid-body slide moves as a non-deformable body with given friction.

(4) The seawater is an incompressible inviscid fluid.

We use standard Cartesian coordinates x, y, z, with z measured vertically upward. For time t, the upper (water) layer consists of seawater with density ρ_1, surface elevation $\eta(x, y; t)$, and horizontal velocity $\mathbf{u}$ with components u and v (Figure 1a). The lower layer consists of viscous sediments (or rigid body) having density ρ_2, dynamic viscosity μ (or friction coefficient k in case of a rigid body), and horizontal velocity $\mathbf{U}$ with components U and V. Both the slope and the slide have small angles, so the motion is essentially horizontal. The slide is bounded by an upper surface $z = -h(x, y; t)$ and the seabed surface $z = -h_s(x, y)$, giving the slide thickness as $D(x, y; t) = h_s(x, y) - h(x, y; t)$.

A schematic of the computational domain for a landslide with a subaerial component is presented in Figure 1. The domain consists of four zones: (1) The *dry* coastal area, D; (2) the *dry* portion of the slide, S_D, corresponding to the subaerial part of the slide; (3) the *wet* portion of the slide, S_W, corresponding to the submarine part of the slide; and (4) the *water*, W. The numerical model must account for the time-varying changes in the areas and locations of these zones.

2.1. VISCOUS SLIDE

Our purpose is to construct the non-linear, vertically integrated Navier-Stokes equations for the landslide. We assume that the landslide occupies a domain from $z = -h_s(x, y)$ to $D(x, y; t) = h_s(x, y) - h(x, y; t)$. Following JLB94, we assume that the landslide rapidly reaches a steady shape so that we can use a locally parabolic approximation in the vertical to describe the horizontal velocities, $U_m(x, y, z; t)$ and $V_m(x, y, z; t)$; specifically,

72

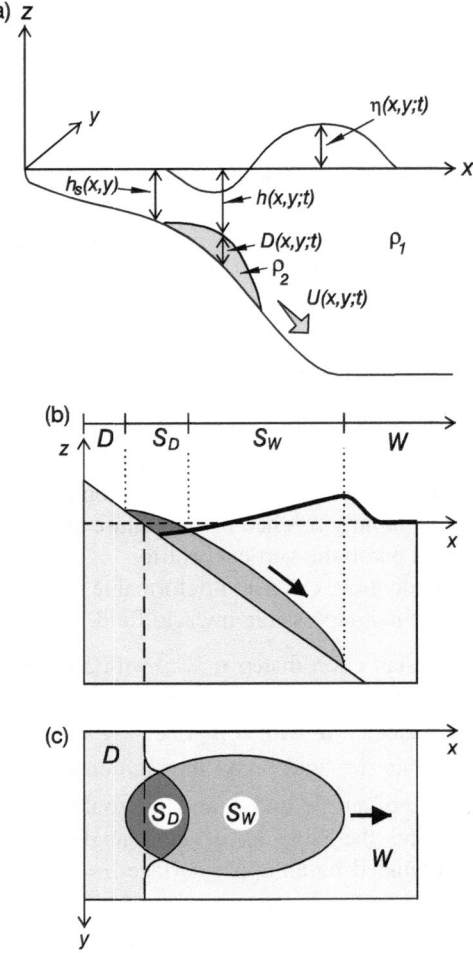

Figure 1. (a) Sketch of a submarine landslide with density ρ_2, thickness D, and water depth h, and associated surface waves of height η. (b) Side view, and (c) plan view of a combined subaerial and submarine slide (see the text for description of the letters).

$$U_m(x, y, z; t) = U(x, y; t)(2\xi - \xi^2), \tag{1a}$$

$$V_m(x, y, z; t) = V(x, y; t)(2\xi - \xi^2), \tag{1b}$$

where $\xi = (z + h_s)/D$. The equations for conservation of mass and momentum for a viscous submarine slide have the form [14]:

$$\frac{\partial D}{\partial t} + \frac{2}{3}\left[\frac{\partial (DU)}{\partial x} + \frac{\partial (DV)}{\partial y}\right] = 0 ; \tag{2}$$

$$\rho_2 \frac{2}{3}\left[\frac{\partial U}{\partial t} - \frac{1}{5}\frac{U}{D}\frac{\partial D}{\partial t} + \frac{4}{5}\left(U\frac{\partial U}{\partial x} + V\frac{\partial U}{\partial y}\right)\right] = -g\left[(\rho_2 - \rho_1)\left(\frac{\partial D}{\partial x} - \frac{\partial h_s}{\partial x}\right) + \rho_1\frac{\partial \eta}{\partial x}\right] - \frac{2\mu U}{D^2} ; \tag{3a}$$

$$\rho_2 \frac{2}{3}\left[\frac{\partial V}{\partial t} - \frac{1}{5}\frac{V}{D}\frac{\partial D}{\partial t} + \frac{4}{5}\left(U\frac{\partial V}{\partial x} + V\frac{\partial V}{\partial y}\right)\right] = -g\left[(\rho_2 - \rho_1)\left(\frac{\partial D}{\partial y} - \frac{\partial h_s}{\partial y}\right) + \rho_1\frac{\partial \eta}{\partial y}\right] - \frac{2\mu V}{D^2} . \tag{3b}$$

The continuum equation (2) is the same as in the JLB94 model. However, the momentum equations (3a) and (3b) are slightly different from those presented by *Jiang and LeBlond* [1] as a result of corrections we have made to several of the constant coefficients in the terms in the square brackets on the left-hand sides of these equations. Numerical experiments we have conducted show that the small errors in these advective terms in the JLB94 model may cause 20-25% errors in computed tsunami heights.

For a subaerial slide, it is useful to introduce a new variable, h_w, the full water thickness $(h_w = \eta + h = \eta - D + h_s)$, and to present equations (3a) and (3b) in the form:

$$\rho_2 \frac{2}{3}\left[\frac{\partial U}{\partial t} - \frac{1}{5}\frac{U}{D}\frac{\partial D}{\partial t} + \frac{4}{5}\left(U\frac{\partial U}{\partial x} + V\frac{\partial U}{\partial y}\right)\right] = -g\left[\rho_2\left(\frac{\partial D}{\partial x} - \frac{\partial h_s}{\partial x}\right) + \rho_1\frac{\partial h_w}{\partial x}\right] - \frac{2\mu U}{D^2} ; \tag{4a}$$

$$\rho_2 \frac{2}{3}\left[\frac{\partial V}{\partial t} - \frac{1}{5}\frac{V}{D}\frac{\partial D}{\partial t} + \frac{4}{5}\left(U\frac{\partial V}{\partial x} + V\frac{\partial V}{\partial y}\right)\right] = -g\left[\rho_2\left(\frac{\partial D}{\partial y} - \frac{\partial h_s}{\partial y}\right) + \rho_1\frac{\partial h_w}{\partial y}\right] - \frac{2\mu V}{D^2} . \tag{4b}$$

For the subaerial zone, S_D, we have the particular case of zero water thickness, $h_w = 0$, for which equations (4a) and (4b) describe slide motion on a dry coast.

The above equations are solved subject to the condition of zero transport through the coastal boundary (G) and require that the slide does not cross the outer (open) boundary (Γ). The condition of no volume transport through the coast gives

$$U_n = 0 \qquad \text{on } G, \tag{5}$$

where U_n is the normal slide velocity. The initial slide has a rectangular bottom periphery oriented at a given angle β and, as noted above, is assumed to have a parabolic cross-section.

2.2. RIGID-BODY SLIDE

The rigid-body model assumes that the shape and dimensions of the initial slide remain invariant during the slide motion. All points of the rigid body move with the same velocity $\mathbf{U} = \mathbf{U}(t)$ and the position of the slide changes with time through the relation:

$$D(x, y; t) = D_0(x - X(t), y - Y(t)), \tag{6}$$

where D_0 is the initial slide distribution, and $X = \int_0^t U dt$, $Y = \int_0^t V dt$. In solving the equations of motion, we further assume that: (1) Bottom friction on the slide is proportional to the normal pressure, P; (2) there are no hydraulic forces ("form drag") on the slide; and (3) the bottom slope is small, $|\nabla h| \ll 1$. Under these assumptions, the momentum equation of the slide becomes

$$\rho_2 \frac{d\mathbf{U}}{dt} \iint_S D ds = \nabla h \cdot P - k \frac{\mathbf{U}}{|\mathbf{U}|} P , \tag{7}$$

where k is the nondimensional coefficient of kinetic friction (the Coulomb friction coefficient), S is the surface area of the slide,

$$P = g \iint_S (\rho_1 \eta + \Delta \rho D) ds , \tag{8}$$

and $\Delta \rho = \rho_2 - \rho_1$ is the density difference between the slide and seawater. The boundary conditions for the rigid slide are the same as for the viscous slide.

2.3. SURFACE WAVES

For surface waves generated by a submarine slide, the water motions are nearly horizontal and the pressure is hydrostatic (long-wave approximation). The nonlinear shallow-water equations then have the form [1,15]:

$$\frac{\partial h_w}{\partial t} + \frac{\partial (h_w u)}{\partial x} + \frac{\partial (h_w v)}{\partial y} = 0; \tag{9}$$

$$\frac{\partial u}{\partial t} + u \frac{\partial u}{\partial x} + v \frac{\partial v}{\partial y} = -g \frac{\partial \eta}{\partial x}; \tag{10a}$$

$$\frac{\partial v}{\partial t} + u \frac{\partial v}{\partial x} + v \frac{\partial v}{\partial y} = -g \frac{\partial \eta}{\partial y}, \tag{10b}$$

which are applicable to wet zones, S_W and W (see Figure 1c). At the shore (boundary G), we assume a vertical wall with zero normal velocity:

$$u_n = 0 \qquad \text{on } G. \tag{11}$$

At the open boundary (Γ), the one-dimensional radiation condition for outgoing waves is:

$$\frac{\partial \eta}{\partial t} = \text{sgn} \left(\frac{\partial \eta}{\partial t} \right) \sqrt{gh - \left[\left(\frac{\partial \eta}{\partial x} \right)^2 + \left(\frac{\partial \eta}{\partial y} \right)^2 \right]} \tag{12}$$

At the initial time, $t = 0$, both the slide and the sea surface are at rest.

2.4. MODEL APPROACH

To solve equations (2)-(4), (7)-(8), and (9)-(10) with boundary conditions (5) and (11)-(12), we used an explicit finite-difference method with the Arakawa C-grid approximation. Velocity computational nodes are shifted by one-half the time and space steps relative to the sea level and slide-water interface computational nodes (the so-called staggered leap-frog scheme [23]). To avoid generation of erroneous small-scale oscillations, the time step (Δt) was taken to be 1/3-1/5 the value required for the Courant stability criterion. To suppress numerical instability, the advective terms in equations (4) and (10) were represented through the upstream approximation scheme [24, 25]. For a detailed description of the present numerical model, the reader is directed to [15].

As previously emphasized, the main problem with numerical simulating subaerial slides is that the wet and dry areas change during the slide/wave motions, creating a variable boundary between the two areas. This is a well-known problem in tsunami run-up studies [23]. The drying of the wet area is not overly complicated. Here, the rule is that if the water thickness becomes equal to or less than zero, the respective point is assumed to be dry. Flooding of the dry area is a more serious problem. To describe the nonlinear interaction between the moving subaerial landslide and the overlying water, we have used the method proposed by *Titov and Synolakis* [26]. In this case, the wet boundary is determined as the intersection of the coastal slope and the horizontal plane of the sea level at the last "wet" point. When sea level at a "dry" point becomes higher than the fixed coastal elevation, this point is assumed to become "wet". This method is more stable to depth and coastline irregularities than other methods. A more detailed discussion of the of the "wetting" and "drying" problem in the area of the landslide using the present numerical algorithm is given by [15]

2.5. NONDIMENSIONAL VARIABLES

Following *Jiang and LeBlond* [1, 13], we have used nondimensional variables in our numerical experiments. We chose the initial maximum slide thickness, D_0, as the vertical length scale, the initial slide length, L, as the horizontal length scale, and $t_0 = (D_0 / g)_{1/2}$ as the time scale. The horizontal velocities are normalized using $U_0 = L_0 / t_0$. Thus, we adopt the following nondimensional variables:

$$(x', y') = (x, y) / L; \tag{13a}$$

$$(h', D', \eta') = (h, D, \eta) / D_0; \tag{13b}$$

$$t' = t / t_0; \qquad t_0 = (D_0 / g)^{1/2}; \tag{13c}$$

$$(u', v', U', V') = (u, v, U, V) / U_0; \qquad U_0 = L / t_0. \tag{13d}$$

The energy of the slide and generated surface waves are normalized as:

$$E' = E / E_0; \qquad E_0 = \frac{\rho_2 g L^2}{D_0}. \tag{14}$$

Nondimensional variables are used in Figures 2-7.

3. Wave Evolution

The two models described above were used to study water waves generated by landslides on a gentle uniform slope in shallow water. We first examined some general properties of tsunami waves generated by both rigid-body and viscous slides. Figures 2 and 3 present results for these models for subaerial and submarine slides. The computations have been made for slide density $\rho_2 = 2.0$ g·cm^{-3} for both slides, Coulomb friction coefficient $k = 0.02$ (for the rigid-body slide), and kinematic viscosity coefficient $v = \mu / \rho_2 = 0.01$ m^{2}s^{-1} (for the viscous slide). The general results are similar for both models and correspond well to those obtained by previous investigators [27, 19, 1, 13]. In particular, the slide moving into deeper water forms a crest wave propagating ahead of the slide with a wave trough following the crest. However, there are some important differences between the rigid-body and viscous models, and between subaerial and submarine landslides. The principal difference is that a rigid-body slide produces much higher waves than a viscous slide, indicating that a rigid-body slide is a much more efficient tsunami wave generator than a viscous slide. Similarly, subaerial slides are much more efficient wave generators than purely submarine slides (Figures 4 and 5).

A subaerial slide entering the water brings an additional volume causing a respective displacement of the sea surface. Therefore, the initial wave crest is sufficiently larger than the following wave trough (Figure 4). *Heinrich* [19] obtained similar results both in laboratory modeling of a subaerial solid triangle block sliding freely downslope into the water and by corresponding numerical computations. *The added volume displacement of the water is one of the reasons why subaerial slides generate significantly larger surface waves than submarine landslides.*

For a submarine slide, three major waves are produced (Figure 5): (1) The leading wave crest propagating rapidly offshore ahead of the moving slide with shallow-water wave speed $c = \sqrt{gh}$; (2) the wave trough propagating offshore with the speed of the slide; and (3) the wave trough propagating shoreward. Similar results were obtained by *Heinrich* [19] and *Jiang and LeBlond* [1, 13]. For a rigid-body slide, the second wave, which is bound to the slide as a forced wave, is significantly larger than two other generated waves (Figure 5a); for a viscous slide all three waves have comparable heights (Figure 5b). An important aspect of this process is that, due to amplitude dispersion, the viscous slide forms a bore-like leading edge with a steep frontal wall as it moves downslope (Figure 3; see also Figures 3-8 in [13]). This frontal bore in the slide gives rise to a corresponding bore-like negative surface wave propagating with the frontal speed of the slide (Figure 5b).

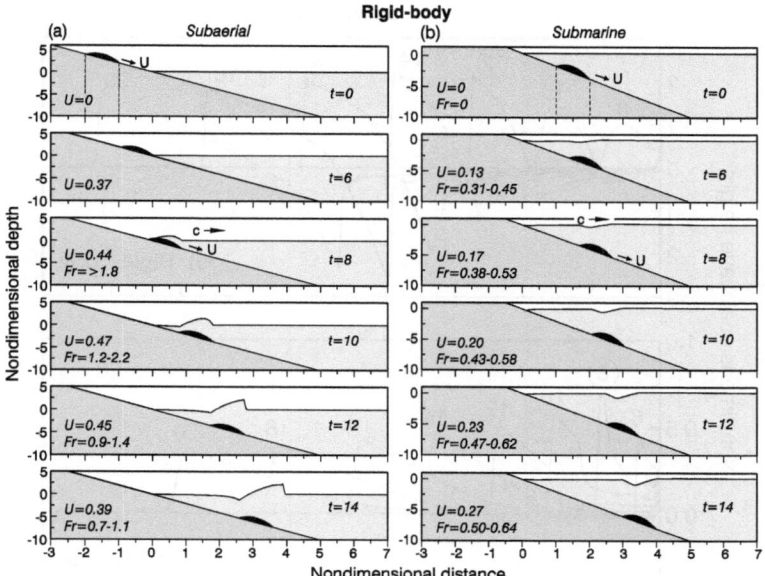

Figure 2. Tsunami waves generated by a rigid-body slide moving downslope at speed U (slope angle ψ = 4°, slide density ρ_2 = 2.0 g·cm^{-3}, friction coefficient k = 0.02). (a) Subaerial slide; (b) submarine slide. Nondimensional time, t, slide speed, U, and Froude number, Fr=U/c, where $c = \sqrt{gh}$ is the long-wave speed, are presented in the plots.

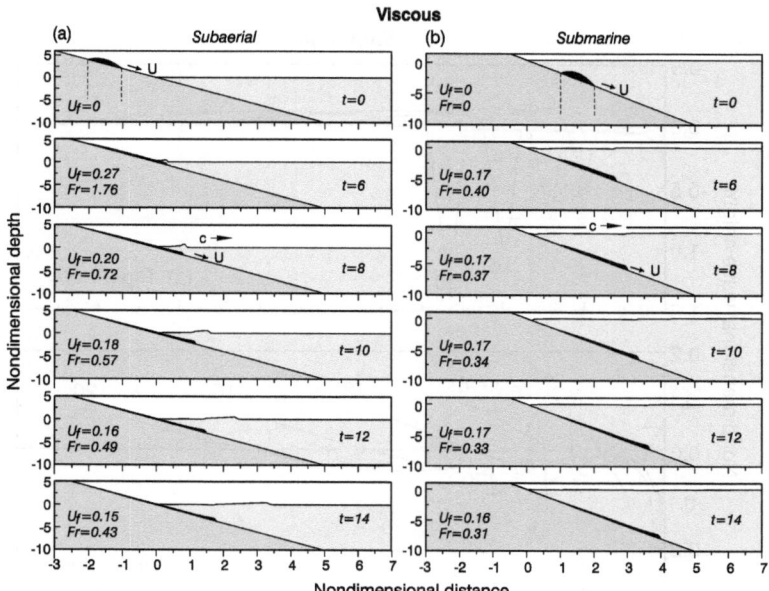

Figure 3. The same as in Figure 2 but for a viscous slide with kinematic viscosity ν = 0.01 m^2·s^{-1}.

78

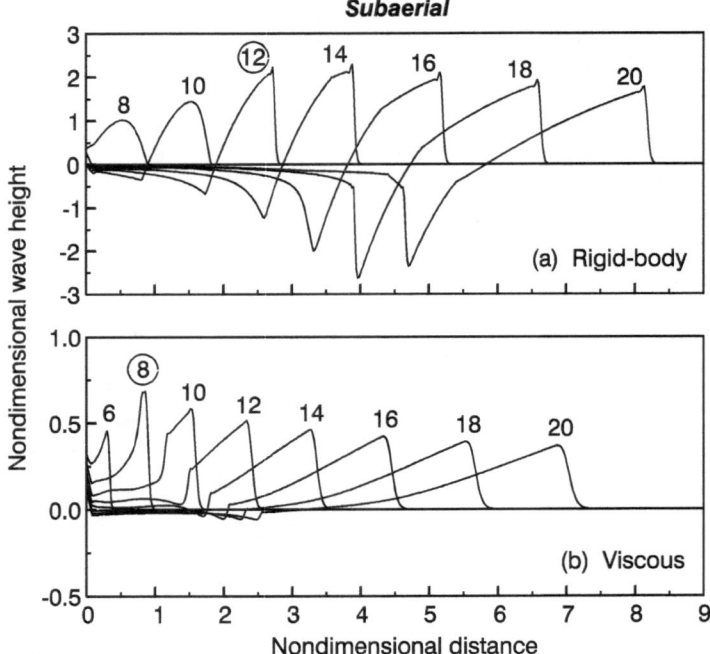

Figure 4. Tsunami waves for fixed times generated by a subaerial slide: (a) A rigid-body slide, and (b) a viscous slide. Parameters of the slides are the same as in Figures 2 and 3. Note that the wave-height scale for the viscous model is four times greater than for the rigid-body model. The circled times are the times closest to resonance (Fr = 1.0).

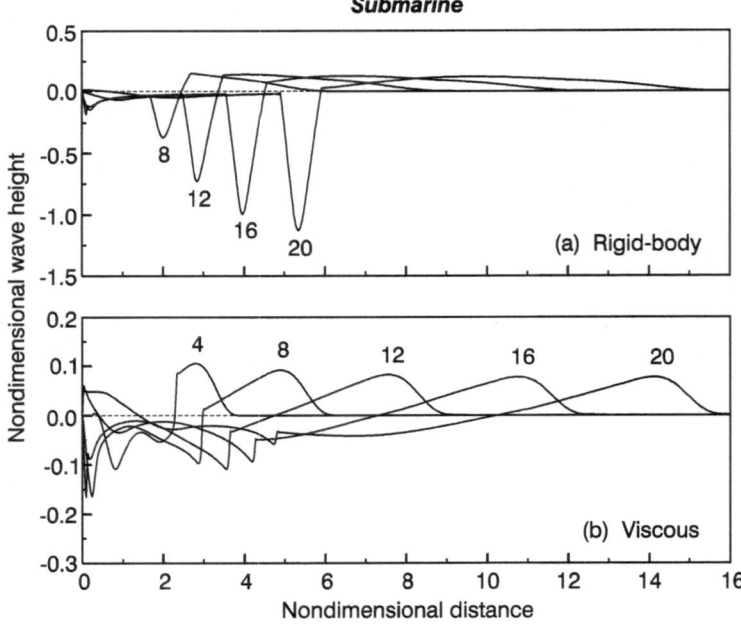

Figure 5. The same as in Figure 4 but for tsunami waves generated by a purely submarine slide.

4. Slide/Wave Speed and Froude Number

The character and intensity of the tsunami waves, as well as the coupling between the slide and the waves, strongly depends on the Froude number, Fr (the ratio between the slide and wave speeds). Resonance occurs when these speeds are equal; i.e. when $Fr = 1.0$.

For a rigid-body, which moves as an entity with the same speed U, the Froude number can be defined as

$$Fr = U/c, \tag{15a}$$

where $c = \sqrt{gh(x, y)}$ is the local long-wave speed. In contrast, different parts of a viscous slide move with different speeds. For the latter case, we may define the Froude number as

$$Fr = U_f/c, \tag{15b}$$

where U_f is the speed of the slide front [13]. Figures 2 and 3 show the values of U (U_f) and corresponding Fr. Because the depth is different near the front and the rear of the rigid-body slide, the wave speed, c, and Froude number are also different (Figure 2). For the viscous slide (Figure 3), the values of U_f and Fr are presented for the front slide point.

The Froude number for submarine landslides plays the same role as the Mach number for high-speed aircraft. As seen from Figure 2a, the character of the water disturbances is significantly different for "super-sonic" ($Fr > 1.0$) and "sub-sonic" ($Fr < 1.0$) slide motions. The super-sonic slide movement ($Fr > 1.0$) does not induce free-propagating water waves so that the water displacement (forced wave) is locked to the slide and moves with the slide speed, almost duplicating its form. When the slide thickness is much smaller than the water depth, we can describe this forced wave roughly as

$$\eta \approx \frac{DU^2}{U^2 - c^2} \tag{16}$$

Expression (16) is similar to well-known "Proudman expression" for cyclone-induced displacements of sea level [28]. For super-sonic slide motion (when $U > c$ and $Fr > 1.0$) the forced wave is positive (crest), whereas for sub-sonic slide motion ($U < c$ and $Fr < 1.0$) it is negative (trough) (Figures 2 and 3).

As shown by our numerical experiments, an initially subaerial rigid-body slide moving underwater downslope spans three consecutive regimes (Figures 2a and 4a):

(1)   *Super-sonic* ($Fr > 1.0$): There is only one wave, the positive forced wave (crest), which repeats the form of the slide and moves together with it.

(2)   *Resonant* ($Fr \sim 1.0$): The height of the positive forced wave is significantly amplified (crest maximum is about $2.5D_0$) and the frontal side of the wave becomes very steep forming a bore. In the rear of the slide an intensified negative (trough) wave forms.

(3)   *Sub-sonic* ($Fr < 1.0$): The first (crest) wave becomes free and propagates away rapidly from the slide with the long-wave speed $c = \sqrt{gh}$, its height decreasing with distance. In contrast, the second (trough) wave is a forced wave, bound to the slide motion, its height slowly decreasing with time.

A subaerial viscous slide moving underwater has the same three regimes as the rigid-body slide above, except that surface wave generation is now less efficient and of a more complex nature (Figures 2b and 4b). The tsunami waves for a viscous slide are induced in three different ways: By the initial water displacement by the slide, by propagation of the slide front, and by underwater changes in the slide configuration.

In all our experiments, *purely submarine slides* (both rigid-body and viscous) were "sub-sonic" (see Figure 3 as an example) and the Froude numbers were always significantly smaller than unity. This result follows from the general properties of underwater slide motions [22, 29]. More specifically, the speed of an underwater rigid-body sliding down a gentle slope may be approximated from the equation:

$$\rho_2 \frac{U^2}{2} + gk(\rho_2 - \rho_1)(x - x_0) = g(\rho_2 - \rho_1)(h - h_0),$$ (17)

where x_0 and h_0 are the initial slide position and water depth, and x and h are the current position and depth. This yield,

$$U = \left[2g \frac{\rho_2 - \rho_1}{\rho_2} [h - h_0 - k(x - x_0)] \right]^{1/2}$$ (18a)

for the slide speed and

$$Fr = \frac{U}{c} = \left[\frac{2g \frac{\rho_2 - \rho_1}{\rho_2} [h - h_0 - k(x - x_0)]}{gh} \right]^{1/2} = \left[2 \frac{\rho_2 - \rho_1}{\rho_2} \left(1 - \frac{h_0}{h} - k \frac{x - x_0}{h} \right) \right]^{1/2}$$ (18b)

for the Froude number.

The fastest slide speed for submarine slides is achieved when a slide starts from the coastline $(h_0 = 0, \quad x_0 = 0)$. Equation (18a) in that case has the form

$$U = \left[2g \frac{\rho_2 - \rho_1}{\rho_2} (h - kx) \right]^{1/2},$$ (19a)

and equation (18b) becomes

$$Fr = \frac{U}{c} = \left[2 \frac{\rho_2 - \rho_1}{\rho_2} \left(1 - k \frac{x}{h} \right) \right]^{1/2}.$$ (19b)

Typical density values for natural alluvial and deluvial sediments are $\rho_2 = 1.2\text{-}2.0 \text{ g·cm}^{-3}$ [13]. Assuming a water density $\rho_1 \approx 1.0 \text{ g·cm}^{-3}$, it follows from expressions (19a) and (19b), that if $\rho_2 \leq 2.0 \text{ g·cm}^{-3}$ then $U < c$, and $Fr < 1.0$ always. Actual Froude numbers will be even smaller because: (1) Expressions (18) and (19) do not take into account hydraulic resistance to slide motion, which would reduce the slide speed[1]; and (2) more realistic deeper initial source areas will reduce the slide speed. We therefore conclude that, for natural submarine landslides, *resonance coupling of slides and surface waves is impossible.* Resonance can occur only if $\rho_2 > 2.0 \text{ g·cm}^{-3}$ (i.e. for strongly consolidated sediments or rock) or if the slide starts above the water. *Existence of the resonance regime for subaerial slides and absence of this regime for purely submarine slides is a second reason why*

---

[1] The influence of water resistance on slide speed and Froude number is discussed by *Harbitz* [22] and *Pelinovsky and Poplavsky* [29], among others.

subaerial slides are much more efficient tsunami generators than submarine slides.

5. Energy Estimates

The numerical experiments described above enable us to derive relative energy estimates of the generated tsunami waves. Multiplying equation (9) by $g\eta$, (10a) by $h_w u$, and (10b) by $h_w v$, and summing all three contributions, gives the time rate of change of total wave energy, $\dfrac{\partial \varepsilon}{\partial t}$:

$$\frac{\partial \varepsilon}{\partial t} = g\eta \frac{\partial D}{\partial t} - \frac{\partial}{\partial x}(h_w u \varepsilon_0) - \frac{\partial}{\partial y}(h_w v \varepsilon_0), \tag{20}$$

where $\varepsilon = \dfrac{1}{2}(h_w u^2 + h_w v^2 + g\eta^2)$ and $\varepsilon_0 = \dfrac{1}{2}(u^2 + v^2 + 2g\eta)$. Integration of (20) over the area, Ω, occupied by the waves and slide yields,

$$\frac{\partial E}{\partial t} = \iint_\Omega g\eta \frac{\partial D}{\partial t} ds - \int_\Gamma \varepsilon_0 h_w u_n dl, \tag{21}$$

where Γ is the open boundary, u_n is the current velocity component normal to this boundary, and $E = \iint_\Omega \varepsilon ds$ is the wave energy. Equation (21) has a simple interpretation: changes in wave energy in the tsunami domain are the sum of the wave generation within the domain and the energy flux through the open boundary. The first term in the right side of (21)

$$W = \iint_\Omega g\eta \frac{\partial D}{\partial t} ds \tag{22}$$

describes the rate-of-energy-generation (energy generated in the area, Ω, per unit time).

Figure 6 presents numerical estimates for the rate-of-energy-generation as a function of time for subaerial and submarine slides moving downslope. Here, the curves (solid lines) give the relative transfer of energy from the slide motion into the surface waves at different time and slides positions/depth. Also shown are the corresponding changes in Froude number. There is clearly good agreement between the two properties for subaerial slides, with sharp resonant peaks in the rate-of-energy transfer corresponding to $Fr = 1.0$ (Figure 6a,b). The small temporal shift in the resonance peak for the viscous subaerial slide arises from the fact that the maximum thickness of the slide lags a little behind the leading edge of the slide (the point at which the Froude number is estimated).

There also is a good correlation between Froude number and rate-of-energy-generation for submarine slides (Figure 6c, d). For example, the time of maximum rate-of-energy-

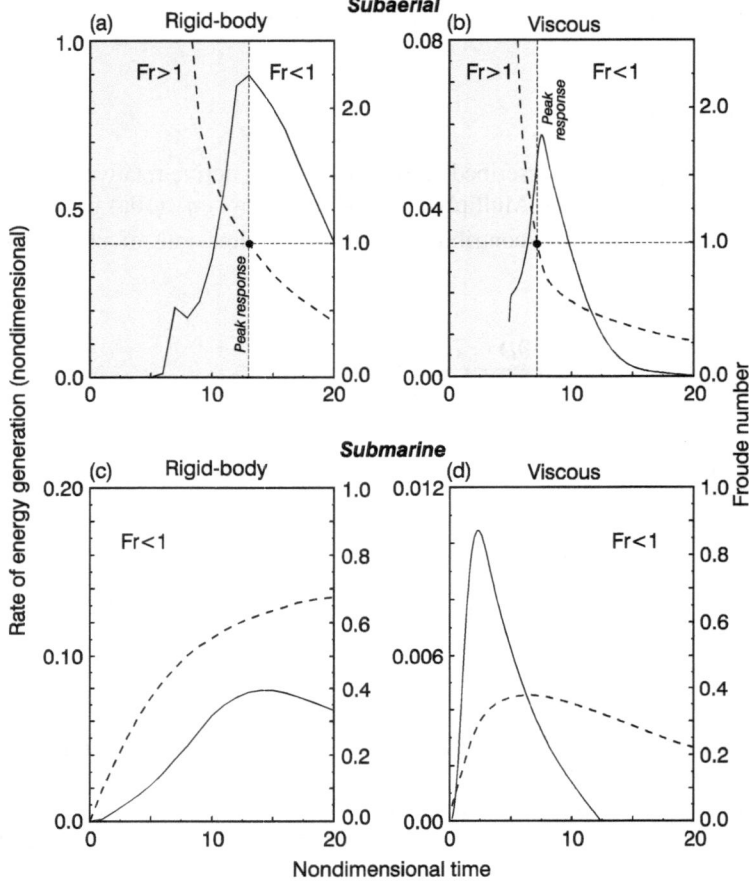

Figure 6. Rate-of-energy-generation (solid lines) as a function of nondimensional time for four slide models: (a) Rigid-body, subaerial; (b) viscous, subaerial; (c) rigid-body, submarine; and (d) viscous, submarine. Dashed lines show variations in the accompanying Froude number. Slope angle is $\psi =$ 4°, slide density $\rho_2 = 2.0$ g·cm^{-3}, friction coefficient k = 0.02 (for a rigid-body slide) and kinematic viscosity $v = 0.01$ m^2·s^{-1} (for a viscous slide).

generation for the viscous submarine slide occurs very close to the time of maximum Froude number (*Fr* = 0.45) (Figure 6d). The rate-of-energy-generation for the rigid-body submarine slide increases with increasing *Fr*. This effect is somewhat hidden by changes in water depth which also have an important impact on wave generation, with wave generation decreasing with increasing depth. As a consequence, the rate-of-energy-generation has a maximum at *t* = 13 (Figure 6c) despite the fact that *Fr* keeps increasing for an infinite slope.

6. Influence of Initial Slide Conditions

It is instructive to examine the effect of initial slide position (i.e. height above sea level for subaerial slides or water depth for submarine slides) on surface wave generation. Results from numerical experiments for various slope angles for rigid-body and viscous models were surprisingly different (Figure 7). For a rigid-body slide, the greater the initial slide height above sea level, the more energetic the generated tsunami waves (Figure 7a).

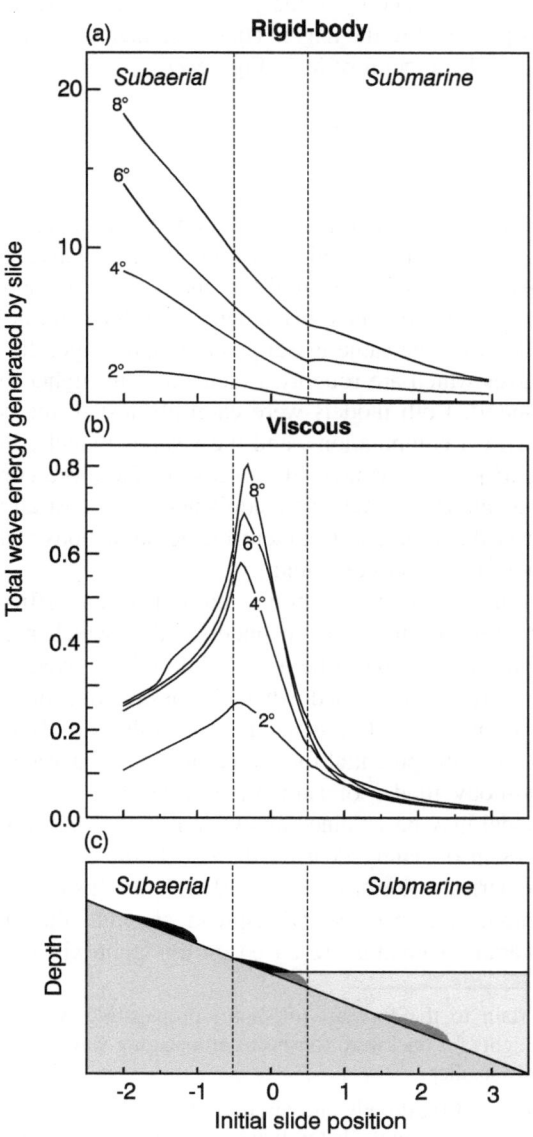

Figure 7. Total wave energy as functions of slope angle and the initial slide position for: (a) A rigid-body slide; and (b) a viscous slide. The initial slide positions are shown in (c). Parameters of the slides are the same as in Figure 6.

This result is in agreement with laboratory and computational results by *Watts* [30, 31] who demonstrated that rigid-body landslide generated tsunamis have amplitudes proportional to their vertical center of mass displacement. For viscous slides, there is an optimal slide position (height), located close to the coastline, which produces the largest tsunami waves. Slides initially located above or below this position generate less energetic waves (Figure 7b).

Our experiments with slope angles in the range from 2° to 10° for both models (rigid-body and viscous) gave identical results: The steeper the slope, the higher the generated surface waves. In general, the energy of the generated tsunami waves for this range of slope angles is roughly proportional to the angle. Thus, a change of angle from 2° to 8° increases the surface wave energy by a factor of four (Figure 7).

7. Discussion

We have conducted a suite of numerical simulations to compare tsunami wave generation by rigid-body and viscous slides and to examine the relative contributions from the subaerial and submarine components of these slides. The numerical experiments reveal significant differences in the computed wave heights for the rigid-body and viscous models. Specifically, for slides with the same initial position and shape, the rigid-body slide model produces surface waves which are roughly a factor of four higher than waves produced by the viscous slide model[2]. Both models were carefully tested, the rigid-body model being verified against analytical computations, and the viscous model compared with the results of *Jiang and LeBlond* [1, 13]. Model differences are therefore considered to be physical rather than mathematical. The models were equivalent in almost all aspects, except that the rigid-body slide treats the slide as an undeformable solid body, while the viscous model allows the slide to deform as a viscous fluid[3].

It is clear from our study that rigid-bodies are much more efficient tsunami generators than viscous slides. What is important to understand is which model is more *physically realistic* and therefore better to describe actual slide failure events. There are at least two key scientific aspects to be addressed when discussing this problem: (1) The accurate numerical modeling of historical slide-generated tsunamis; and (2) the estimation of tsunami risk for coastal areas with the potential for submarine and subaerial slides.

Use of the rigid-body model of submarine slides has a long history. Some simple estimates for this model may be obtained analytically [29, 30, 31]. Laboratory modeling of slide-generated waves also mainly deals with rigid bodies, which can then be compared with analytical or numerical solutions [27, 19, 30, 31]. On this basis, it makes sense that use of the rigid-body model has been so widespread and that the majority of publications examining slide-generated tsunamis are based on this approximation. Unfortunately, most

---

[2] These estimates pertain to the forward (offshore) propagating waves in Figures 4 and 5; the differences in wave heights for backward (onshore) propagating waves (i.e. run-up heights) for these models are considerably smaller.

[3] *Watts et al.* [32] compared rigid-body and viscous slide models and found that the former model generates higher surface waves than the latter; however, in their case the difference was much smaller than in ours, probably because parameters of their models were fitted to reduce this difference. In addition, their rigid-body model was simply a boundary element model for potential flow, and their viscous model a shallow-water two-layer model similar to JLB94.

geotechnical information for land and marine slides, slumps, avalanches, and rock falls indicate that the rigid-body approximation is too simplistic and that the viscous fluid model better describes these processes [33]. For this reason, the viscous fluid slide model, first rigorously formulated by *Jiang and LeBlond* [1, 13], is now widely used to simulate catastrophic tsunamis arising from submarine landslides. Examples include Nice, France (1979) [18], Skagway (1994) [14, 8, 15], and PNG (1999) [4, 16, 17]. In all of the above events, the viscous model gives reasonable agreement with the existing empirical data.

The consistency between the observed tsunami waves and numerical simulations of these waves using the viscous slide model implies that this is the preferred model for determining landslide-generated tsunami risk along the coast (i.e. possible tsunami heights which could be caused by potential submarine or subaerial landslides). Long-term tsunami prediction in coastal regions ("tsunami-zoning") is a key problem for hazard mitigation and long-term planning, and is also a necessary element of any planning for new construction in the coastal zone [34,35]. A common argument is that a rigid-body model produces higher tsunami waves, so this model should be used for estimation of tsunami risk. We disagree with this notion. As the present results indicate, the rigid-body model strongly overestimates actual tsunami heights. If accepted for engineering design, such overestimation could greatly increase construction costs. There are no perfect models. Nevertheless, it is important to use the *best available science* for tsunami predictions [*Frank González*, Pers. Comm.]. In our opinion, the results from the viscous slide model are more realistic than those from the rigid-body model, making the viscous model the model of choice.

Another important consideration is the influence of the subaerial component of landslides on tsunami generation. Although most existing models omit this component of the slide, our study demonstrates that it is of primary importance. There are two fundamental effects associated with this component: (1) The subaerial component of the slide adds an additional volume to the ocean, causing a corresponding displacement of the sea surface; and (2) there is always a point (in time and space) where $Fr = 1.0$ for which there is a resonant transfer of energy from the slide into the surface waves. In contrast, for purely submarine slides with $\rho_2 \leq 2.0$ g·cm^{-3}, the Froude number is always less than unity and resonance coupling of slides and surface waves is physically impossible (at least on Earth). Due to these two effects, subaerial slides are much more efficient tsunami generators than submarine slides. The catastrophic consequences and destructive waves associated with a relatively small slide failure in 1994 in Skagway Harbor were apparently initiated by the subaerial component of the slide. Incorporating this component into existing models is essential for correct representation of natural events.

For subaerial slides, the energy of the generated tsunami waves strongly depends on the initial slide height above sea level. Rigid-body subaerial slides with higher initial elevations generate larger tsunami waves because their potential energy is greater, resulting in more energy being transferred into the surface waves (for similar reasons, earthquakes of larger magnitude generally produce higher tsunamis). However, for viscous slides there is an additional opposing effect. Because different parts of a viscous slide move with different velocities, the slide stretches and spreads during its movement downslope, forming a long sediment tail (see Figure 3). As a consequence, the tsunami-generating efficiency of the slide is reduced. The greater the distance traveled by the slide, the more pronounced is this effect. The existence of an optimal subaerial point (slide height) associated with maximum generated tsunami waves is apparently the direct consequence of these two opposing effects.

8. Conclusions

The findings of our numerical simulations can be summarized in point form as follows:

(1) A rigid-body slide has greater tsunami-generating efficiency and produces much higher tsunami waves than a viscous (liquid) slide. The choice of "most realistic" slide model should be a physical, rather than a mathematical, consideration. We argue that, in most cases, the viscous slide model is the model of choice.

(2) The maximum wave height and energy of generated surface waves depend on various slide parameters and factors, including: Slide volume, type of slide (viscous or rigid), slide density, slide position (relative height or depth), and slope angle. For a rigid-body slide, the higher the initial slide above sea level, the higher the generated waves. For a viscous slide, there is an optimal slide position (height), located close to the shore, which produces the largest tsunami waves. Slides initially located above or below this position generate less energetic waves. An increase in slide volume, density, and slope angle always increases the energy of the generated waves.

(3) The added volume that occurs when a subaerial slide enters the water results in a displacement of the sea surface and a significant increase in height of the leading wave crest.

(4) The critical parameter determining the generation of surface waves is the Froude number (the ratio between the slide and wave speeds). The most efficient generation occurs near resonance when $Fr = 1.0$. For purely submarine slides with water density $\rho_1 \sim 1.0$ g·cm^{-3} and slide density $\rho_2 \leq 2.0$ g·cm^{-3}, the Froude number is always less than unity and resonance coupling of slides and surface waves is physically impossible. For subaerial slides there is always a resonant point (in time and space) where $Fr = 1.0$ for which there is a significant transfer of energy from a slide into surface waves. This effect, combined with the displacement of the sea surface (point 3, above), are the two main reasons that subaerial slides are much more important for tsunami generation than submarine slides.

Acknowledgements

The authors thank Patricia Kimber for drafting the figures. This research was partially sponsored by INTAS project 99-1600.

References

1. Jiang, L. and LeBlond, P.H.: 1994, Three-dimensional modeling of tsunami generation due to a submarine mudslide, *J. Phys. Oceanogr.* **24** (3), 559-572.
2. Imamura, F., and Gica, E.C.: 1996, Numerical model for tsunami generation due to subaqueous landslide along a coast, *Science of Tsunami Hazards* **14** (1), 13-28.
3. Tappin, D. et al.: 1999, Sediment slump likely caused 1998 Papua New Guinea tsunami, *EOS* **80**, 329, 334, 340.
4. Heinrich, P., Piatensi, A., Okal, E., and Hébert, H.: 2000, Near-field modeling of the July 17, 1998 tsunami in Papua New Guinea, *Geophys. Res. Let.* **27**, 3037-3040.
5. Altinok, Y., Alpar, B., Ersoy, S. and Yalciner, A.C.: 1999, Tsunami generation of the Kocaeli Earthquake (August 17th, 1999) in the Izmit Bay: Coastal observations, bathymetry and seismic data, *Turkish J. Marine*

Sciences **5** (3), 131-148.

6. Kulikov, E.A., Rabinovich, A.B., Thomson, R.E., and Bornhold, B.D.: 1996, The landslide tsunami of November 3, 1994, Skagway Harbor, Alaska, *J. Geophys. Res.* **101** (C3), 6609-6615.

7. Kowalik, Z.: 1997, Landslide-generated tsunami in Skagway, Alaska, *Science of Tsunami Hazards* **15** (2), 89-106.

8. Rabinovich, A.B., Thomson, R.E., Kulikov, E.A., Bornhold, B.D., and Fine, I.V.: 1999, The landslide-generated tsunami of November 3, 1994 in Skagway Harbor, Alaska: A case study, *Geophys. Res. Let.* **26**, (19), 3009-3012.

9. LeBlond, P.H. and Jones, A.T.: 1995, Underwater landslides ineffective at tsunami generation, *Science of Tsunami Hazards* **13** (1), 25-26.

10. Miller, D.J.: 1960, The Alaska Earthquake on July 10, 1958: Giant wave in Lituya Bay, *Bull. Seism. Soc. America* **50** (2), 253-266.

11. Lander, J.F.: 1996, *Tsunamis Affecting Alaska, 1737-1996*. Boulder, US Dep. Comm., 195 p.

12. Raichlen, F., J.J. Lee, C. Petroff, and P. Watts, 1996: The generation of waves by a landslide: Skagway, Alaska: A case study, *Proc. 25th Coastal Eng. Conf.*, ASCE, Orlando, Florida, 1478-1490.

13. Jiang, L. and LeBlond, P.H.: 1992, The coupling of a submarine slide and the surface waves which it generates, *J. Geophys. Res.* **97** (C8), 12,731-12,744.

14. Fine, I.V., Rabinovich, A.B., Kulikov, E.A., Thomson, R.E., and Bornhold, B.D.: 1998, Numerical modelling of landslide-generated tsunamis with application to the Skagway Harbor tsunami of November 3, 1994, *Proc. Int. Conf. on Tsunamis*, Paris, 211-223.

15. Thomson, R.E., Rabinovich, A.B., Kulikov, E.A., Fine, I.V., and Bornhold, B.D.: 2001, On numerical simulation of the landslide-generated tsunami of November 3, 1994 in Skagway Harbor, Alaska, in *Tsunami Research at the End of a Critical Decade*, edited by G. Hebenstreit, Kluwer, Dordrecht, 243-282.

16. Titov, V.V., and González, F.: 2001, Numerical study of the source of the July 1998 PNG tsunami, in *Tsunami Research at the End of a Critical Decade*, edited by G. Hebenstreit, Kluwer, Dordrecht, 197-207.

17. Imamura, F., Hashi, K., and Imteaz, Md.M.A.: 2001, Modeling for tsunamis generated by landsliding and debris flow, in *Tsunami Research at the End of a Critical Decade*, edited by G. Hebenstreit, Kluwer, Dordrecht, 209-228.

18. Assier-Rzadkiewicz, S., Heinrich, P., Sabatier, P.C., Savoye, B., and Bourillet, J.F.: 2000, Numerical modelling of landslide-generated tsunami: The 1979 Nice event, *Pure Appl. Geophys.* **157**, 1707-1727.

19. Heinrich, P.: 1992, Nonlinear water waves generated by submarine and aerial landslides, *J. Waterways, Port, Coastal and Ocean Eng.*, ASCE, **118** (3), 249-266.

20. Heinrich, P., Mangeney, A., Guibourg, S., Roche, R., Boudon, G., and Vheminée, J.-L.: 1998, Simulation of water waves generated by a potential debris avalanche in Montserrat, Lesser Antilles, *Geophys. Res. Let.* **25** (19), 3697-3700.

21. Norem, H., J. Locat, and B. Schieldrop: 1991, An approach to the physics and modeling of submarine flowslides, *Mar. Geotechnol.* **9**, 93-111.

22. Harbitz, C.B.: 1992, Model simulations of tsunamis generated by the Storegga slides, *Marine Geology* **105**, 1-21.

23. Imamura, F.: 1996, Review of tsunami simulation with finite difference method, in *Long-Wave Runup Models,* edited by H. Yeh, P. Liu, and C. Synolakis, World Scientific, Singapore, 25-42.

24. Roache, P.J., 1976: *Computational Fluid Dynamics*, Hermosa Publ., Albuquerque, N.M., 446 p.

25. Mader, C.L.: 1988, *Numerical Modeling of Water Waves*, Univ. California Press, Berkeley.

26. Titov, V.V., and Synolakis, C.E.: 1998, Numerical modeling of tidal wave runup, *J. Waterw., Port, Coastal and Ocean Eng.*, ASCE, **124**, (3), 157-171.

27. Wiegel, R.L.: 1955, Laboratory studies of gravity waves generated by the movement of a submerged body, *Trans. Am. Geophys. Union* **36** (5), 759-774.

28. Proudman, J.: 1953, *Dynamic Oceanography*, Methuen and Co., London, 409 p.

29. Pelinovsky, E., and Poplavsky, A.: 1996, Simplified model of tsunami generation by submarine landslides, *Phys. Chem. Earth* **21** (12), 13-17.

30. Watts, P.: 1998, Wavemaker curves for tsunamis generated by underwater landslides, *J. Waterways, Port, Coastal and Ocean Eng.*, ASCE, **124** (3), 127-137.

31. Watts, P.: 2000, Tsunami features of solid block underwater landslides, *J. Waterways, Port, Coastal and Ocean Eng.*, ASCE, **126**, (3), 144-152.

32. Watts, P., Imamura, F., and Grilli, S.: 2000, Comparing model simulations of three benchmark tsunami generation cases, *Science of Tsunami Hazards* **16** (2), 107-123.

33. Andresen, A., and Bjerrum, L.: 1967, Slides in subaqueous slopes in loose sand and silt, in *Marine Geotechnique*, edited by A.F. Richards, Univ. Illinois Press, Urbana, 221-239.

88

34. Chaudhry, M.H., Mercer, A.G., and Cass, D.: 1983, Modeling of slide-generated waves in a reservoir, *J. Hydr. Engin.*, ASCE, **109** (11), 1505-1520.
35. Mofjeld, H.O., González, F.I., and Newman, J.C.: 1999, Tsunami prediction in U.S. coastal regions, in *Coastal Ocean Prediction, Ch. 14, Coastal and Estuarine Studies 56* edited by C.N.K. Mooers, AGU, Washington, 353-375.

TSUNAMI SIMULATION TAKING INTO ACCOUNT SEISMICALLY INDUCED DYNAMIC SEABED DISPLACEMENT AND ACOUSTIC EFFECTS OF WATER

TATSUO OHMACHI
Tokyo Institute of Technology, Tokyo, Japan

1. Introduction

Recently, we have presented a new technique to simulate generation and propagation of tsunamis [1]. In contrast with convention [2] where the initial sea surface is assumed to be the same as the seismically induced static displacement of the seabed and propagation (tsunamis) is simulated using the long-wave approximation, the new technique takes into account effects of dynamic seabed displacement resulting from seismic faulting, as well as of acoustic water waves.

Due to an assumption of the new technique that the seawater-seabed system can be regarded as a weakly coupled system, our numerical simulation is made up of two steps. The first is earthquake ground motion simulation for which the boundary element method (BEM) is used, and the second is tsunami simulation for which the finite difference method (FDM) is used. Considering a rupture mechanism of seismic faulting, the dynamic seabed displacement is first simulated, including the static displacement in near-fault area. The velocity associated with the seabed displacement is input accordingly at the bottom of the seawater, and the resulting seawater disturbance is simulated by solving the Navier-Stokes equations without using the long wave approximation, introducing a height function. To secure reliable solutions, the CFL stability criterion [3] is used.

2. Simulation in two Dimensions

An analytical model used in a 2-D simulation is shown in Fig. 1. The fault model is a thrust-faulting with dipping at 30 degrees, 30km wide and 5km deep under the seabed. Dislocation is 10m, rupture velocity is 3km/s, and rise time is 2sec. The fault rupturing starts at the bottom and propagates upwards. The water depth is 3km, and simulated area extends to 100km on both sides of the epicenter

Snapshots from the simulation are shown in Fig. 2, in which the lower and upper surfaces represent the seabed and the sea surface, respectively. The time indicated in the snapshots is elapsed time from the initial fault rupturing. At about 10 seconds after the fault rupturing, the seabed shows the maximum uplift, which is 1.6 times as large as the static displacement and a few seconds later, the sea surface shows the maximum water wave height. At 15 seconds, the Rayleigh wave appears mainly on the right side of the peak, and travels to the right along the seabed. A single peak on the sea surface is divided into two peaks. One of them is propagated to the right at a same velocity as the Rayleigh wave,

A. C. Yalçıner, E. Pelinovsky, E. Okal, C. E. Synolakis (eds.),
Submarine Landslides and Tsunamis 89-99.
©2003 Kluwer Academic Publishers.

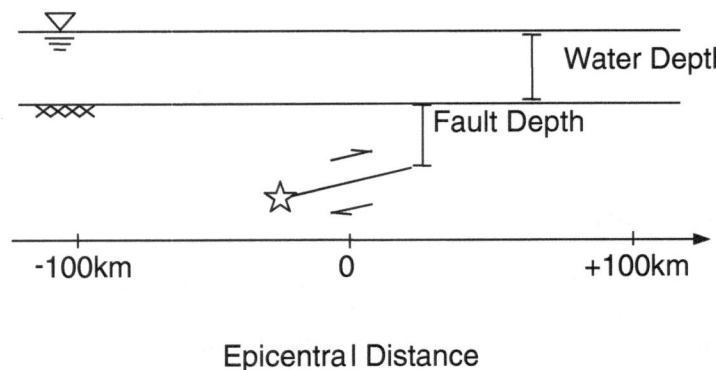

Figure1 . A 2D tsunami simulation model.

producing several later phases. The water wave is called the oceanic Rayleigh wave. After 40 seconds, the seabed in this area stops its motion, and shows the static displacement. After 50 seconds, a single peak on the sea surface is flattened a little, divided into two and continue propagation as tsunami waves. The height of the water wave and velocity of the tsunami are just the same as those of the long waves

Thus, the height of the water wave in near-field is found to be remarkably larger than that of the static seabed displacement. In this case, it is almost twice. This is mainly due to superposition of two types of waves. One is the oceanic Rayleigh wave and the other is tsunami.

3. Simulation in Three Dimension the 1998 Sanriku-oki Earthquake Tsunami.

At the bottom of the Pacific Ocean, off Sanriku coast, Japan are deployed monitoring systems for earthquakes and tsunamis by researchers of Tohoku University and University of Tokyo [4]. On May 30, 1998, the systems recorded time histories of an earthquake (M_w 6.1) followed by a tsunami at stations shown in Fig. 3, in which OBS1, OBS2 and OBS3 are ground motion stations, and TM1 and TM2 are tsunami stations. Fault parameters of the earthquake are shown in Table 1.

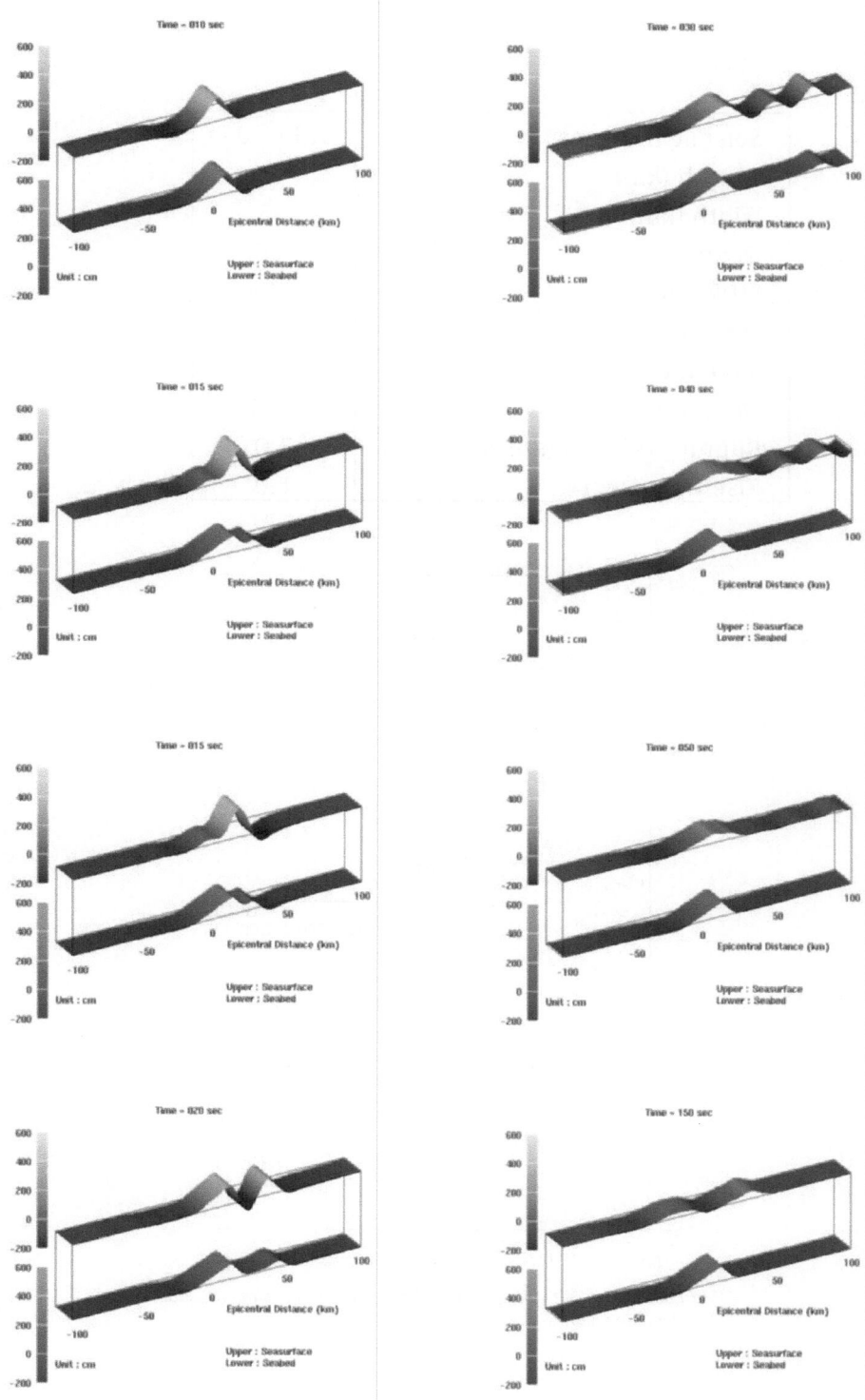

Figure 2. 2D tsunami simulation by the present technique.

TABLE 1. Fault parameters.

Seismic moment (dyne·cm)	1.9×10^{25}
Length (km)	15
Width (km)	15
Depth (km)	10
Strike (degree)	204
Dip (degree)	13
Rake (degree)	95
Dislocation (m)	0.3
Rupture velocity (km/sec)	3.0
Rise time (sec)	1.0

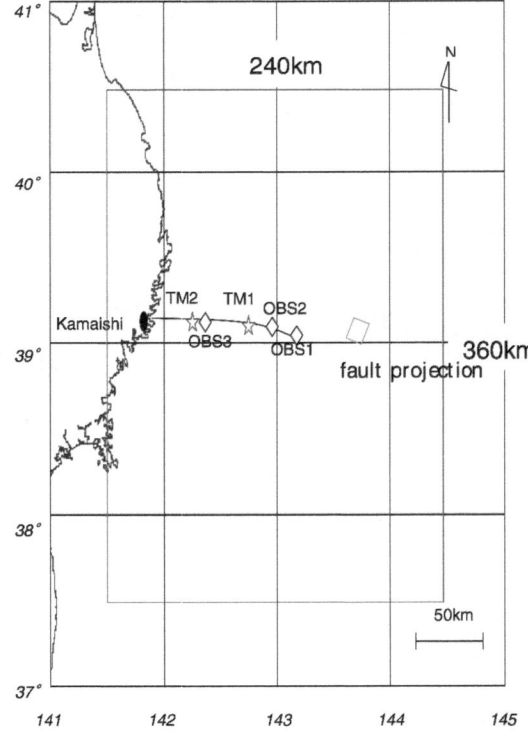

Figure 3. Fault plane projection, monitoring stations and calculation area.

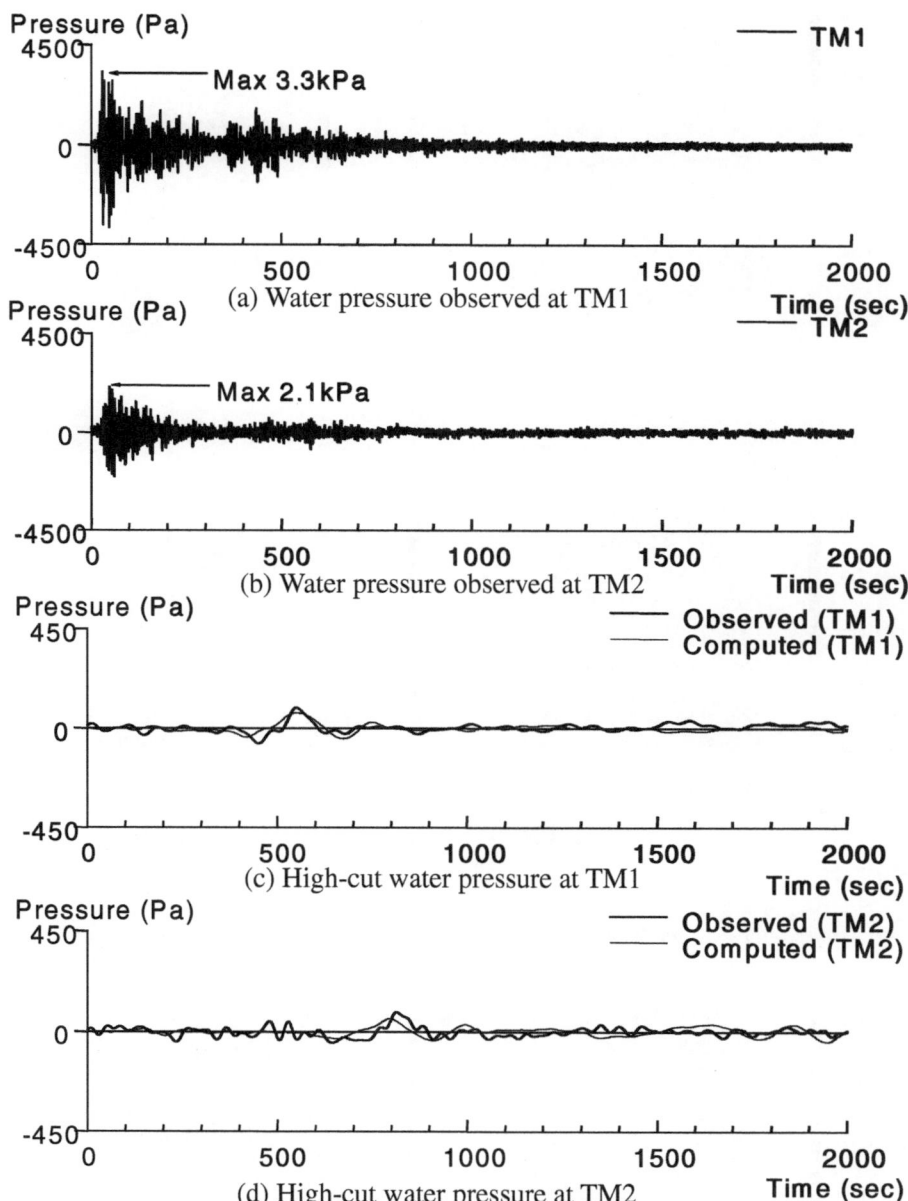

Figure 4. Comparison of the observations and the results of the simulation.

94

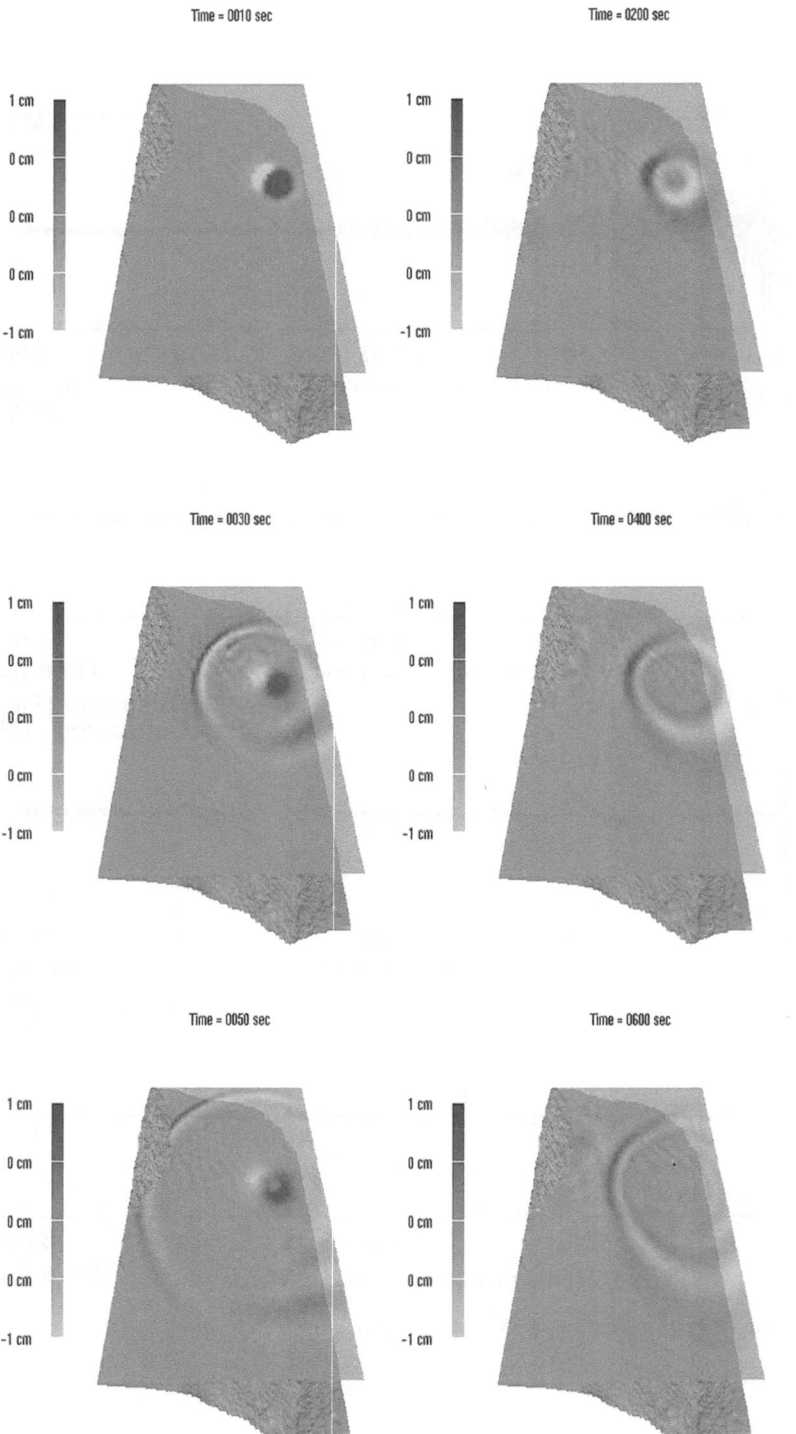

Figure 5. 3D tsunami simulation of the 1998 Sanriku-oki, Japan earthquake tsunami.

The simulation area is 360km long and 240km wide as shown in Fig. 3. In Fig. 4, time histories of the water pressure at TM1 and TM2 are shown. Since high frequency components are apparently predominant in the original records shown in (a) and (b), they are reduced by a low-pass filter, with the results shown in (c) and (d) in which the simulated time histories are also shown. Although the water level change equivalent to the pressure change, which can be stated as a small tsunami, is as small as 1cm, the simulated pressure agrees well with the observation, demonstrating validity of our technique. According to snapshots shown in Fig. 5, sea surface disturbance generated by the oceanic Rayleigh waves passes the stations TM1 and TM2 before 50seconds, from which the high frequency components in the original pressure records in Fig. 4(a) and (b) are found to be associated with seismic waves such as the oceanic Rayleigh waves.

4. The 1983 Nihonkai-chubu Earthquake Tsunami

The Nihonkai-chubu earthquake (M7.7) occurred on May 26, 1983. Among many fault models, Sato's model [5] estimated from seismic data is used in the present simulation, in an attempt to reduce the difference between fault models from tsunami data and those from seismic data. As shown in Fig. 6 and Table 3, the fault model consists of three sub-faults. The first fault rupture supposedly started at the south tip of the southern sub-fault, propagating in NE direction at 3km/s. After 10seconds interval, the second fault rupture developed on the central sub-fault at 2km/s in NE direction, and finally the third rupture propagated on the northern sub-fault in NNW direction at 1.5km/s, with the total rupture time of 63sec, as shown in Figs. 7 and 8. From snapshots shown in Fig. 9, apparently the patterns of the sea surface disturbance including tsunamis are somewhat different from that simulated by the conventional (static) technique, especially in the near-fault area during some time immediately after the fault rupturing. Because of the difference, to the author's belief, the present technique will help us to characterize near-field tsunamis with accuracy, and to reduce the difference between fault models from tsunami data and those from seismic data.

96

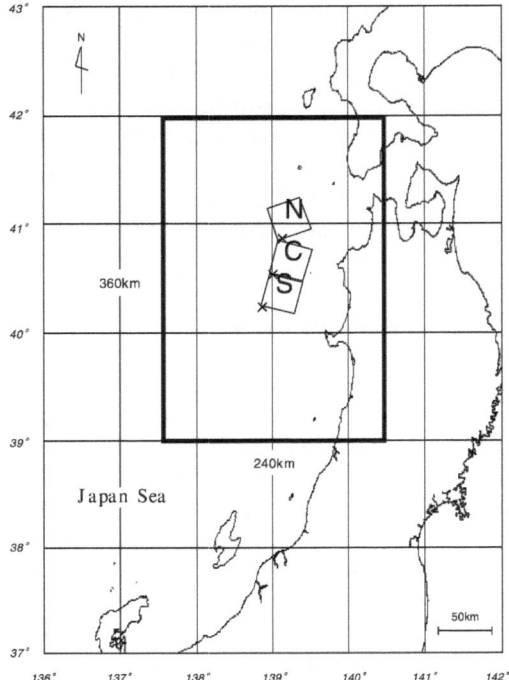

Figure 6. Projection of the fault planes projections and calculation area of the 1983 Nihonkai-chubu earthquake tsunami.

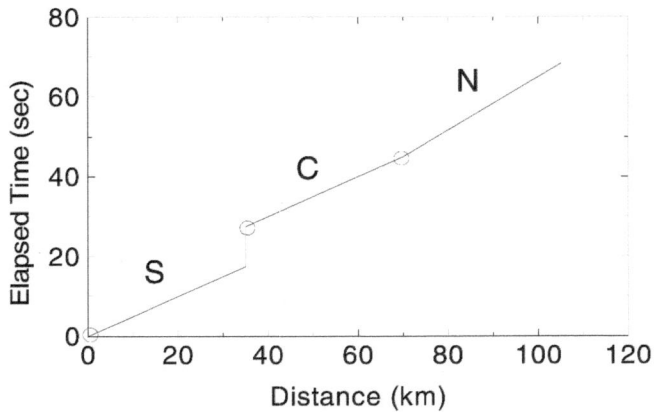

Figure 7. Relation between elapsed time and rupture distance.

TABLE 2. *Fault parameters of the 1983 Nihonkai-chubu earthquake.*

	South	Central	North
Seismic moment (dyne·cm)	3×10^{27}	2×10^{27}	3×10^{27}
Length (km)	35	35	35
Width (km)	35	35	35
Depth(km)	0	0	0
Strike (degree)	15	15	345
Dip (degree)	20	20	20
Rake (degree)	90	90	90
Dislocation(m)	6.8	4.6	6.8
Rupture velocity(km/sec)	2.0	2.0	3.0
Rise time(sec)	3.5	3.5	3.0

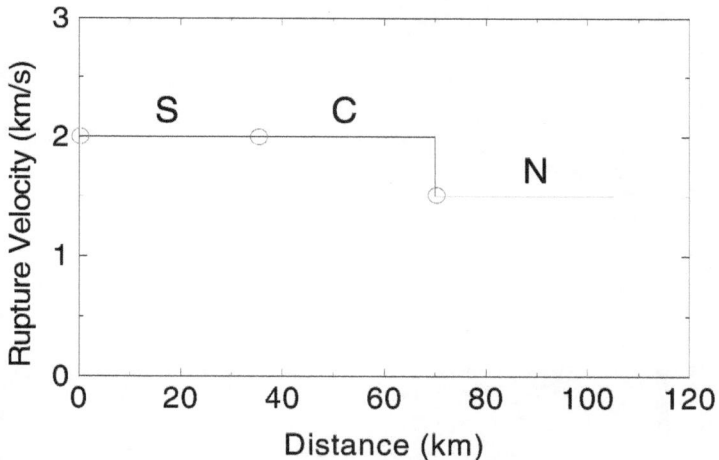

Figure 8. *Relation between rupture velocity and rupture distance.*

98

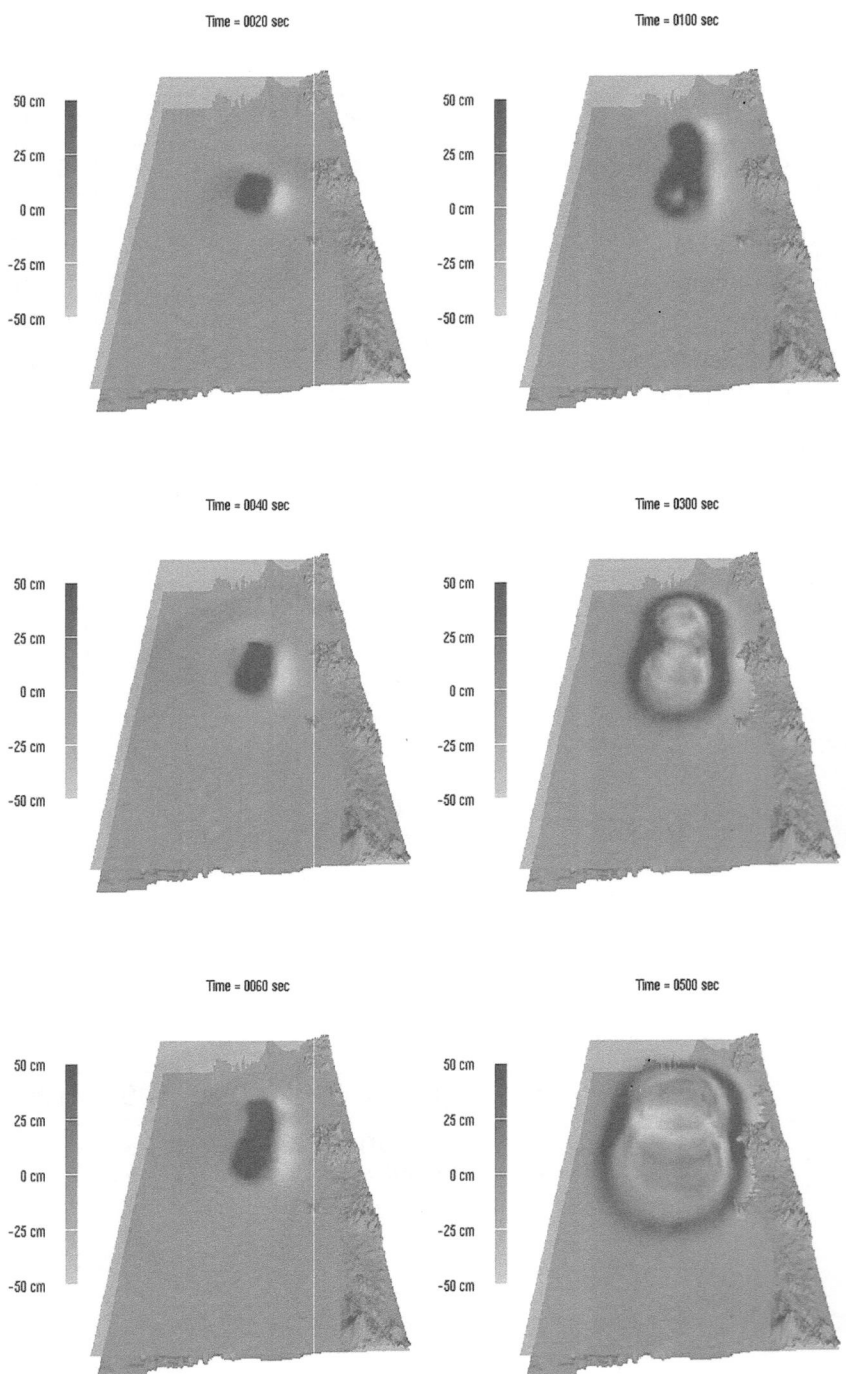

Figure 9. Three dimensional tsunami simulation about the 1983 Nihonkai-chubu earthquake tsunami.

5. Conclusions

A series of dynamic simulation of tsunami were conducted, taking into account the effects of dynamic seabed displacement caused by seismic faulting and acoustic waves of the seawater. Although there are several points to be improved in the present dynamic simulation technique, findings drawn from the present study are the followings;

1. In comparison with the seismically induced static seabed displacement, the dynamic seabed displacement makes a remarkable contribution to increase water wave height, especially in the near-fault area.

2. The increase is mainly due to superposition of two types of water waves. One is the tsunami that travels as a long wave, and the other is the oceanic Rayleigh wave that travels much faster than the tsunami.

3. In the far-field, there is little difference between the wave height from dynamic simulation and that from static simulations.

4. From case studies on the 1983 Nihonkai-chubu Earthquake tsunami and the 1989 Sanriku-oki Earthquake tsunami, the present technique has proved the validity of the simulation results and advantage over the conventional techniques.

Acknowledgements

The author is grateful to Dr. H. Tsukiyama, Tsukiyama Research Inc., and Dr. H. Matsumoto, Japan Marine Science and Technology Center, for their collaboration in the present study.

References

1. Ohmachi, T., Tsukiyama, H, and Matsumoto, H. (2001): Simulation of Tsunami Induced by Dynamic Displacement of Seabed due to Seismic Faulting, Bulletin of the Seismological Society of America, 91, 6, pp. 1898-1909.
2. Aida, I. (1984): A Source Model of the Tsunami Accompanying the 1983 Nihonkai-Chubu Earthquake, Bull. Res. Inst., Vol. 59, pp. 93-104. (in Japanese)
3. Courant, R., K. O. Friendrichs, and H. Lewy (1967): On the partial difference equations of mathematical physics, IBMJ Res. Dev., Vol. 11, pp. 215-234.
4. Hino, R., Tanioka, Y., Kanazawa, T., Sakai, S. and Nishino M. (2001): Micro-tsunami from a local interplate earthquake detected by cabled offshore tsunami observation in northeastern Japan, Geophysical Research Letters, Vol. 28, No. 18, pp.3533-3536, September 15..
5. Sato, T. (1985): Rupture Characterisitics of the 1983 Nihonkai-chubu (Japan sea) Earthquake as Inferred from Strong Motion Accelerograms, Journal of Physics of the Earth, Vol. 33, pp.525-557.

IMPULSIVE TSUNAMI GENERATION BY RAPID BOTTOM DEFLECTIONS AT INITIALLY UNIFORM DEPTH

P. A. TYVAND[1], T. MILOH[2] and K. B. HAUGEN[1]
[1]*Department of Agricultural Engineering*
Agricultural University of Norway, 1432 Aas Norway
[2]*Faculty of Engineering Tel Aviv University,*
Ramat Aviv 69978 Israel

Abstract

An analytical small-time expansion is developed for impulsive tsunami generation at initially uniform depth in two and three dimensions. The flow is due to impulsive bottom deflections that are rapid, with duration shorter than the gravitational time scale for the given depth. The bottom deflection is assumed separable in space and time. A third-order small-time expansion is given. The first and second-order solutions are general, while only the gravitational effect is included. The solutions are given as Fourier transforms and Green function integrals. The solutions are integrated numerically for the simple case of a rising block.

1. Introduction

The present paper is concerned with tsunami generation due to impulsive bottom deflections which has a shorter duration than the gravitational time $(h/g)^{1/2}$. Here h is the ocean depth, which is assumed initially constant, and g is the gravitational acceleration. For typical ocean depths of 3000 meters, this means that the significant bottom motions must have stopped after 15 seconds. This constraint may apply to tsunamis generated by earthquakes, but not to tsunamis due to landslides or volcanic eruptions.

The present work is only concerned with the evolution of the wave shape during a rapid bottom deflection, not with the subsequent wave propagation. After the rapid bottom motion has ceased, the wave will evolve as a Cauchy-Poisson problem with a rigid bottom.

Traditional numerical modelling of tsunami generation due to rapid bottom motion has often taken the surface deflection to be the same as that of the bottom [1]. This piston-type hydrostatic modelling is not supported by Laplace's equation. Analytical descriptions of tsunami generation and propagation have been made on the basis of Laplace's equation ([2], [3], [4]).

Detailed analyses of the flow during a rapid impulsive generation process are still lacking. Tyvand and Storhaug [5] and Miloh, Tyvand and Zilman [6] have developed Green functions for the initial flow for various bottom topographies in two and three dimensions. This covers only the first-order theory in a small-time expansion. It has the disadvantage that it disregards nonlinearity as well as the early gravitational flow during the generation. In the present work a third-order small-time expansion will be developed, which accounts for early nonlinearity and early gravitational effects.

A. C. Yalçıner, E. Pelinovsky, E. Okal, C. E. Synolakis (eds.),
Submarine Landslides and Tsunamis 101-109.

2. Formulation of the problem

We consider inviscid flow in an incompressible fluid, which is initially at rest with constant depth h. The gravitational acceleration is g. The free surface of the fluid is subject to constant atmospheric pressure. An impulsive flow due to rapid bottom deflections starts at time $t=0$. From Kelvin's circulation theorem the flow will be irrotational and obey Laplace's equation:

$$\nabla^2 \Phi = 0 \tag{1}$$

The velocity field is the gradient of the velocity potential $\Phi(x,y,z,t)$. The surface elevation is $\eta(x,y,t)$. We apply dimensionless variables with unit of length h, unit of time $(h/g)^{1/2}$, unit of velocity $(gh)^{1/2}$. The z axis points vertically upwards, while the x, y plane is in the undisturbed free surface. We will investigate the early surface disturbance generated by a rapid deflection $\xi(x,y,t)$ of an initially horizontal and impermeable bottom. The deflection starts impulsively at $t=0$, which gives the initial conditions:

$$\Phi(x,y,0,0)=0, \quad \eta(x,y,0)=0 \tag{2}$$

The bottom deflection is assumed to be separable in space and time:

$$\xi(x,y,t) = Z(x,y)\,T(t) \tag{3}$$

Apart from this separation constraint, the functions $Z(x,y)$ and $T(t)$ can be quite arbitrary. But in order for our small-time expansion to be valid we must put three additional restrictions on $\xi(x,y,t)$:

(i) The maximum value of ξ must be small compared with the constant unit depth. Otherwise nonlinear effects would be too strong to be captured by a second-order theory.

(ii) The forced bottom velocity $\partial \xi/\partial t$ must be negligible for times greater than one. More precisely: $T'(t)$ must be approximately zero for $t>t_0$ where t_0 is somewhat smaller than one.

(iii) The forced bottom motion should start with a nonzero velocity: $T'(0)\neq 0$. If this is not the case, the orders of approximation cannot be classified the way this is done in the following.

Surface conditions:

- Kinematic: $\quad \partial\eta/\partial t + \nabla\Phi\bullet\nabla\eta = \partial\Phi/\partial z, \qquad z=\eta(x,y,t) \tag{4}$

- Dynamic: $\quad \partial\Phi/\partial t + (1/2)|\nabla\Phi|^2 + z = 0, \qquad z=\eta(x,y,t) \tag{5}$

Bottom condition:

- Kinematic: $\quad \partial\xi/\partial t + \nabla\Phi\bullet\nabla\xi = \partial\Phi/\partial z, \qquad z = -1+\xi(x,y,t) \tag{6}$

We develop an asymptotic series approximating the surface elevation $\eta(x,y,t)$. We seek to separate each term in the time and space-dependencies in order to get an asymptotic series with n terms:

$$\eta(x,y,t)= H(t) \sum_{i=1}^{n} \eta_i (x,y) \, T_i(t) \tag{7}$$

Here $H(t)$ denotes the Heaviside unit step function, which is defined equal to zero for $t \leq 0$ and equal to one for $t>0$.

Each term is governed by the given bottom deflection $\zeta(x,y,t)=Z(x,y)T(t)$ and the boundary conditions. The kinematic free-surface condition (4) will now be linearized. This is compatible with full second-order theory, and includes the third-order gravitational flow. This linearized kinematic condition makes the velocity potential separable to each order:

$$\Phi(x,y,z,t)= H(t) \sum_{i=1}^{n} \Phi_i (x,y,z) \, T_i'(t) \tag{8}$$

Here the time dependencies are the derivatives of the time dependencies of the corresponding surface elevations so that each $\Phi_i(x,y,z)$ meets the requirement:

$$\eta_i (x,y) = [\partial \Phi_i (x,y,z)/\partial z]_{z=0} \tag{9}$$

Boundary value problems to each order

With the restrictions we have put on $\zeta(x,y,t)$, the asymptotic series (7) may be useful up to $t=t_0$. Below the boundary value problems for the space dependent parts of the velocity potentials are listed along with the corresponding surface elevations. Laplace's equation is valid to each order. Also listed are the differential equations the time dependent parts must satisfy. There are three possible agents driving the flow to each order: The forced bottom motion, nonlinearity at the free surface, and gravity. The two former agents are active in second order, and their time dependencies are in general different. Therefore the second-order problem must be split into two sub-problems (subscripts $2B$ and $2S$).

First-order bottom contribution

$$\nabla^2 \Phi_1 = 0$$

$$\Phi_1 \big/_{z=0} = 0$$

$$[\partial \Phi_1/\partial z]_{z=-1} = Z(x,y)$$

$$\eta_1 (x,y) = [\partial \Phi_1/\partial z]_{z=0}$$

$$T_1'(t) = T'(t)$$

Second-order bottom contribution

$$\nabla^2 \Phi_{2B} = 0$$

$$\Phi_{2B}\,|_{z=0} = 0$$

$$[\partial\Phi_{2B}/\partial z]_{z=-1} = [\,\nabla\bullet(Z(x,y)\;\nabla\,\Phi_1)]_{z=-1}$$

$$\eta_{1B}\,(x,y) = [\partial\Phi_{1B}/\partial z]_{z=0}$$

$$T_{2B}{'}(t) = T(t)\;T'(t)$$

Second-order surface contribution

$$\nabla^2 \Phi_{2S} = 0$$

$$\Phi_{2S}\,|_{z=0} = -\;\eta_1^{\,2}\,/2$$

$$[\partial\Phi_{2S}/\partial z]_{z=-1} = 0$$

$$\eta_{1S}\,(x,y) = [\partial\Phi_{1S}/\partial z]_{z=0}$$

$$T_{2S}{''}(t) = (T'(t))^2$$

Third-order gravity contribution

$$\nabla^2 \Phi_3 = 0$$

$$\Phi_3\,|_{z=0} = -\;\eta_1$$

$$[\partial\Phi_3/\partial z]_{z=-1} = 0$$

$$\eta_3\,(x,y) = [\partial\Phi_3/\partial z]_{z=0}$$

$$T_3{''}(t) = T(t)$$

On the time factors

Let us summarize the conditions for the time factors to each order:

$$T_1(t) = T(t), \;\; T_{2B}(t) = (1/2)\,T(t)^2, \;\; T_{2S}{''}(t) = (T'(t))^2, \;\; T_3{''}(t) = T(t) \tag{10}$$

We have here taken into account the following constraints for an initially flat bottom:
$T_i(0)=T(0)=0$. In this note we do not give examples of integrating up the different functions $T_i(0)$ when $T(t)$ is given. But we note that $T_{2S}(t)$ is governed by a differential equation, which will usually give a solution different from the explicit formula for $T_{2B}(t)$. This shows the necessity of working with two second order sub-problems.

3. General solutions in two dimensions

In the present work we give only results in two dimensions. There are two different exact methods for constructing exact solutions. Below we show the complete results according to the two-dimensional Fourier transform method. First we show the Green function solutions for the linearized first and third order problems:

Green function approach

First-order contribution:

$$\eta_1(x) = \int_{-\infty}^{\infty} Z(x') \, sech \, ((\pi(x-x')/2) \, dx' \tag{11}$$

Third-order gravity contribution:

$$\eta_3(x) = \int_{-\infty}^{\infty} Z(x') \, [(\pi(x-x')/2) \, tanh \, ((\pi(x-x')/2) - 1] \quad sech \, ((\pi(x-x')/2) \quad dx' \tag{12}$$

Fourier transform approach

First-order contribution:

$$\eta_1(x) = (2 \, \pi)^{-1} \int_{-\infty}^{\infty} sech \, (k) \, \Xi \, (ik) \, e^{ikx} \, dk \tag{13}$$

Second-order contributions:

$$\eta_{2B}(x) = - \, \eta_{2S}(x) = (4 \, \pi^2)^{-1} \int_{-\infty}^{\infty} (k+K) \, sech \, (k+K) \, tanh \, (k) \, \Xi \, (ik) \, \Xi \, (iK) \,) \, e^{i(k+K)x} \, dK \, dk \tag{14}$$

Third-order gravity contribution:

$$\eta_3(x) = - \, (2 \, \pi)^{-1} \int_{-\infty}^{\infty} k \, tanh \, (k) \, sech \, (k) \, \Xi \, (ik) \, e^{ikx} \, dk \tag{15}$$

In the equations above, Z is the position dependent part of the bottom deflection, and Ξ is the corresponding Fourier transform:

$$\Xi \, (ik) = \int_{-\infty}^{\infty} Z(x) \, e^{-ikx} \, dx$$

106

In Figures 1-4 we present some results from numerical evaluation of these Fourier integrals.

We consider a rising block of dimensionless width L:

$$Z(x) = \varepsilon, \ |x| < L/2$$

$$Z(x) = 0, \ |x| > L/2$$

(16)

Figures 1 and 2 show numerical results for the case $L=10$. This block is wide enough to produce an almost flat region in the middle. In Figure 1 the first and third order solutions are shown. The third-order solution shows that the early influence of gravity vanishes in the middle of the block. Gravity is important only near the edges of the rising block. Figure 2 shows the second order elevations, which are equal in magnitude but with opposite sign.

Figures 3-4 show the corresponding results for a narrower block: $L=2$. There is no longer a flat middle region to first order. Gravity is not merely an edge effect, but it will influence the rising heap of fluid as a whole. The negative third-order elevation will in fact have largest amplitude above the middle of the block. From Figures 2 and 4 we see that the second order elevations, being equal but with opposite sign, will have their zero-elevation points slightly outside the block. The block displacements at the bottom are included in Figures 1 and 3 as dotted lines. The first-order curve alone takes care of the mass flux due to the rising block.

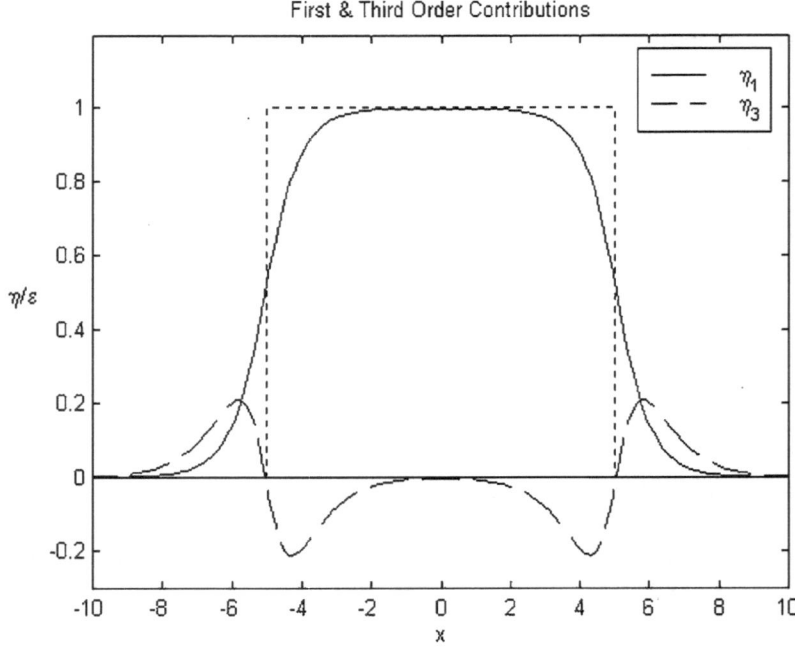

Figure 1. First and third order contributions to the surface elevation for a two-dimensional, rectangular bottom deflection. The width is ten times the water depth.

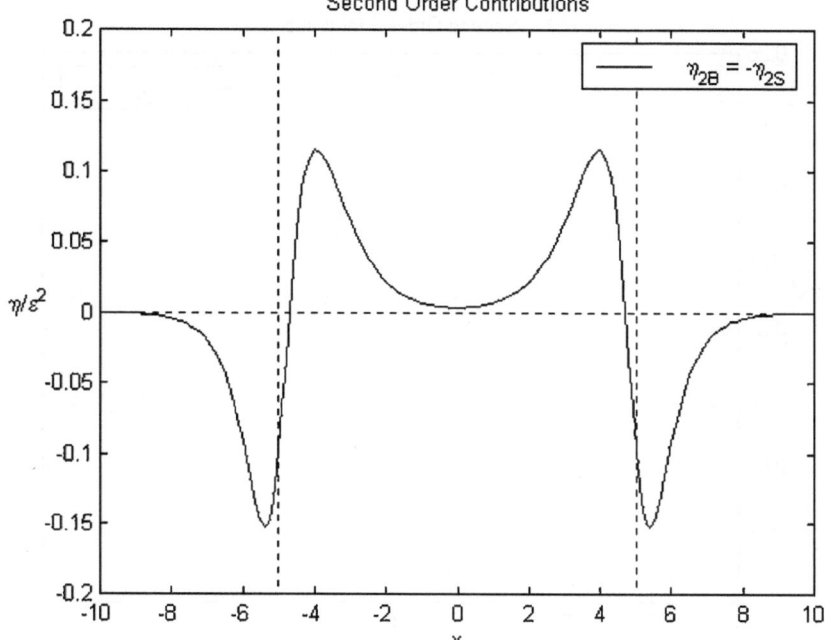

Figure 2. Second order contributions to the surface elevation for a two-dimensional, rectangular bottom deflection. The width is ten times the water depth.

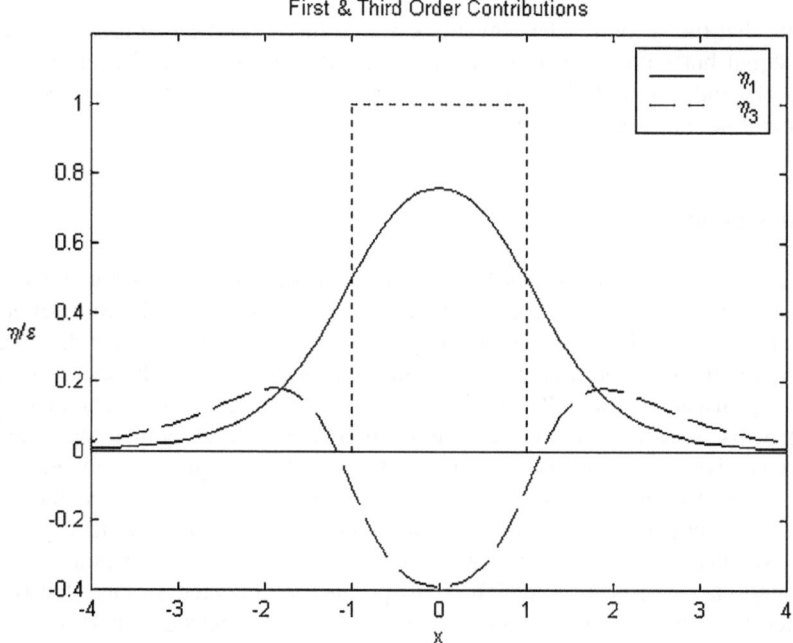

Figure 3. First and third order contributions to the surface elevation for a two-dimensional, rectangular bottom deflection. The width is twice the water depth.

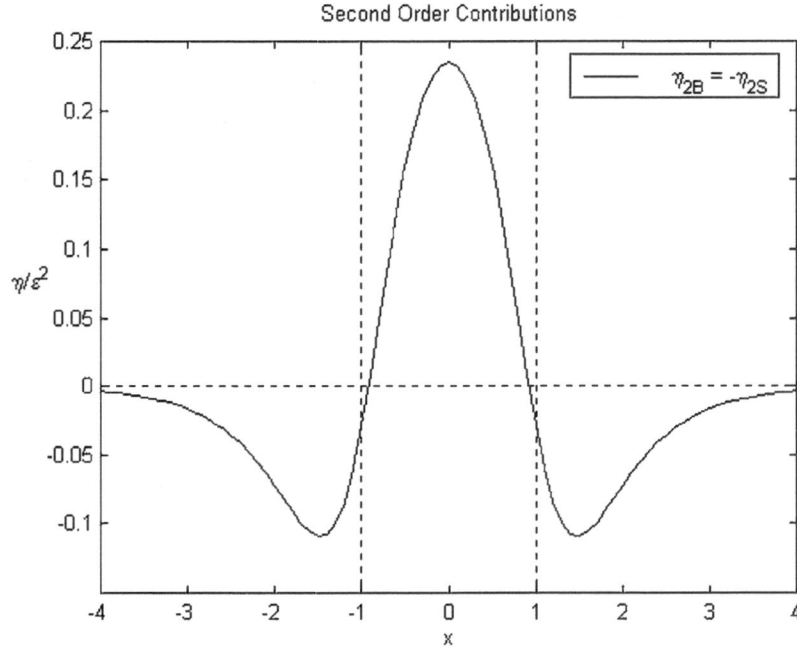

Figure 4. Second order contributions to the surface elevation for a two-dimensional, rectangular bottom deflection. The width is twice the water depth.

We note that the second-order elevations due to the bottom effect and the free-surface effect are equal but with opposite sign. In general they do not cancel each other because their time dependence will be different. In three dimensions these two second-order elevations will not be equal in magnitude.

4. Concluding remarks

The scope of the present work has been to improve the modelling of a special type of tsunami generation. The initial build-up of surface elevation by rapid bottom deflections on initially uniform water depth. The simplest possible model for the surface displacement is to let it be equal to the bottom displacement. This piston-type hydrostatic modelling is physically inconsistent, but still quantitatively acceptable if the gradient in the bottom displacement is small. A more consistent model, which smoothes out the surface deflections is given by impulsive Green functions. This is equivalent to our first-order theory. It shares the disadvantage with the hydrostatic piston model that the surface is assumed to be completely at rest after the bottom motion has ceased. Our theory is a type of third-order small-time expansion that accounts for the early nonlinearities as well as early gravity effects in a consistent way. This will produce a deflected surface that remains in motion after the generation process stops. The later wave propagation can be described as a Cauchy-Poisson problem with nonzero initial velocity as well as nonzero initial elevation.

The results presented here are only two-dimensional. The corresponding theory in three dimensions will be published elsewhere.

References

1. Y. Tanioka and K. Satake, Tsunami generation by horizontal displacement of ocean bottom, Geophys. Res. Letters 23, 861-864 (1996).
2. J.L. Hammack, A note on tsunamis: their generation and propagation in an ocean of uniform depth, J. Fluid Mech. 60, 769-799 (1973).
3. C.C. Mei, The Applied Dynamics of Ocean Surface Waves, Wiley, New York (1983).
4. P.C. Sabatier, On water waves produced by ground motions. J. Fluid Mech. 126, 27-58 (1983).
5. P.A. Tyvand and A.R.F. Storhaug, Green functions for impulsive free-surface flows due to bottom deflections in two-dimensional topographies, Phys. Fluids 12, 2819-2833 (2000).
6. T. Miloh, P.A. Tyvand and G. Zilman, Green functions for initial free-surface flows due to three-dimensional impulsive bottom deflections. J. Eng. Math., v. 43, 57-74 (2002)

References

ANALYTICAL MODELS OF TSUNAMI GENERATION BY SUBMARINE LANDSLIDES

EFIM PELINOVSKY
Laboratory of Hydrophysics and Nonlinear Acoustics, Institute of Applied Physics, 46 Uljanov Street, Nizhny Novgorod, 603950 Russia

Abstract:

The simplified models of tsunami generation by underwater landslides are derived and their analytical solutions are discussed. The shallow-water model describes the properties of the generated tsunami waves usually well for all regimes expect the resonance case. The nonlinear-dispersive model based on the forced Korteweg-de Vries equation is developed to describe the resonant mechanism of the tsunami wave generation by the landslide moving with near-critical speed (long wave speed).

1. Introduction

The last catastrophic events: Papua New Guinea (July 17, 1998) and Izmit (August 17, 1999) confirm again an extreme importance of the landslides in the process of the tsunami wave generation. The series of symposia to study landslides and tsunamis were organised in last years. Many presentations made at these meetings are published in the special issue [1]. The main purpose of the NATO ARW workshop organised in Turkey is to discuss the fundamental studies of underwater ground failures related to tsunami generation. The various numerical models (mainly shallow-water models) are applied to simulate the tsunami waves in the basins with real bathymetry taking into account the properties of the landslide body [2-20]. An applicability of the analytical methods for real events of course is very limited. However, analytical results are extreme useful to understand the physical processes of the tsunami generation and evaluate the significant role of different physical mechanisms. Simplified shallow-water and linear potential models of the tsunami generation are described in [21-27]. The present paper gives the review of theoretical approaches to study tsunami waves induced by the moving bottom disturbances like landslides.

A. C. Yalçıner, E. Pelinovsky, E. Okal, C. E. Synolakis (eds.),
Submarine Landslides and Tsunamis 111-128.

2. Shallow Water Model of the Tsunami Generation

The basic hydrodynamic model of tsunami generation by the underwater landslides is based on the well-known Euler equations of ideal fluid on the non-rotated Earth:

$$\frac{\partial \vec{u}}{\partial t} + (\vec{u}\nabla)\vec{u} + w\frac{\partial \vec{u}}{\partial z} + \frac{1}{\rho}\nabla p = 0 \tag{1}$$

$$\frac{\partial w}{\partial t} + (\vec{u}\nabla)w + w\frac{\partial w}{\partial z} + \frac{1}{\rho}\frac{\partial p}{\partial z} = -g \tag{2}$$

$$div\,\vec{u} + \frac{\partial w}{\partial z} = 0 \tag{3}$$

with corresponding boundary conditions at the bottom and ocean surfaces (geometry of the problem is shown on Figure 1):
at the solid bottom ($z = -h(x,y,t)$),

$$w - (\vec{u}\nabla)h = W_n(x, y, t) \tag{4}$$

at the free surface ($z = \eta(x,y,t)$), the kinematic condition is

$$w = \frac{d\eta}{dt} = \frac{\partial \eta}{\partial t} + (\vec{u}\nabla)\eta \tag{5}$$

and the dynamic equation,

$$p = P_{atm} \tag{6}$$

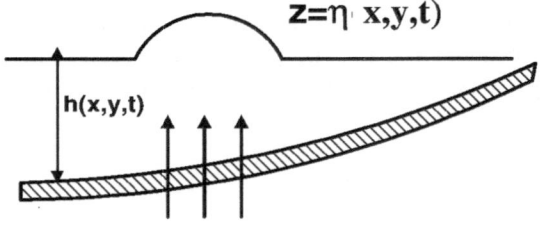

Figure 1. Problem geometry

Here $\eta(x,y,t)$ is an elevation of the water surface, $\vec{u}$ and w are horizontal ($\vec{u} = (u,v)$) and vertical components of the velocity field, x and y are coordinates in the horizontal plane; z-axis is directed upwards vertically, ρ is water density, p is pressure and p_{atm} is atmospheric pressure, which is assumed constant, g is gravity acceleration, $h(x,y,t)$ is variable ocean depth due to the bottom displacement in the source area or moving solid landslides, and W_n is a velocity of the bottom motion on the perpendicular direction (when the bottom motion is vertical only, W coincides with dh/dt). Differential operators ∇ and div act in the horizontal plane.

Usually, tsunami waves are long (as compared to the ocean depth). Therefore, it is natural first to consider the long-wave (or shallow-water) approximation of the model of tsunami generation, and then to estimate conditions of its applicability. The theory of long waves is based on the main assumption that vertical velocity and acceleration are low as compared to horizontal ones and can be calculated from the initial system with the help of an asymptotic procedure. As a small parameter it uses the ratio of the vertical velocity to the horizontal one or the ocean depth to the characteristic wavelength and will be given further. And here a simpler algorithm is used, which consists in neglecting vertical acceleration dw/dt in (2). In this case equation (2) is integrated and, with dynamic boundary condition (6) taken into account, determines the hydrostatic pressure:

$$p = p_{atm} + \rho g(\eta - z)$$ (7)

Substituting (7) into (1) and neglecting the vertical velocity once again, we obtain the first equation of the long-wave theory:

$$\frac{\partial \vec{u}}{\partial t} + (\vec{u}\nabla)\vec{u} + g\nabla\eta = 0$$ (8)

The second equation is yielded by integration of (3) over the depth from the bottom ($z = h(x,y,t)$) to the surface ($z = \eta(x,y,t)$), taking into account boundary conditions (4) - (5), as well as the fact that horizontal velocity $\vec{u}$ does not depend on vertical coordinate, z:

$$\frac{\partial \eta}{\partial t} + div[(h+\eta)\vec{u}] = W_n$$ (9)

Equations (8) - (9) are closed as related to functions η and $\vec{u}$. They are *nonlinear* (the so-called nonlinear shallow-water theory), *inhomogeneous* (the right-hand part is non-zero), and contain a preset function h that is variable in time and space.

Most generally used within the tsunami problem is the linear version of the shallow-water theory. In this case variations of depth are assumed weak, as well as the velocity of the bottom motion. As a result, we obtain a linear set of equations:

$$\frac{\partial \vec{u}}{\partial t} + g\nabla\eta = 0 ,$$ (10)

$$\frac{\partial \eta}{\partial t} + div\left[(h\vec{u}\right] = W_n \tag{11}$$

with the right-hand part $W_n(x,y,t)$. It is convenient to exclude $\vec{u}$ and pass over to the wave equation for the surface elevation:

$$\frac{\partial^2 \eta}{\partial t^2} - div(c^2 \nabla \eta) = \frac{\partial W_n}{\partial t} \tag{12}$$

where

$$c(x, y,t) = \sqrt{gh(x, y,t)} \tag{13}$$

is the long-wave speed.

Equation (12) is the basic one within the linear theory of tsunami generation and must be supplemented by initial conditions. It is natural to believe that at the initial moment the ocean is quiet, i.e.

$$\eta = 0, \quad \vec{u} = 0 \quad \text{or} \quad \partial \eta / \partial t = 0, \tag{14}$$

though due to linearity of (12) a more general case can be also considered.

From the point of view of mathematical physics the wave equation (12) is too well studied to discuss the details of its solution here. Note only that since the function W_n is "switched on'" at the time moment $t = 0$, we formulate here the Cauchy problem for the wave equation in a generalized sense, i.e., we include into the consideration generalized functions and do not require that the level and flow velocity should be differentiated twice.

The initial and long-wave systems of equation presented here are basic hydrodynamic models of tsunami generation by underwater landslides.

3. Shallow Water Wave Generation in a Sea Of Constant Depth

The simplest model here uses the linear wave equation (12) with the preset right-hand part $W_n(x,y,t)$. We can identify W_n with the vertical velocity of the bottom motion (dz_b/dt). In this case (12) reduces to the classical wave equation

$$\frac{\partial^2 \eta}{\partial t^2} - c^2 \Delta \eta = \frac{\partial^2 z_b}{\partial t^2} \tag{15}$$

with constant long-wave speed, c and zeroth initial conditions.

We will first consider a source as the simplest 2D (x, z) bottom displacement as the model of the horizontally moved (with constant speed V) landslide. Actually, in this case we have a uni-dimensional problem; the solution for this problem, which satisfies the zeroth initial conditions, is easily found explicitly [23], [27]

$$\eta(x,t) = \frac{V^2}{V^2 - c^2} z_b(x - Vt) -$$

$$- \frac{V^2}{2c(V - c)} z_b(x - ct) + \frac{V^2}{2c(V + c)} z_b(x + ct) \quad . \tag{16}$$

This solution is a superposition of three waves: one of them is bounded, and the two others, free. After some time they become separated in space: the first wave stays in the source zone (moved with a landslide), and the two others leave it. Let us discuss first the field in the source for sufficiently long times, when the waves become separated in space; it is described by the first term in (16). We see that in the case of the large speed ($V \to \infty$) the surface elevation answers the bottom displacement ($\eta \approx z_b$), and when the motion is slow due to small vertical acceleration the surface level is practically not perturbed ($\eta \approx V^2 z_b/c^2$). Of a special interest is the case of synchronism between the landslide motion and excited wave, when even small bottom displacement cause strong water elevation (formally, infinite within this model). The free wave moving in the same direction with the landslide is similarly amplified (but it is trough, not crest). The wave leaving the landslide zone in the opposite direction is limited with respect to the amplitude at any velocity (that is caused by great difference in motion velocities) and this wave is not resonant. These conclusions, certainly, are valid for relatively extended (long) landslides; otherwise one should take into account relative retardation of the elevation, which excites the nonlinear process. Recently, these results were generalised for a basin of variable depth [27]. Also, the analytical solution for gravitational flow over the inclined plane should be mentioned [28].

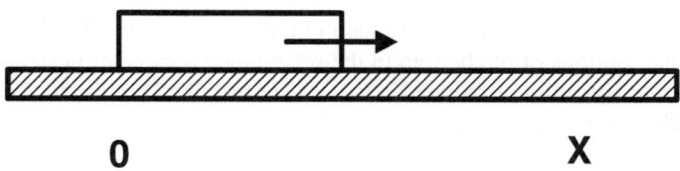

0   **X**

Figure 2. Moving landslide

Let us consider a model of the horizontally "growing" landslide, when its right end moves the finite distance X, such that velocity of the bottom motion, V differs from zero only in the region ($0 - X$), see, Figure 2,

$$W = H_b \delta(t - x/V), \quad 0 < x < X . \tag{17}$$

Solution of (12) is easily found in the form of the Duhamel integral

$$\eta(x,t) = \frac{1}{2c} \int_0^t d\tau \int_{x-c(t-\tau)}^{x+c(t-\tau)} dy \frac{\partial W_n}{\partial \tau}(y,\tau)$$

. (18)

Out of the source region the wave is constituted of two rectangular pulses moving in the opposite directions. The amplitude and duration of the pulse moving towards $(x > 0)$ are

$$H = \frac{H_b}{2} \frac{V}{|V-c|}, \qquad T = \frac{X}{c} \frac{|V-c|}{V},$$

(19)

and in the opposite direction,

$$H = \frac{H_b}{2} \frac{V}{V+c}, \qquad T = \frac{X}{c} \frac{V+c}{V}.$$

(20)

We see that here the resonance also takes place. When speed of the bottom displacement motion V nears velocity of wave propagation c, the amplitude of the resonance wave goes to infinity, and the pulse duration becomes very small. Such short pulses break applicability of the long-wave theory, and the resonance should be considered within more accurate models.

Similar calculations can be made also for the plane (x,y) problem, when the bottom displacement runs within a rectangular region. Detailed calculations for this case were performed in [31], and later confirmed within numerical modelling [29-30]. It should be emphasized that the resonance is retained in the plane problems, wherein all the displacements with velocities higher than c are of the resonance character, and maximum radiation occurs along directions $\theta = \arccos(c/V)$, determined through the so-called Mach angle. Such relations are well-known in the theory of the wave radiation. Wave amplitude stays finite at $c \neq V$, and it is proportional to factor $(1-c^2/V^2)^{-1/2}$ for the Mach direction.

Thus, horizontal motions of the bottom in the seismo-active zone or horizontally moved landslides cause intensification of radiated waves. "Catastrophic" within the linear shallow-water theory is the case of the resonance $c = V$, by this, the wave amplitude becomes formally unbounded.

To estimate the accuracy of the linear model, let us consider the problem of tsunami generation by a moving bottom displacement within the nonlinear formulation. We will limit ourselves to the uni-dimensional variant of the nonlinear shallow-water theory (8) and (9),

$$\frac{\partial u}{\partial t} + u \frac{\partial u}{\partial x} + g \frac{\partial \eta}{\partial x} = 0,$$

(21)

$$\frac{\partial \eta}{\partial t} + \frac{\partial}{\partial x} \left[(h - z_b + \eta) u \right] = \frac{\partial z_b}{\partial t}.$$

(22)

Assuming that the elevation is the function of only the running coordinate $\xi = x - Vt$, we will try to find solutions for this system in the same form, i.e., depending on the same coordinate ξ. Then the system (21) and (22}) yields nonlinear algebraic equations,

$$(h - z_b + \eta)u = V(\eta - z_b), \quad g\eta = Vu - u^2/2,$$
(23)

where it is assumed that in the outside region $z_b = 0$ there are no perturbations of the sea level and velocity. If bottom displacement is small and $V \neq c$, this algebraic system yields linear relation, the first term in (16). However, there is no unlimited growth of the amplitude under resonance $V = c$, and at small z_b the following asymptotic formula is valid [23]

$$\eta \approx V\sqrt{2z_b/3g}, \quad u \approx \sqrt{2gz_b/3}.$$
(24)

Thus, a sufficiently small moving bottom displacement excited waves of the same order as in the case of $V \neq c$; in the vicinity of resonance $V \approx c$ the wave amplitude is essentially larger, it is proportional to $\sqrt{z_b}$. Dependence of the amplitude on the velocity difference $V-c$ reminds of the known resonance curves of a nonlinear oscillator [32].

4. Linear Dispersive Model of Tsunami Generation

Within the linear shallow-water theory we obtained very short duration of the generated waves in the resonance case. All this shows that substantiation of the long-wave model is necessary. Let us consider the exact model based on the potential form of the hydrodynamic equations. In this case it is convenient to introduce a flow potential,

$$\vec{u} = \nabla\varphi, \quad w = \partial\varphi/\partial z,$$
(25)

and pass over from system (1) - (6) to the Laplace equation,

$$\Delta\varphi + \frac{\partial^2\varphi}{\partial z^2} = 0,$$
(26)

with the following boundary conditions at the bottom, $z = - h(x, y, t)$,

$$\partial\varphi/\partial z - \nabla\varphi\nabla h = W_n(x, y, t),$$
(27)

and at the free surface, $z = \eta(x, y, t)$, kinematic and dynamic conditions,

$$\frac{\partial\varphi}{\partial z} = \frac{d\eta}{dt} = \frac{\partial\eta}{\partial t} + \nabla\varphi\nabla\eta,$$
(28)

$$\frac{\partial \varphi}{\partial t} + \frac{1}{2}\left(\frac{\partial \varphi}{\partial z}\right)^2 + g\eta = 0$$

(29)

In particular, within the linear version of the potential theory only the boundary conditions at the free surface ($z = 0$) are changing. They permit one to exclude surface elevation η and take into consideration only the flow potential φ,

$$\frac{\partial^2 \varphi}{\partial t^2} + g\frac{\partial \varphi}{\partial z} = 0$$

(30)

The Laplace equation (26) together with boundary conditions (27) and (30) determines the boundary-value problem for the potential within the linear theory. Steady-state solutions of the linear potential model for the non-resonant motion of the landslide were studied by Pelinovsky & Poplavsky [24]. The unsteady solution of the linear potential model can be expressed in the integral form for the surface elevation [33, 34]

$$\eta(\vec{r},t) = \frac{1}{2\pi h^2}\int d\vec{\rho}\int_0^t d\tau W_n(\vec{\rho},\tau)\times$$

$$\times\int_0^\infty \frac{mI_0(m|\vec{r}-\vec{\rho}|/h)}{\cosh m}\cos[\gamma\sqrt{g/h}(t-\tau)]dm$$

(31)

where $\gamma = m\,\tanh(m)$ and I_0 is the Bessel function. For the sake of simplicity we will consider here the bottom elevation of the following form (Figure 2),

$$W_n = z_b(\vec{r})\delta(t - x/V)$$

(32)

such that the bottom displacement is instantaneous behind the front, $x = Vt$. In this case one of the integrals in (30) is calculated,

$$\eta(\vec{r},t) = \frac{1}{2\pi h^2}\int z_b(\vec{\rho})d\vec{\rho}\int_0^\infty \frac{mI_0(m|\vec{r}-\vec{\rho}|/h)}{\cosh m}\cos[\gamma\sqrt{g/h}(t-\rho_x/V)]dm$$

(33)

It is easily seen that the presence of multiplier cosh (m) results in the converging integral with respect to m, and, consequently, in the wave field being finite. Thus, rejection of the long-wave-approximation is of principal importance and permits us to remove unboundedness of the wave field under resonance within the linear theory.

The wave amplitude within (32) cannot be evaluated analytically. Roughly, it can be estimated at the resonance in the following way. The shallow-water theory is valid at wavelengths, $\lambda \gg h$, or wave duration, $T \gg h/c$. If (19) is transformed as follows,

$$H = \frac{H_b L}{2cT},$$

(34)

then, taking into account limitations for the wave duration within the shallow-water theory, we find the upper boundary for the wave amplitude under resonance [23]

$$H \leq \frac{H_b L}{2h}.$$

(35)

For sufficiently long sources ($L \gg h$) the wave amplification can be rather noticeable.

5. Nonlinear Dispersion Model of Resonant Generated Waves

The formulas and data presented above yield that out of the resonance $V = c$ the wave field is well described within the linear shallow-water theory, and in the case of the resonance our rejection of long-wave approximation and account of nonlinearity is of the decisive importance for explanation of finiteness of the wave field energy. Let us consider here a simplified model for resonance generation of tsunami waves by the horizontally moving landslide. During this consideration we will apply the potential form of hydrodynamic equations (26) - (29). The boundary condition at the bottom (27) can be changed for constant mean depth

$$\frac{\partial z_b}{\partial t} + \nabla \varphi \nabla z_b = \frac{\partial \varphi}{\partial z} (z = -h + z_b).$$

(36)

Since the flow potential is a harmonic function, it can be differentiated with respect to all of its arguments and represented as a Taylor series over the vertical coordinate

$$\varphi(x, y, z, t) = \sum_{n=0}^{\infty} q_n(x, y, t)(z + h - z_b)^n.$$

(37)

Substitution of (37) into the Laplace equation (26) yields recurrent correlations for unknown functions q_n,

$$(n+2)(n+1)q_{n+2} + \Delta q_n - 2(n+1)\nabla q_{n+1}\nabla z_b - $$
$$- (n+1)q_{n+1}\Delta z_b + (n+2)(n+1)q_{n+2}(\nabla z_b)^2 = 0,$$

(38)

such that only two of then (namely, q_0 and q_1) are independent. Specifically, q_2 equals

$$q_2 = \frac{\Delta q_0 - 2\nabla q_1 \nabla z_b - q_1 \Delta z_b}{2[1 + (\nabla z_b)^2]}.$$

(39)

Now, having substituted series (37) to the boundary condition at the bottom (36) we find the connection between q_1 and q_0,

$$q_1 = \frac{\partial z_b / \partial t + \nabla q_0 \nabla z_b}{[1 + (\nabla z_b)^2]}.$$

(40)

Thus, series (37) is completely determined by one function, $q_0(x,y,t)$. Boundary conditions at the free surface (28) and (29) determine equations to be found for η and q_0. They can be conveniently rewritten in terms of surface values of flow velocities,

$$\frac{\partial \eta}{\partial t} + (\vec{u}\nabla)\eta = W_n \tag{41}$$

$$\frac{\partial \vec{u}}{\partial t} + (\vec{u}\nabla)\vec{u} + w\frac{\partial \vec{u}}{\partial z} + g\nabla\eta = 0 \tag{42}$$

where

$$\vec{u} = \sum_{n=0}^{\infty}[\nabla q_n - (n+1)q_{n+1}\nabla z_b](h+\eta-z_b)^n \tag{43}$$

$$w = \sum_{n=0}^{\infty}(n+1)q_{n+1}(H+\eta-z_b)^n \tag{44}$$

Equations (41) and (42) together with (39), (40), (43) and (44) are completely equivalent to the initial boundary problem for the potential. They can be written in a very elegant form for fully nonlinear waves [37], [36]. Here we simplify them with the use of a series of approximations:

• slowness of spatial and temporal changes of the bottom displacement (as compared to the source size L and characteristic time L/c);

• great excess of the source size over the basin depth ($L \gg h$);

• smallness of the bottom displacement (as compared to the depth) providing smallness of nonlinear effects.

The first two approximations permit us to limit ourselves to only three terms in series (37), the third one, to retain only the first term of this series in all nonlinear terms. Having omitted the technical details of using these approximations we can write the final form of simplified equations (41) and (42) (see, e.g., [35]),

$$\frac{\partial \eta}{\partial t} + div[(h-z_b+\eta)\vec{u}] = \frac{\partial z_b}{\partial t} \tag{45}$$

$$\frac{\partial \vec{u}}{\partial t} + (\vec{u}\nabla)\vec{u} + g\nabla\eta = -\frac{h}{2}\frac{\partial^2}{\partial t^2}\nabla z_b + \frac{h^2}{3}\frac{\partial}{\partial t}\Delta\vec{u} \tag{46}$$

Simplified equations (usually called in literature "*systems of the Boussinesq type*") differ from the shallow-water equations only by the presence of the right-hand part in (46), wherein the first term determines the correction for the gravity acceleration due to the oscillations of the bottom displacement. The system obtained describes the tsunami generation by inhomogeneous and unsteady moving bottom displacements. Taking into account that the maximum efficiency of excitation is achieved in the resonance case, when the landslide propagates with the velocity close to $c = (gh)^{1/2}$, we can simplify the problem further assuming it to be uni-dimensional. Besides, it is reasonable to keep z_b only in the right-hand part of (45) and neglect it in other places (their influence, as one can show, is essentially weaker). This results in the following system,

$$\frac{\partial \eta}{\partial t} + div[(h+\eta)\vec{u}] = \frac{\partial z_b}{\partial t} , \tag{47}$$

$$\frac{\partial \vec{u}}{\partial t} + (\vec{u}\nabla)\vec{u} + g\nabla\eta = \frac{h^2}{3}\frac{\partial}{\partial t}\Delta\vec{u} . \tag{48}$$

Taking into account the resonant character of tsunami generation, this system can be simplified. Let us re-write (47) and (48) in the form of the nonlinear wave equation for the surface elevation η,

$$\frac{\partial^2 \eta}{\partial t^2} - c^2\frac{\partial^2 \eta}{\partial x^2} = \Pi\{\eta, u\} + \frac{\partial^2 z_b}{\partial t^2} , \tag{49}$$

$$\Pi = -\frac{\partial}{\partial x}\left(\eta\frac{\partial u}{\partial t}\right) + \frac{h}{2}\frac{\partial^2 u^2}{\partial x^2} - \frac{h^3}{3}\frac{\partial^4 u}{\partial t\partial x^2} . \tag{50}$$

Here the right-hand part of (49) can be treated as proportional to the small parameter, characterized the weak nonlinearity, dispersion and forcing. Using the linear relation, $u = \eta g/c$ in (50), equation (49) can be reduced to [38]

$$\frac{\partial \eta}{\partial t} + c\frac{\partial \eta}{\partial x} + \alpha\eta\frac{\partial \eta}{\partial x} + \beta\frac{\partial^3 \eta}{\partial x^3} = \frac{\partial f}{\partial x} , \tag{51}$$

where

$$\alpha = 3c/2h, \quad \beta = ch^2/6, \quad f = -cz_b/2 . \tag{52}$$

Equation (51) is the known Korteweg - de Vries equation that has been repeatedly studied with its right-hand part; it is often called the *forced Korteweg - de Vries equation* and currently it is considered as one of the basic equations of the nonlinear mathematical physics.

Let us assume that the landslide moves with constant velocity, V close to the linear velocity of propagation (resonance). Therefore, right-hand part in (52) is a function of x -

Vt. Then it is convenient to pass over to the system of coordinates connected with the landslide

$$x' = x - Vt, \qquad t' = t, \tag{53}$$

and rewrite (52) in its final form (subscripts omitted),

$$\frac{\partial \eta}{\partial t} + (c - V)\frac{\partial \eta}{\partial x} + \alpha \eta \frac{\partial \eta}{\partial x} + \beta \frac{\partial^3 \eta}{\partial x^3} = \frac{\partial f(x)}{\partial x}. \tag{54}$$

At first, the steady-state solutions of the forced Korteweg - de Vries equation will be analysed. In this case (54) reduces to the nonlinear ordinary differential equation

$$\beta \frac{d^2 \eta}{dx^2} + (c - V)\eta + \frac{\alpha \eta^2}{2} = f(x). \tag{55}$$

Simplified model here is the source of a very small length (but larger than water depth for applicability of shallow water theory). In this case the surface elevation outside of source area is described by the "normal" Korteweg - de Vries equation and corresponds to the steady-state waves of solitary or cnoidal form. In the source area both solutions should be matched using the boundary conditions derived from (55),

$$\eta\Big|_-^+ = 0, \qquad \frac{d\eta}{dx}\Big|_-^+ = \int_{-\infty}^{\infty} f(x)dx. \tag{56}$$

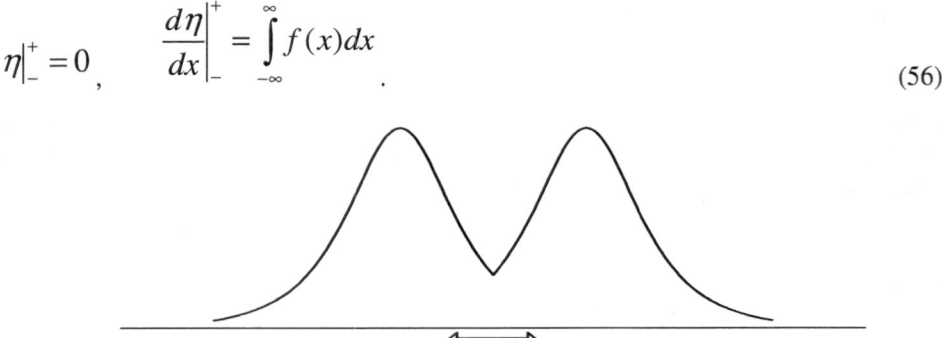

Figure 3. Steady-state nonlinear wave generated by the moving short source (its location is shown by arrow).

As a result, two pieces of solitons with different amplitudes can propagate together with a force (Figure 3), or two different cnoidal waves, depending from the velocity of the landslide [39]. The ambiguity of steady-state solutions requires proofing the theorem of uniqueness.

6. Soliton Interaction with the Moving Source

For demonstration the possible effects of the unsteady solution of the forced Korteweg - de Vries equation, let us assume that the soliton has been formed already and consider the process of its amplification under the effect of a moving force. We will suppose that the forcing is sufficiently weak, so that the soliton retains its form in the process of interaction (the condition for adiabatic interaction is slowness of variation of soliton parameters on distances of the order of the nonlinearity length). The solution for the forced Korteweg - de Vries equation is seeking in the form of the asymptotic series

$$\eta(x,t) = a(t)\operatorname{sech}^2\left(\Gamma(t)[x-\Psi(t)]\right)+..., \quad \Gamma=(\alpha a/12\beta)^{1/2},$$
$$\Psi(t) = \int(p_0+...)dt.$$

(57)

The procedure for obtaining the equations for the soliton amplitude and velocity in the framework of the asymptotic method is well-known and will not be considered here. Specifically, in the first approximation we obtain the energy balance equation for the amplitude,

$$\frac{d}{dt}\int_{-\infty}^{\infty}\frac{\eta^2}{2}dx = \int_{-\infty}^{\infty}\eta\frac{\partial f}{\partial x}dx,$$

(58)

and or the position (phase) of the soliton we have the following unperturbed relation,

$$\frac{d\Psi}{dt} = c - V + \frac{\alpha a}{3}.$$

(59)

These equations describe the simplified model of the process of adiabatic interaction of the soliton with the moving force. In the case of the landslide of large length the soliton acts as the delta function, which makes it possible to calculate the integral in (58). As the result, the equation (58) has the differential form,

$$\frac{da}{dt} = 2\frac{df}{d\Psi}.$$

(60)

This system of differential equations was obtained by Warn & Brasnett, [40] for the problems of atmosphere dynamics. It is evident that system (59) and (60) is reduced to the equation of nonlinear oscillator,

$$\frac{d^2\Psi}{dt^2} = \frac{2\alpha}{3}\frac{df}{d\Psi},$$

(61)

and is easily studied on the phase plane. Note that all the integral trajectories can be found in the explicit form,

$$\frac{4\alpha}{3}f(x) = \left(c - V + \frac{\alpha a}{3}\right)^2 + \text{const}.$$

(62)

We will consider symmetric perturbations $f(x) = f(-x)$ in the form of one positive or negative pulse with one extremum. For the sake of certainty we assume that

$$f(x) = b\,\text{sech}^2(x/l),$$

(63)

and make some substitutions,

$$\theta = \Psi/l, \quad A = a/a_0, \quad a_0 = 3|V - c|, \quad Q = \frac{4ab}{3(V-c)^2}.$$

(64)

Then equation (62) can be written in the dimensionless form,

$$Q\,\text{sech}^2\theta = [1 - A\,\text{sign}(V - c)]^2 + \text{const}.$$

(65)

The phase pattern of system (59) and (60), all the trajectories of which are determined by (65), depends on the sign of Q and $(c - V)$. First we will consider the case of $c > V$. The phase pattern for that case is shown on Figure 4 for $Q > 0$. The main regime here is the regime of the trajectory passage corresponding to fast motion of solitons through the source region. The soliton amplitude grows at the moment of interaction (and decreases at the opposite sign of forcing) and recovers after interaction. At the phase plane one can also see (in the region of low amplitudes) trajectories that correspond to generation of virtual solitons. These solitons are generated behind the source, and then they grow, take over the source and dissipate. The same regime exists for $Q < 0$; in that case the solitons situated behind the source dissipate during interaction, and a part of the solitons is generated in front of the perturbation.

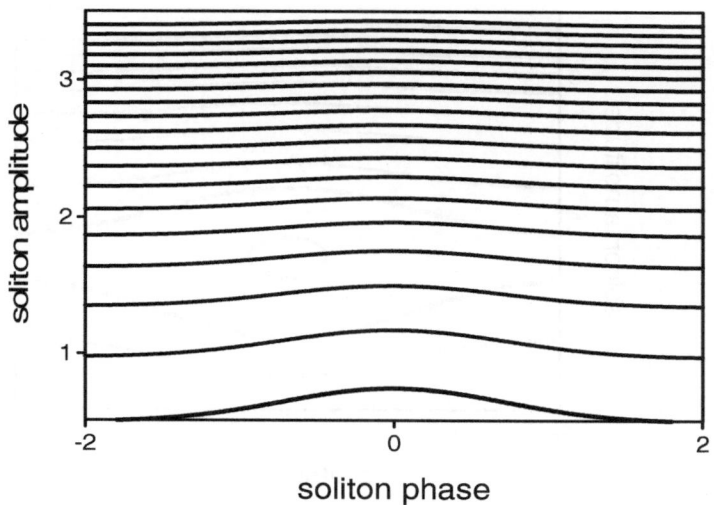

Figure 4. Phase plane of system (59) and (60) for the "slow" source.

Let us consider now the opposite case, $c < V$. Dynamics of solitons is much richer here. First, here exists the equilibrium state,

$$\alpha a_0 = 3(V - c), \quad \theta = 0. \tag{66}$$

It is equivalent to existence of the steady-state soliton propagating with the same velocity as the source. Naturally, it is possible only at certain amplitude of the soliton. The character of the equilibrium depends on the sign of Q. Specifically, at $Q < 0$ it is a saddle, and at $Q > 0$, it is the center. In the first case the solitons are reflected from the source region, and in the second case, they oscillate over their amplitudes with frequency

$$\omega = \sqrt{4\alpha b / 3l^2}. \tag{67}$$

This regime corresponds to the soliton capture by the moving source. The phase plane for $Q > 0$ is displayed on Figure 5. Here there are also regimes of the trajectory passage and virtual solitons, and they are clearly visible on the phase plane. Note the important formula yielded by (65)

$$A_{\max} + A_{\min} = 2. \tag{68}$$

It connects variation of the soliton amplitude as it reflects from the perturbation at $Q > 0$ or as it is trapped when $Q > 0$.

126

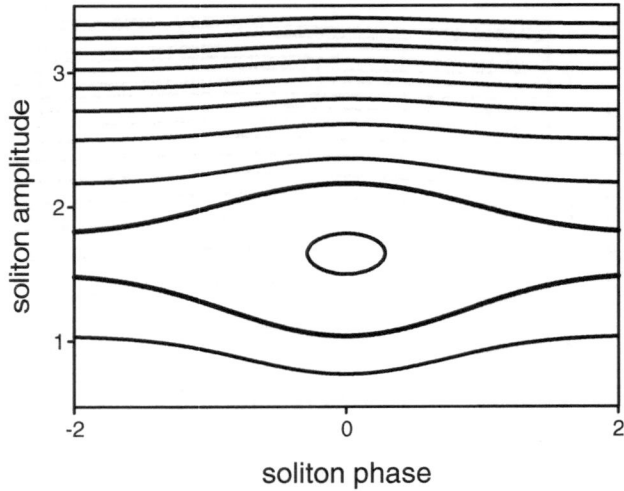

Figure 5. Phase plane of system (59) and (60) for the "fast" source.

The same pattern should be retained for the landslides with arbitrary length. As numerical solutions of the forced Korteweg - de Vries equation showed, the simplified model given here yields a correct physical representation of the interaction between the soliton and the moving source [41]. The generalisation of this theory for a case of landslide moved with variable speed, and also with taking into account the wave dissipation, can be found in papers by Grimshaw et al [42, 43]. If the moving source generates several waves, they may intersect and interact between them forming sometimes large-amplitude waves, called the *freak* waves [44].

7. Conclusion

The paper gives a short review of the analytical solutions used in the theory of tsunami generation by the by underwater landslides. The simplified linear and nonlinear shallow water model is derived and their analytical solutions for a basin of constant depth are discussed. The shallow-water model describes the properties of the generated tsunami waves usually well for all regimes expect the resonance case. The nonlinear-dispersive model based on the forced Korteweg-de Vries equation is developed to describe the resonant mechanism of the tsunami wave generation by the landslide moving with near-critical speed (long wave speed). Some analytical solutions of the forced Korteweg – de Vries equation are obtained; they illustrate the steady-state and unsteady regimes of tsunami wave generation.

Acknowledgement
This work was supported by the grants from INTAS (99-1068, 01-2156) and RFBR (02-05-65107).

References

1. Keating, B.H., Waythomas, C.F., and Dawson, A.G. (2000) Landslides and Tsunamis. *Pageoph* Topical Volumes, **157**, No. 6/7/8.
2. Assier-Rzadkiewicz, S., Mariotti, C., and Heinrich, P. (1996) Modelling of submarine landslides and generated water waves. *Phys. Chem.Earth.*, **21**, 7 -12.
3. Assier-Rzadkiewicz, S., Mariotti, C., and Heinrich, P. (1997) Numerical simulation of submarine landslides and their hydraulic effects. *J. Waterway, Port, Coastal, and Ocean Eng.*, **124**, 149-157.
4. Assier-Rzadkiewicz, S., Heinrich, P., Sabatier, P.C., Savoye, B., and Bourillet, J.F. (2000) Numercial modelling of a landslide-generated tsunami: the 1979 Nice event. *Pageoph*, **157**, 1707 – 1727.
5. Harbitz, C.B. (1992) Model simulation of tsunami generated by Storegga slides. *Marine Geology*, **104**, 1 – 21.
6. Harbitz, C.B., Pedersen, G., and Gjevik, B. (1993) Numerical simulations of large water waves due to landslides. *J.HydraulicEngineering*, **119**, 1325 - 1342
7. Heinrich, P. ((1992) Nonlinear water waves generated by submarine and aerial landslides. *J. Waterway, Port, Coastal, and Ocean Eng.*, **118**, 249 – 266.
8. Henrich, P., Guibourg, S., Mangeney, A., and Roche, R. (1999) Numerical modelling of a landslide-generated tsunami following a potential explosion of the Montserrat volcano. *Phys. Chem. Earth*, **24**, 163 – 168.
9. Imamura, F., and Gica, E.C. (1996) Numerical model for tsunami generation due to subacqueous landslide along a coast. *Science of Tsunami Hazards*, **14**, 13 – 28.
10. Jiang, L., and LeBlond, P.H. (1994) Three-dimensional modelling of tsunami generation due to a submarine mudslide. *J. Phys. Oceanography*, **24**, 559 - 572.
11. Johnsgard, H., and Pedersen, G. (1996) Slide-generated waves in near-shore regions. A Lagrangian description. *Phys. Chem. Earth*, **21**, 45 – 49.
12. Kienle, J., Kowalik, Z., Troshina, E. (1996) Propagation and runup of tsunami waves generated by St. Augustine volcano, Alaska. *Science of Tsunami Hazards*, **14**, 191 – 206.
13. Kowalik, Z. (1997) Landslide-generated tsunami in Skagway, Alaska. *Science of Tsunami Hazards*, **15**, 89 – 106.
14. Kulikov E.A., Rabinovich, A.B., Thomson, R.E., and Bornhold, B.D. (1996) The landslide tsunami of November 3, 1994, Skagway Harbor, Alaska. *J. Geophys. Res.*, **C101**, 6609 – 6615.
15. Mader, C.L. (1997) Modeling of the 1994 Skagway tsunami. *Science of Tsunami Hazards*, **15**, 41 – 48.
16. Tanioka, Y., and Satake, K. (1996) Tsunami generation by horizontal displacement of ocean bottom. *Geophys. Res. Letters*, **23**, 861 – 864.
17. Tinti, S., Bortolucci, E., and Armigliato, A. (1999a) Numerical simulationof the landslide-induced tsunami of 1988 in Vulcano island, Italy. *Bull. Volcanol.*, **61**, 121 – 137.
18. Tinti, S., Romagnoli, C., and Bortolucci, E. (1999b) Modeling a possible holocenic landslide-induced tsunami at Stromboli volcano, Italy. *Phys. Chem. Earth*, **24**, 423 – 429.
19. Watts, P. (1998) Wavemarker curves for tsunami generated by underwater landslides. *J. Waterway, Port, Coastal, and Ocean Eng.*, **124**, 127 – 137.
20. Watts, P., Imamura, F., and Grilli, S. (2000) Comparing model simulations of three benchmark tsunami generation cases. *Science of Tsunami Hazards*, **18**, 107 – 123.
21. Garder, O., Dolina, I., Pelinovsky, E., Poplavsly, A., and Fridman, V. (1993) Generation of tsunami by gravity meta-dynamical processes. *Tsunami Researches*, **5**, 50 - 60.
22. Noda, E.K. (1970) Water waves generated by a local surface disturbance. *J. Waterway, Port, Coastal, and Ocean Eng.*, **76**, 7389 – 7400.
23. Pelinovsky, E.N. (1996) *Hydrodynamics of tsunami waves*. Institute of Applied Physics, Nizhny Novgorod.
24. Pelinovsky, E., and Poplavsky, A. (1996) Simplified model of tsunami generation by submarine landslides. *Phys. Chem. Earth*, **21**, 13 – 17.
25. Tinti, S., and Bortolucci, E. (2000a) Energy of water waves induced by submarine landslides. *Pageoph*, **157**, 281 – 318.
26. Tinti, S., and Bortolucci, E. (2000b) Analytical investigation of tsunamis generated by submarine slides, *Annali di Geofisica*, **43**, 519 – 536.
27. Tinti, S., Bortolucci, E., and Chiavettieri, C. (2001) Tsunami excitation by submarine slides in shallow-water approximation, *Pageoph*, **158**, 759 – 797.
28. Mangeney, A., Heinrich, P., and Roche, R. (2000) Analytical solution for testing debris avalanche numerical models. *Pageoph*, **157**, 1081 – 1096.
29. Marchuk, A.G., Chubarov, L.B., and Shokin, Yu.I. (1983) *Numerical modelling of tsunami waves*. Novosibirsk, Nauka.
30. Nosov, M.A., and Shelkovnikov, N.K. (1995) Tsunami generation by moving bottom displacement. *Vestnik of Moscow University*, Ser. 3., **36**, 96 - 101.

128

31. Novikova, L.E., and Ostrovsky, L.A. (1978) Excitation of tsunami waves by a travelling displacement of the ocean bottom. *Marine Geodesy*, **2**, 365 – 380.
32. Thompson, J.M.T., and Stewart, H.B. (1993) *Nonlinear dynamics and chaos*. Wiley.
33. Kajiura, K. (1963) The leading wave of a tsunami. *Bull. Earthq. Res. Inst.*, **41**, 535 – 571.
34. Stoker, J.J. (1957) *Water waves: the mathematical theory with applications*. Wiley.
35. Dorfman, A.A. (1977) The equations of the approximated nonlinear – dispersive theory of the long gravity waves induced by the bottom displacement. *Theoretical and experimental investigations of tsunami problems*. Moscow, Nauka, 18 – 25.
36. Madsen, P.A., and Schaffer, H.A. (1998) High-order Boussinesq-type equations for surface gravity waves – derivation and analysis. *Phil. Trans. Royal Soc. London*, **356**, 1 – 59.
37. Wei, G., Kirby, J.T., Grilli, S., and Subramanya, R. (1995) A fully nonlinear Boussinesq model for surface waves. 1. Highly nonlinear unsteady waves. *J. Fluid Mech.,* **294**, 71 – 92.
38. Cole, S.L. (1985) Transient waves produced by flow past a bump. *Wave Motion*, **7**, 579 – 587.
39. Malomed, B.A. (1988) Interaction of a moving dipole with a soliton in the KdV equation. *Physica D*, **32**, 393 - 408.
40. Warn, T., and Brasnett, B. (1983) The amplification and capture of atmospheric solitons by topography: a theory of the onset of regional blocking. *J. Atmos. Sci.*, **40**, 28 – 38.
41. Grimshaw R., Pelinovsky E., Tian X. (1994) Interaction of solitary wave with an external force. *Physica D*, **77**, 405 - 433.
42. Grimshaw R., Pelinovsky E., Sakov P. (1996) Interaction of a solitary wave with an external force moving with variable speed. *Stud. Applied Mathematics*, **97**, 235 - 276.
43. Grimshaw R., Pelinovsky E., Bezen A. (1997) Hysteresis phenomena in the interaction of a damped solitary wave with an external force. *Wave Motion*, **26**, 253 - 274.
44. Pelinovsky, E., Talipova, T., and Kharif, C. (2000) Nonlinear dispersive mechanism of the freak wave formation in shallow water. *Physica D*, **147**, 83-94.

TSUNAMI GENERATION IN COMPRESSIBLE OCEAN OF VARIABLE DEPTH

M.A. NOSOV, S.V. KOLESOV
*Department of Marine Physics, Physical Faculty, Moscow State
University, Vorobjevy Gory, Moscow 119899 Russia*

Abstract

This paper presents results of numerical modelling of tsunami waves generation by small bottom displacements in compressible ocean of variable depth. The problem is considered within the framework of the linear potential theory. Free surface disturbance, dynamic pressure distribution, and energy of compressible water layer for different bottom profiles are calculated and analyzed.

1. Introduction

The role of water compressibility in the tsunami problem has been discussed many times [1-5]. It is well-known that submarine earthquakes can radiate not only gravitational but hydroacoustic waves (T-phase) [6]. However, in most cases, perhaps except very few studies [7, 8], tsunami is considered as a process in incompressible fluid.

From the physical point of view [9], a fluid can be regarded as incompressible if $\Delta\rho/\rho << 1$ (ρ is the fluid's density). In case of time-independent motion, this condition is fully equivalent to the following one: (1) $v << c$, where **v** is the mass velocity of fluid and **c** is the acoustic velocity in fluid. It is obvious that the problem of tsunami generation is time-dependent; therefore, an additional condition must be satisfied. As related to the tsunami generation problem [13], the second condition is as follows: (2) $\tau >> \{H/c, L/c\}$. Here, H is the depth of the ocean, L is a horizontal size of the source, and τ is the duration of bottom motion. Note that, in justifying the applicability of the theory of incompressible fluid to the description of tsunami, the second condition is usually ignored. The characteristic values of the above parameters are as follows: $v \sim 1\,m/s$, $c \sim 1500\,m/s$, $H \sim 4000\,m$, $L \sim 10^4 - 10^5\,m$, and $\tau \sim 1-100\,s$. One can see that the first condition is well satisfied, while the second condition can be broken in many cases. However, if a fast traveling motion, such as seismic waves or rupture formation (2000-6000 m/s), is taken as a tsunami source, then the first condition is also broken.

It should be noted that, in problems of tsunami propagation and run-up, the first condition is well satisfied. The second condition assumes the form: $T >> \{H/c, \lambda/c\}$, where T is the period and λ is the wavelength of tsunami ($T \sim 10^2 - 10^4\,s$, $\lambda \sim 10^4 - 10^5\,m$). Thus, the second condition is also satisfied.

In framework of linear potential theory of compressible fluid *the problem of wave generation by small bottom displacements of finite duration* was examined by us analytically in case of ocean of constant depth [10-13]. Major results for ocean of constant depth are as follows: independently on the time-spatial history of bottom displacements, the behavior of a compressible ocean differs from that of an incompressible one mostly by the

A. C. Yalçıner, E. Pelinovsky, E. Okal, C. E. Synolakis (eds.),
Submarine Landslides and Tsunamis 129-137.

formation of "fast" surface oscillations with the dominating period $4H/c$. In the source domain, the oscillations amplitude can be several times greater than the amplitude of bottom displacements.

In this paper, particular features of *the problem* in case of ocean of variable depth are examined numerically.

2. Mathematical model

Let us consider a layer of an ideal compressible homogeneous fluid in the field of gravity. The layer is bounded by free surface above and by absolutely rigid bottom below. The origin of the Cartesian coordinate system OXZ finds itself at the unperturbed free surface, and the OZ-axis is oriented vertically upward. The bottom position is set by function

$$z_b(x,t) = -H(x) + \eta(x,t),$$

where $H(x)$ is the depth, and $\eta(x,t)$ is the bottom displacements of small amplitude ($|\eta| \ll H$). It is assumed that a given point of the bottom moves in a given direction $\vec{\alpha}_0$ ($|\vec{\alpha}_0| = 1$, $\vec{\alpha}_0 \neq \vec{f}(x)$) with velocity $U(x,t)$. To find acoustic and gravitational waves generated by bottom displacements the following linear problem for the current velocity potential $\varphi(x,z,t)$ is solved:

$$\varphi_{xx} + \varphi_{zz} = c^{-2}\varphi_{tt}, \tag{1}$$

$$\varphi_{tt} = -g\varphi_z, \quad z = 0, \tag{2}$$

$$\frac{\partial \varphi}{\partial \bar{n}} = (\vec{\alpha}_0, \bar{n}(x))U(x,t), \quad z = -H(x), \tag{3}$$

$$c\varphi_{xt} - \varphi_{tt} + 0.5c^2\varphi_{zz} = 0, \quad x = x_{min}, x_{max}, \tag{4}$$

where g is the acceleration of gravity, $\bar{n}(x)$ is the normal to the bottom surface at given point x. All characteristics required to describe fluid behavior can be calculated using the potential [9]: the dynamic pressure $p = -\rho\varphi_t$, the fluid velocity $\vec{v} = \vec{\nabla}\varphi$, the fluid particles displacements $\vec{\xi} = \int \vec{v}dt$, and the fluid surface displacement $\xi = -g^{-1}\varphi_t|_{z=0}$.

3. FD scheme

The following traditional explicit Finite Difference scheme (rectangular grid) is used for equation (1):

$$\varphi_{x,z}^{t+1} = 2\varphi_{x,z}^{t} - \varphi_{x,z}^{t-1} + c^2 \frac{\Delta t^2}{\Delta x^2}(\varphi_{x+1,z}^{t} - 2\varphi_{x,z}^{t} + \varphi_{x-1,z}^{t}) +$$

$$+ c^2 \frac{\Delta t^2}{\Delta z^2}(\varphi_{x,z+1}^{t} - 2\varphi_{x,z}^{t} + \varphi_{x,z-1}^{t}) \tag{5}$$

where Δt is time increment, Δx and Δz are space increments. The condition for stability of the scheme is Courant criterion $\Delta t < \min(\Delta x, \Delta z)/c$. In practice, the time increment was computed as follows: $\Delta t = 0.68 \min(\Delta x, \Delta z)/c$.

Formula (5) allows computing values of the current velocity potential in interior grid points. The values on the boundary are computed in accordance with equations (2)-(4) using the following schemes:

- *free surface*

$$\varphi_{x,z}^{t+1} = 2\varphi_{x,z}^{t} - \varphi_{x,z}^{t-1} - g\Delta t^2 \frac{\varphi_{x,0} - \varphi_{x,1}}{\Delta z},$$

- *left and right boundaries (free pass condition)*

$$\varphi_{0,z}^{t+1} = \frac{c\Delta t}{c\Delta t + \Delta x}\left(\varphi_{1,z}^{t+1} - \varphi_{1,z}^{t} + \varphi_{0,z}^{t}\right) + \frac{\Delta x}{\Delta t + \Delta x}\left(2\varphi_{0,z}^{t} - \varphi_{0,z}^{t+1}\right) +$$

$$+ \frac{c^2\Delta t^2\Delta x}{2\Delta z^2\left(\Delta t + \Delta x\right)}\left(\varphi_{0,z+1}^{t} - 2\varphi_{0,z}^{t} + \varphi_{0,z-1}^{t}\right),$$

$$\varphi_{N,z}^{t+1} = \frac{c\Delta t}{c\Delta t + \Delta x}\left(\varphi_{N,z}^{t} - \varphi_{N-1,z}^{t} + \varphi_{N-1,z}^{t+1}\right) + \frac{\Delta x}{\Delta t + \Delta x}\left(2\varphi_{N,z}^{t} - \varphi_{N,z}^{t+1}\right) +$$

$$+ \frac{c^2\Delta t^2\Delta x}{2\Delta z^2\left(\Delta t + \Delta x\right)}\left(\varphi_{N,z+1}^{t} - 2\varphi_{N,z}^{t} + \varphi_{N,z-1}^{t}\right),$$

- *bottom (5 cases)*

$$\varphi_{x,z} = \varphi_{x,z-1}U_x \cos\gamma \, \Delta z \quad \text{(horizontal bottom)},$$

$$\varphi_{x,z} = \varphi_{x+1,z} + U_x \sin\gamma \, \Delta x \quad \text{(left vertical wall)},$$

$$\varphi_{x,z} = \varphi_{x-1,z} + U_x \sin\gamma \, \Delta x \quad \text{(right vertical wall)},$$

$$\varphi_{x,z} = \frac{\varphi_{x,z-1} + \varphi_{x+1,z}tg^2\beta - U_x \cos(\gamma-\beta)\sqrt{\Delta x^2 + \Delta z^2}tg\beta}{tg^2\beta + 1} \quad \text{(left slope)},$$

$$\varphi_{x,z} = \frac{\varphi_{x,z-1} + \varphi_{x-1,z}tg^2\beta - U_x \cos(\gamma+\beta)\sqrt{\Delta x^2 + \Delta z^2}tg\beta}{tg^2\beta + 1} \quad \text{(right slope)},$$

where $tg\,\beta = \Delta z/\Delta x$, γ is the angle between the normal $\vec{n}(x)$ and the bottom movement direction $\vec{\alpha}_0$.

4. Verification of model

Before practical calculations the numerical model was tested in respect of the following points: physical adequacy of results, efficiency of the free pass boundary condition, fulfillment of the energy conservation law, etc.

Since the problem (1)-(3) have been solved analytically for ocean of constant depth [10-13], in verification of the numerical model we could rely on exact analytical solutions. The comparison of free surface displacements calculated analytically and numerically give us ground to state that numerical model compute amplitudes of both gravitational and acoustic waves with accuracy better than 1% if number of grid points between surface and bottom is more than 20.

In case of variable depth the numerical model was tested in respect of gravitational waves only. A few numerical experiments show that in a basin with slightly sloping bottom, the wave amplitude changes exactly in accordance with the Green's law $A \sim H^{-1/4}$.

5. Discussion of results

In order to reduce number of initial parameters we choose rather simple bottom topography. Two domains of fixed depths H_1 and H_2 are connected by a slope of length L=80km. The depths H_1 and H_2 have range within 0.5 - 8.5 km ($H_1 \leq H_2$). Waves in fluid are generated by bottom displacements of duration τ. The displacements are in the direction of the normal $\bar{n}$. The displacements duration range is $1-100 s$. The increments for the FD scheme are determined as follows: $\Delta x = 800 \, m$, $\Delta z = H_1/20$, and $\Delta t = 0.68 \Delta z/c$. The calculating area and the governing law for bottom displacements are shown on Fig. 1.

Fig. 2 demonstrates free surface disturbance generated by bottom displacements of duration τ=10 s. The disturbance is calculated at time t=1000 s. It incorporates both slow gravitational and fast acoustic modes (tsunami precursor). It is seen that the acoustic mode radiation directivity is strongly depended on bottom topography, whereas the gravitational mode is hardly sensitive to changes of bottom topography. Anyway the gravitational disturbance is also asymmetrical; the wave of larger amplitude propagates toward the shallow domain. Nevertheless, larger part of gravitational waves energy radiates in the deep domain.

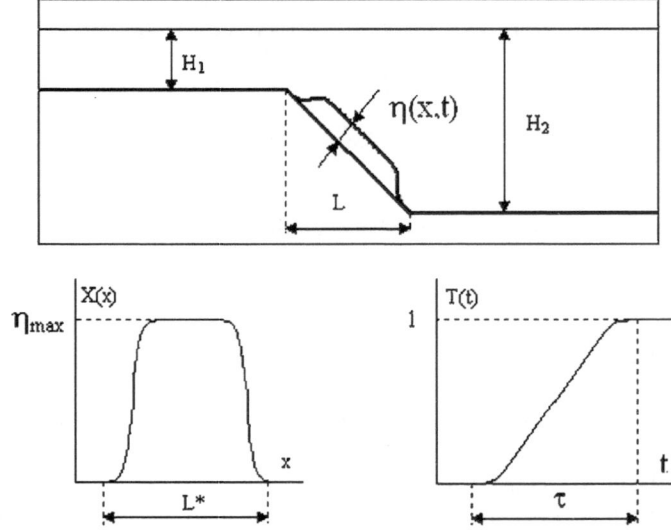

Figure 1. Bottom topography and time-spatial history of bottom displacements.

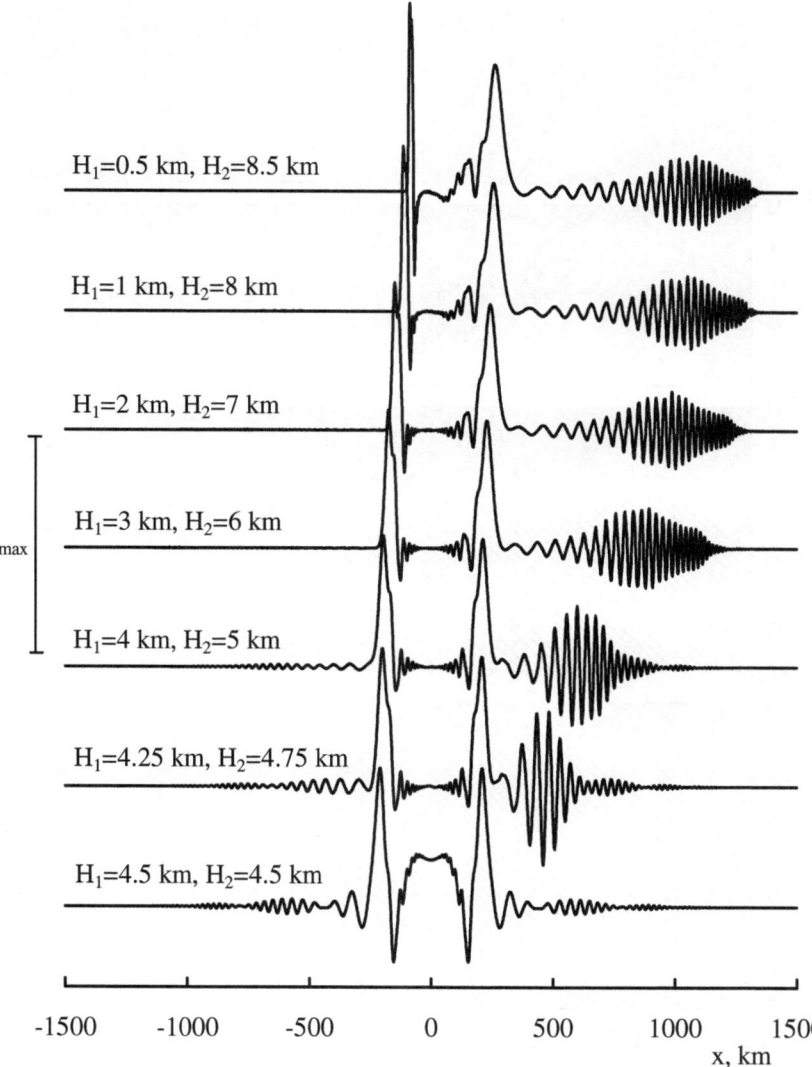

Figure 2. Free surface disturbance calculated for different bottom topography.

Being a wave of significant amplitude in the deep part of the basin, acoustic tsunami precursor is not observed at the water surface in the shallow domain. It means that acoustic waves generated by bottom motions can not penetrate in the shallow regions. This point can be explained in terms of the normal mode theory. Compressible water layer bounded above by free surface and below by absolutely rigid bottom is a sound waveguide. Any disturbance in such waveguide can be considered as a superposition of the normal modes. The longest one has wavelength $\lambda_{max} = 4H$, where H is the water layer thickens (depth).

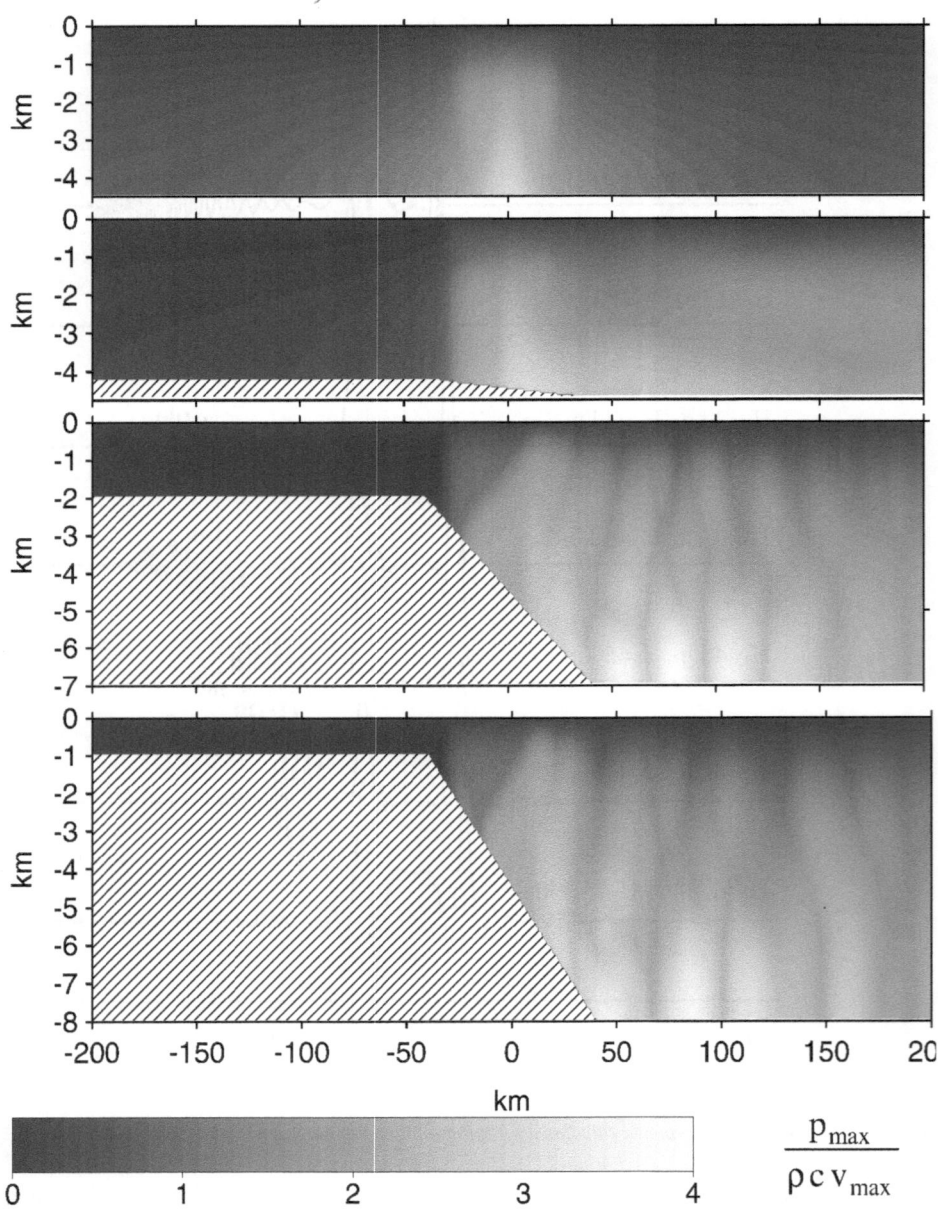

Figure 3. Spatial distribution of maximum pressure amplitude.

Bottom displacements of duration τ radiates wavelength $\lambda = c\tau$. Thus penetration of the acoustic waves in region where depth H is smaller than $c\tau/4$ is suppressed. Actually bottom motions generate a wide spectrum of acoustic waves, including components which are short enough to reach near shore regions, where these waves can be registered as the T-phase.

Examples of spatial distributions of the maximum dynamic pressure are shown on Fig. 3. In case of constant depth the distribution is symmetrical. The dynamic pressure reaches its maximum values near bottom, just above the source. Even for slightly sloping bottom (1:160) the distribution changes significantly. The maximum pressure area is tended to shift outside the source toward the deep domain, where the maximum pressure reaches value of $4\rho c v_{max}$ (v_{max} is the maximum bottom velocity). At the same time the maximum dynamic pressure in the shallow region does not exceed relatively small value $\sim 0.02\rho c v_{max}$, which is mostly a contribution of the gravitational wave. This confirms the reasoning above that shallow domains are closed for penetration of the acoustic waves.

During tsunami (or tsunami earthquakes) fish migration is often observed [14-16]. In particular, it was reported that some deepwater species have been found on beach or watched in the vicinity of water surface. Such fish behavior can be easily explained in terms of the maximum dynamic pressure distribution. Avoiding large pressure variations, fish migrates toward shallow regions or water surface.

At calculation of the energy transferred from the moving bottom to the compressible water layer, the following components should be considered:

$$W_k = \rho \sum \frac{v_x^2 + v_z^2}{2} dxdz$$
- kinetic energy;

$$W_p = \frac{1}{c^2\rho} \sum \frac{p^2}{2} dxdz$$
- elastic potential energy;

$$W_g = \rho g \sum \frac{\xi^2}{2} dx$$
- gravitational potential energy.

Also we determine the energy imparted to the water layer [9]:

$$W_0 = c\rho \sum \left(\frac{\partial \eta}{\partial t}\right)^2 dxdt$$

An example of the time-history of the energy components is shown on Fig 4. After finishing of the bottom motions, total energy $W_\Sigma = W_k + W_p + W_g$ remains a constant. It proves, that gravitational and elastic (acoustic) waves exist in the system inseparably.

The normalized total water layer energy (W_Σ / W_0) is plotted as a function of the bottom displacement duration in Figure 5. If the duration τ is shorter than acoustic wave propagation time along the distance "bottom-surface-bottom" $2H/c$ the value W_Σ is equal to W_0. Otherwise $W_\Sigma < W_0$, moreover W_Σ depends on the duration τ non-monotonically. It is important to emphasize, that in the realistic range of the duration values ($\tau \sim 10$ s) the energy captured by water layer can changes considerably (more than one order of magnitude). Oscillatory character of the dependence in questions is a consequence of resonance features of compressible fluid with free surface on rigid bottom (quarter-wave resonator). The increase of the bottom slope "detunes" the resonator, and the dependence becomes smoother. In case of large values of τ the ratio W_Σ / W_0 tends to go up. It can be explained in the following way. Value W_0 is calculated as integrated acoustic energy flux,

136

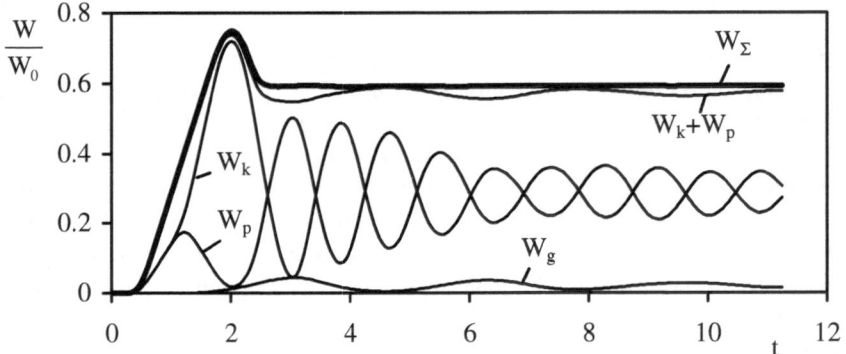

Figure 4. Example of the time-history of energy components.

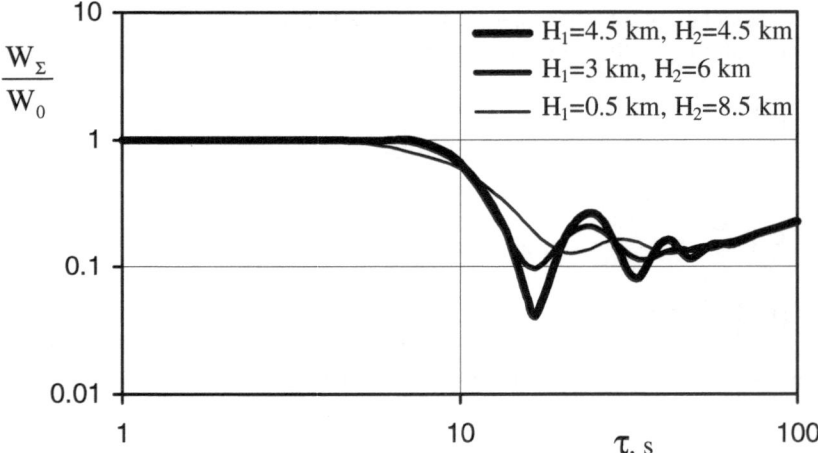

Figure 5. Total water layer energy as a function of the bottom displacement duration.

missing gravitational potential energy of "initial elevation". At large τ, this gravitational potential energy becomes not negligible compared to the value W_0. This is why the right part of the curve deviates upward.

Acknowledgements

This work was supported by the Russian Foundation for Basic Research, project 01-05-64547.

References

1. Sells, C.C.H. (1965), The effect of a sudden change of shape of the bottom of a slightly compressed ocean. Philos. Trans. R. Soc. A (London), no.1092, pp. 495-528.

2. Kajiura, K. (1970), Tsunami Source, Energy and the Directivity of Wave Radiation. Bull. Earthquake Res. Inst. Univ. Tokyo, Vol. 48, pp. 835-869.

3. Podyapolsky, G.S. (1970), Generation of the tsunami wave by earthquake. In: Tsunamis in the Pacific Ocean, Honolulu, pp.19-32.

4. Gusiakov, V.K. (1972), Excitation of Tsunami and Oceanic Rayleigh Waves under a Submarine Earthquake (in Russ.). Mathematical Problem in Geophysics, Novosibirsk, issue 3, pp.250-272.

5. Yanushaushkas, A.I. (1981), Cauchy-Poisson Theory for a Compressible Fluid (in Russ.). Tsunami propagation and run-up, Moscow, pp.41-55.

6. Fox, C.G. and Hammond, S.R. (1994), The VENTS Program T-Phase Project and NOAA's role in ocean environmental research. MTS Journal, 27(4), pp. 70-74.

7. Panza, F.G., Romanelli, F., Yanovskaya, T.B. (2000), Synthetic Tsunami Mareograms for Realistic Oceanic Models, Geophys. J. Int., 141, pp.498-508.

8. Ohmachi, T., Matsumoto, H., Tsukiyama, H. (2001), Seawater Pressure Induced by Seismic Ground Motions and Tsunamis. ITS 2001 Proceedings, Session 5, Number 5-4, pp.595-609.

9. Landau, L.D., Lifshitz, E.M. (1982), Fluid Mechanics, Oxford: Pergamon, 2nd ed.

10. Nosov, M.A. (1999), A Model for Tsunami Generation by Bottom Movements Incorporating Water Compressibility, *Volcanology and Seismology*, **20**, 731-741.

11. Nosov, M.A. and Sammer, K. (1998), Tsunami Excitation by a Moving Bottom Displacement in Compressible Water, *Moscow University Physics Bulletin,* **53**, 67-70.

12. Nosov, M.A. (1999), Tsunami Generation in Compressible Ocean, *Phys. Chem. Earth (B)*, **24**, 437-441.

13. Nosov, M.A. (2000), Tsunami Generation in a Compressible Ocean by Vertical Bottom Motions, *Izvestiya, Atmospheric and Oceanic Physics,* **36**, 718-726.

14. Soloviev, S.L. and Go, Ch.N. (1974), Catalog of Tsunamis on the Western shore of the Pacific Ocean (in Russian). Nauka, Moscow, 308 p. [Translated in 1984 by the Canadian Institute for Scientific and Technical Information, National Research Council, Ottawa, Canada, K1A 0S2].

15. Soloviev, S.L. and Go, Ch.N. (1975) Catalog of Tsunamis on the Eastern Shore of the Pacific Ocean (in Russian). Nauka, Moscow, 203 p. [Translated in 1984 by the Canadian Institute for Scientific and Technical Information, National Research Council, Ottawa, Canada, K1A 0S2].

16. Soloviev, S.L., Solovieva, O.N., Go, Ch.N., Kim, Kh.S., Shchetnikov, N.A. (2000), Tsunamis in the Mediterranean Sea 2000 B.C. - 2000 A.D. Dordrecht/Boston/London, Kluwer Academic Publishers, 237 p.

TSUNAMI WAVE EXCITATION BY A LOCAL FLOOR DISTURBANCE

I.T. SELEZOV
Department of Wave Processes
Institute of Hydromechanics, Nat. Acad. Sci.
8/4 Sheliabov Str., Kiev 03680, Ukraine

Abstract

This paper presents the modeling of tsunami wave generation by a bottom displacement in three-dimensional settlement taking into account elastic floor. The corresponding initial boundary value problem is solved and analyzed. The IBV-problem based on the shallow water wave model and the problem of nonlinear-dispersive waves over an excitable bottom are discussed and first order asymptotic approximations describing the problem are presented. A new type earthquake wave model is presented and analyzed. The existence of the solutions of jump type is established which can lead to triggering phenomena. Undeterminacy to state the initial boundary value problems for tsunami generation is discussed.

1. Introduction

Generation of tsunami waves is a problem of great importance and many researches have been focused on this problem for a long time [2, 7, 9, 12]. Up to now there are no perfectly clear explanations of the earthquake triggering mechanism and as a consequence there are no perfect models to formulate initial conditions. Nevertheless, in any case the problem of tsunami wave generation is stated as initial boundary value (IBV) problem with the given initial conditions corresponding to some idealized situations.

Considering the problem to state initial boundary value problem for tsunami wave generation shows that this problem is undeterminate. The main reason is the undeterminancy of earthquake triggering mechanism in general, including underwater earthquakes particularly, in spite of having been investigated for a long time [1-3, 7-20]. As a result, till now there are no completed satisfactory achievements and explanation of the causes of this phenomenon. As far as a triggering mechanism is concerned it is difficult to make this problem to perfectly determine, in the mathematical sense.

For example, in 1998 three huge tsunami waves of 10 m elevation run up on lagoon in Papua, New Guinea. Tsunami waves were generated by the underwater earthquake due to underground shocks. When should these appear? With what time intervals between them? The question arises how to state the IBV-problem? Later on we would present some particular models from this point of view.

A. C. Yalçıner, E. Pelinovsky, E. Okal, C. E. Synolakis (eds.),
Submarine Landslides and Tsunamis 139-150.
©2003 Kluwer Academic Publishers.

It was noted by Braddock et al [2] and this situation has not changed that the actual bottom motion is not completely understood. For example, seismic investigations show that actual flow conditions in the mantle are complex and undeterminate varying from cracks of whole mantle to the layered structure [4]. As has been noted by Papanicolaou in [11], the energy localization in random elastic media is possible due to such causes as the mode conversion, the transfer of energy from compression to shear waves, and polarization. Brevdo [3] considered an elastic layer in 3-D and instead of a convenient purely normal wave mode treatment investigated the asymptotic behaviour of wave packet. On this basis, a possible resonant triggering mechanism of certain earthquakes is shown due to localized low amplitude oscillatory forcing at resonant frequencies. Considering the energy budget of deep-focus earthquakes suggests that they may be slip-sliding away [8], leading to undeterminate triggering mechanisms. One possible triggering mechanism can be initiated by the re-polarization of elastic waves at an interface where the perfect matching can be violated by enough strong tension stresses [20].

A new earthquake model of wave type has been proposed in [18]. It is based on the hydrodynamic flow of geomaterial along the ray tube of tectonic stream and on the evolution of the medium damage governed by a kinetic equation for the damage ratio $\lambda \in [0,1]$. The analysis of dispersion equation shows that wave propagation soliton-like disturbances can be expected for the narrow wave beams in the vicinity of the critical Morse points. In the case $\lambda \to 1$ the singular degeneration takes place and this leads to the jumps in structural parameters which can cause earthquake. The triggering time can not be established exactly if the above presented procedure is followed.

This paper considers the problem of excitation of the surface gravity waves in ocean due to the underwater source. The concentrated source is placed at the interface between water and elastic half-space. The original problem is essentially simplified and can be useful for analysis of tsunami waves. The IBV-problem based on the shallow water wave model and the problem of nonlinear-dispersive waves over an excitable bottom are discussed and first order asymptotic approximations describing the problem are presented.

2. Wave Generation by a Source on Elastic Floor

Corresponding initial boundary value (IBV) problem is stated for the fluid of finite depth in $\Omega_1 = \{(r,\theta,z) : r \in [0,\infty), z \in [0,1]\}$ over an elastic half-space in Ω_2 when at the interface $z=0$ a source is switched on at the initial time $t=0$ which sharply increases up to the maximum and then exponentially decreases with time.

The motions of fluid and elastic solid are governed by the potential flow equations for incompressible inviscid fluid and by the elastodynamic equations for isotropic homogeneous medium respectively

$$\beta^2 \nabla^2 \varphi_1 + \frac{\partial^2 \varphi_1}{\partial z^2} = 0 \qquad \text{in} \quad \Omega_1, \tag{1}$$

$$\beta^2 \frac{\partial^2 \varphi_1}{\partial t^2} + \frac{\partial \varphi_1}{\partial z} = 0 \quad \text{at} \quad z=1, \qquad \frac{\partial \varphi_1}{\partial z} = \beta^2 \frac{\partial u_z}{\partial t} \quad \text{at} \quad z=0 \tag{2}$$

$$\nabla^2 \vec{u} + \left(\gamma_0^2 - 1\right)\vec{\nabla}\left(\vec{\nabla} \cdot \vec{u}\right) = \gamma_2^2 \frac{\partial^2 \vec{u}}{\partial t^2} \quad \text{in} \quad \Omega_2, \tag{3}$$

$$\beta\left(\gamma_0^2 - 2\right)\left(\frac{\partial u_r}{\partial r} + \frac{u_r}{r}\right) + \gamma_0^2 \frac{\partial u_z}{\partial z} - \alpha \frac{\partial \varphi_1}{\partial t} = -\frac{A_0}{\pi} F(t) H(t) \frac{\delta(r)}{r}, \quad z=0, \tag{4}$$

$$\frac{\partial u_r}{\partial z} + \beta \frac{\partial u_z}{\partial r} = 0 \quad \text{at} \quad z = 0 \tag{5}$$

$$\varphi_1 = \frac{\partial \varphi_1}{\partial t} = u_z = \frac{\partial u_z}{\partial t} = u_r = \frac{\partial u_r}{\partial t} = 0 \quad \text{under } t = 0, \tag{6}$$

$$u_r, \ u_z \to 0 \quad \text{under} \quad z \to -\infty, \tag{7}$$

where $\nabla^2 = 1/r\, \partial/\partial r\, (r\partial/\partial r)$; φ_1 is the velocity potential; $\vec{u}$ is the elastic displacement vector, $\vec{v} = (u_r, u_z)$; $H(t)$ is the Heaviside function; $\delta(r)$ is the delta-function; $F(t)$ is the excitation function. The equations (1)-(7) are written in dimensionless form according to the formulas

$$r^* = \frac{r}{R}, \quad z^* = \frac{z}{h}, \quad t^* = t\frac{\sqrt{gh}}{R}, \quad \eta^* = \frac{\eta}{h}, \quad \vec{u}^* = \frac{u}{h}, \quad \varphi_1^* = \frac{\varphi_1}{R\sqrt{gh}}, \quad \beta = \frac{h}{R},$$

$$\gamma_0 = \frac{c_e}{c_s}, \quad \gamma_1 = \frac{\sqrt{gh}}{c_e}, \quad \gamma_2 = \frac{\sqrt{gh}}{c_s}, \quad \alpha = \frac{\rho_0\, gh}{G}.$$

where R is a characteristic horizontal scale.

It is necessary to find the potential φ_1 as the solution of the Laplace equation (1) and the radial and vertical displacements u_r, u_z as the solutions of the Lame equation (3), $\vec{u} = \{u_r, 0, u_z\}$, satisfying the conditions (2) at the free surface $z=1$ and at the interface $z = 0$, the boundary conditions (4), (5) for normal and shear stresses σ_{zz}, σ_{rz} at $z = 0$, the initial conditions (6) and the regularity conditions (7).

The problem is solved by using the Laplace transform in time

$$\varphi_1^L = \int_0^\infty \varphi_1 e^{-pt}\, dt, \quad u_r^L = \int_0^\infty u_r e^{-pt}\, dt \quad u_z^L = \int_0^\infty u_z e^{-pt}\, dt \tag{8}$$

The values (8) are presented in the form

$$\varphi_1^L = \int_0^\infty \Phi(p, s, z) J_0(sr)\, ds \quad u_r^L = \int_0^\infty U(p, s, z) J_1(sr)\, ds,$$

$$u_z^L = \int_0^\infty W(p, s, z) J_0(sr)\, ds \tag{9}$$

Taking into account the formula $\dfrac{\delta(r)}{r} = \int\limits_0^\infty s\,J_0\,(rs)\,ds$ and the expressions

(8), (9) reduces the problem (1)-(7) to the boundary-value problem for amplitudes Φ, U, W

$$\Phi'' - s^2\beta^2\,\Phi = 0,\tag{10}$$

$$\Phi' - p^2\beta^2\,\Phi = 0 \quad\text{at}\quad z = 1,\quad \Phi' = p\,\beta^2\,W \quad\text{at}\quad z = 0,\tag{11}$$

$$\gamma_0^2\left(W'' + s\beta\,U'\right) - s^2\beta^2 W - s\beta\,U' = p^2\beta^2\gamma_2^2\,W\tag{12}$$

$$-\gamma_0^2\left(s^2\beta^2 U + s\beta\,W'\right) + U'' + s\beta\,W' = p^2\beta^2\gamma_2^2\,U,\tag{13}$$

$$U' - s\beta\,W = 0,\quad \left(\gamma_0^2 - 2\right)s\beta\,U + \gamma_0^2\,W' - \alpha\,p\,\Phi = -\frac{A_0\,s}{\pi}F^L(p) \quad\text{at}\quad z = 0\tag{14}$$

The resultant equation describing the free surface elevation is given in Eq. 15

$$\eta^L = F^L(p)\frac{A_0}{\pi}\,p^2\beta\int\limits_0^\infty f(s)\frac{s\,J_0\,(rs)}{\Delta_0}\,ds\tag{15}$$

where

$$f(s) = \frac{2s\,e^{-s\beta}}{s + p^2\beta + (s - p^2\beta)\,e^{-s\beta}},$$

$$\Delta_0 = \frac{s(\lambda_1 - \lambda_2)(\gamma_0^2 + \gamma_0^2\,c_1\,c_2 - 2) - (c_2 - c_1)(\gamma_0^2\,\lambda_1\,\lambda_2 + s^2(\gamma_0^2 - 2))}{c_2\,\lambda_1 - c_1\,\lambda_2}\times$$

$$\times s\frac{p^2\beta + s\,th\,s\beta}{s + p^2\beta\,th\,s\beta} - \alpha\,p^2.$$

Using the Cagniard approach [8] the exact analytical solution for the epicenter elevation can be obtained. For arbitrary values of r (radial coordinate) calculations are carried out on the basis of a numerical Laplace transform inversion. The approach developed is based on the expansions of desired functions with respect to the orthonormal system of Fourier-Bessel functions

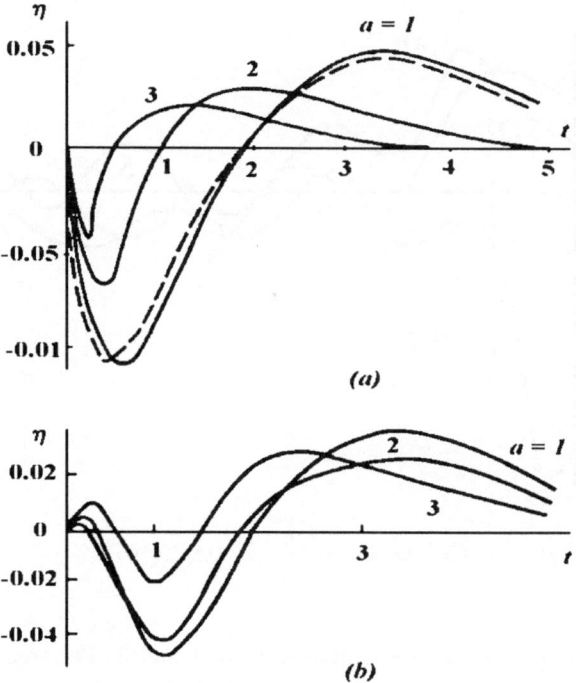

Figure 1. Free surface elevation generated by the excitation function $F(t) = t^2 \exp(-at)$ *for* $a = 1, 2, 3$:
a) at the epicenter $r \,/\, R = 0$; *b) over the edge of seismic center* $r \,/\, R = 1$.

$$f(t) = \sum_{n=1}^{\infty} c_{n\nu} J_\nu (k_{n\nu} e^{-\sigma t}) \quad, \quad J_\nu (k_{n\nu}) = 0,$$

(16)

$$c_{n\nu} = \frac{2\sigma}{J_{\nu+1}^2 (k_{n\nu})} \sum_{m=0}^{\infty} \frac{(-1)^m}{m! \Gamma(m+\nu+1)} \left(\frac{k_{n\nu}}{2}\right)^{2m+\nu} F((2m+2+\nu)\sigma)$$

(17)

and on the Tikhonov regularization procedure to improve the series convergence

$$f(t) = \sum_{n=1}^{\infty} c_{n\nu} \Theta_n \phi_{n\nu} (t) \quad,$$

(18)

$$\Theta_n = \frac{1}{1 + \delta k_{n\nu}^2}$$

(19)

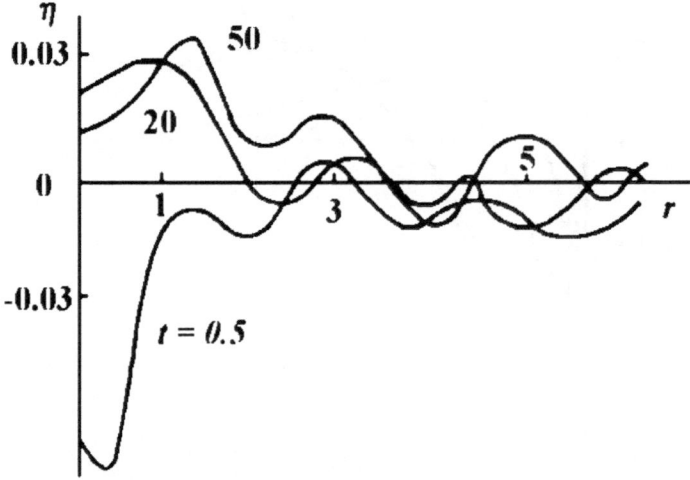

Figure 2. Free surface shape at different times: $\dfrac{\sqrt{gh}}{R} t = 0.5; 20; 50$.

The results of calculations are presented in Figs. 1 and 2. The free surface elevation depending on time at two places: $r=0$ (epicenter) and $r=1$ (the edge of seismic center), for different values of a are presented in Fig. 1. The parameter a characterizes the sharpness of the excitation impulse, so that increasing a increases the impulse sharpness, as well as the sharpness of the free surface response, but in this case the magnitude decreases due to decreasing the time to transmit the energy from seismic center to the free surface.

Solid curves in Figure 1a correspond to exact solution, while dotted curve shows the results of numerical inversion for evaluation of the exactness of numerical transform inversion. The calculations were also carried out for the free surface elevations at different ratios of the propagation velocities of shallow water waves and shear waves in solid, γ_1 and γ_2, and for different relative depths.

Approximate analysis of tsunami wave generation in water of the variable depth can be carried out on the basis of shallow water wave equations. In this case the corresponding IBV-problem is essentially simplified. At the same time, this approximate model is applicable with sufficient exactness at some distances from an epicenter. The corresponding IBV-problem is stated as follows

$$\vec{\nabla}\cdot\left[H(r)\vec{\nabla}\eta_1\right]-\frac{\partial^2\eta_1}{\partial t^2}=0, \qquad r\in[0,a],\quad t\in[0,\infty), \tag{20}$$

$$\eta_1\big|_{t=0}=\eta_0(r),\qquad \frac{\partial\eta_1}{\partial t}\bigg|_{t=0}=0,\qquad \eta_1\to 0 \ \text{under}\ r=0 \tag{21}$$

$$\nabla^2 \eta_2 - \frac{\partial^2 \eta_2}{\partial t^2} = 0,$$
$$r \in (a, \infty), \quad t \in [0, \infty), \tag{22}$$

$$\eta_2 \big|_{t=0} = 0, \quad \frac{\partial \eta_2}{\partial t} \big|_{t=0} = 0, \quad \eta_1 \big|_{r=a} = \eta_2, \quad \frac{\partial \eta_1}{\partial r} \big|_{r=a} = \frac{\partial \eta_2}{\partial r} \big|_{r=a}. \tag{23}$$

On the basis of (20)-(23) the effect of the initial bottom elevation (H_0) which varies as $H(r) = H_0 + (1 - H_0)r^2$ on the tsunami wave generation has been investigated in [15, 17].

3. Excitation of nonlinear water waves

The investigation of nonlinear-dispersive effects during tsunami wave propagation is a problem of great importance [12]. The problem of particular interest is the excitation of nonlinear water waves by a bottom surface. The original problem of nonlinear water wave propagation over inhomogeneous moving bottom is stated as follows

$$\beta \nabla^2 \varphi + \frac{\partial^2 \varphi}{\partial z^2} = 0 \qquad \text{in} \quad \Omega \tag{24}$$

$$\eta_t + \alpha \vec{\nabla} \varphi \cdot \vec{\nabla} \eta = \frac{1}{\beta} \varphi_z \qquad \text{at} \quad z = \alpha \eta \tag{25}$$

$$\eta + \varphi_t + \frac{\alpha}{2\beta} \varphi_x^2 + \frac{\alpha}{2} (\vec{\nabla} \varphi)^2 = 0 \qquad \text{at} \quad z = \alpha \eta \tag{26}$$

$$\gamma(\xi_t + \alpha \vec{\nabla} \varphi \cdot \vec{\nabla} \xi) - \alpha \vec{\nabla} \varphi \cdot \vec{\nabla} H = \frac{\alpha}{\beta} \varphi_z \qquad \text{at} \quad z = -H(x, y) + \gamma \xi(x, y, t), \tag{27}$$

where ∇^2 and $\vec{\nabla}$ are horizontal operators. In (24)-(25) nondimensional values are used according to the formulas (asterisks are omitted): $(x^*, y^*) = (x, y)l$, $(z^*, H^*) = (z, H)/h_0$, $\xi^* = \xi/\xi_0$, $\eta^* = \eta/a$, $\varphi^* = \varphi\sqrt{g h_0}/gla$, $t^* = t\sqrt{g h_0}/l$, where l and h_0 are the characteristic length and depth, a and ξ_0 are the amplitudes of free surface and bottom elevations, respectively.

As we can see from (24)-(27), the nonlinear parameter $\alpha = a/h_0$, the dispersion parameter $\beta = (h_0/l)^2$ and the parameter of nonstationary bottom state $\gamma = \xi_0/l$ are responsible for the phenomena under consideration.

The problem (24)-(27) after some considerations and with the assumption $\alpha\beta \ll 1$ is reduced to the following simplified form

$$\beta\frac{\partial^2\varphi}{\partial t^2}+\frac{\partial\varphi}{\partial z}=0,\qquad z=\alpha\eta, \qquad \eta=-\frac{\partial\varphi}{\partial t}\bigg|_{z=\alpha\eta},$$

$$\beta\left[\frac{\partial\xi}{\partial t}-\vec{\nabla}\varphi\cdot\vec{\nabla}H\right]=\frac{\partial\varphi}{\partial z},\qquad z=-H+\gamma\xi.$$

(28)

(29)

Considering the case $H=1$ and using power series expansion $\varphi=\sum\limits_{n=0}^{\infty}\varphi_n(x,y,t)(z+1)^n$ also simplifies the problem (28), (29) to a recurrence system of equations.

Assuming $\alpha\ll1$, $\beta\ll1$, $\gamma=O(\alpha)$ and applying asymptotic analysis, this system is reduced to the exactness up to the order $O(\alpha,\beta,\gamma)$ of the terms in evolution equations.

$$\frac{\partial^2\varphi_0}{\partial t^2}-c_0^2(\eta,\xi)\nabla^2\varphi_0-\frac{\beta}{2}\frac{\partial^2\nabla^2\varphi_0}{\partial t^2}+\frac{\beta}{6}\nabla^4\varphi_0=\frac{\partial F}{\partial t},$$

(30)

$$F=-\xi-\beta\frac{\partial^2\xi}{\partial t^2}+\frac{\beta^2}{2}\nabla^2\xi,\quad \eta_0=-\frac{\partial\varphi_0}{\partial t},$$

(31)

$$\vec{\nabla}\cdot\left(H\vec{\nabla}\eta_0\right)-\frac{\partial^2\eta_0}{\partial t^2}=-\frac{\partial^2\xi}{\partial t^2},\qquad c_0^2(\eta,\xi)=1+\alpha\eta_0-\gamma\xi$$

(32)

In the dispersion-free case but of variable depth, $H\ne const$, the system (30), (31) is reduced to the following equation

$$\vec{\nabla}\cdot\left(H\vec{\nabla}\eta_0\right)-\frac{\partial^2\eta_0}{\partial t^2}=-\frac{\partial^2\xi}{\partial t^2}$$

(33)

As we can see from the equation (30), the presence of moving bottom leads the appearance of excitational force and changing the propagation velocity c_0.

4. A new earthquake model of wave type

There are several seismic models but two of them are in the interest of this paper. First of all, the diffusion theory developed by Elsasser [6] is well known but not capable to explain the migration of seismicity to large distances. Unlike this theory the model of Nikolaevsky et al. [10] predicts propagation of tectonic stress disturbances which are like triggers to initiate earthquakes. He considers the bending and compression of lithosphere plates contacted with tectonic streams neglecting inertial forces in these streams.

We developed a new model to predict the possibility of solitary wave propagation. This model is based on a geological concept of tectonic streams as hydrodynamic structures. The tectonic streams appear at the boundaries of plates as a result of their interactions. These streams are characterized by small velocities (1 cm/year), nonlinear effects, laminar flows, and small Reynolds' numbers $\left(10^{-10} - 10^{-20}\right)$. Typical behaviour of such tectonic streams is observed in Carpathian region.

In a Cartesian coordinate system $(x_1, x_2, x_3) \leftrightarrow (x, y, z)$ we consider 2-D flow of the medium the state of which is characterized by the vector

$$\bar{Q}^t = \{u_i, \rho, p, \lambda\} \qquad (i = 1,3) \tag{33}$$

where $\vec{u} = (u_1 = u_x, u_3 = u_z)$ is the velocity vector, ρ is the density, p is the pressure, λ is the damage ratio of medium.

Vector $\bar{Q}t$ is presented as a superposition of undisturbed and disturbed states

$$\bar{Q}^t = \bar{Q}^0(x, z, t) + \alpha \bar{\bar{Q}}\left(x, z, t'\right) \tag{34}$$

corresponding to slow (tectonic) time t and rapid time t', so that $t' \ll t$. In the expression (34) $\bar{Q}0$ is independent of t' and it can be considered as a "frozen" background field, α is a small parameter. The field $\bar{\bar{Q}}$ is considered in the thin layer of a unit width (thin ray tube).

Let us introduce the non-dimensional axial and transverse coordinates $\eta = l/l_0$, $\chi = r/h$, where $\eta \in [-\infty, \infty]$, $\chi \in [-1,1]$. Hereinafter it is assumed that the dynamic viscosity has a local minimum on the axial line, so that $\mu_r = 0$ at $\chi = 0$.

The governing equations are written as follows

$$\frac{D\rho}{Dt} + \rho u_{k,k} = 0 \qquad\qquad k = 1,3 \tag{35}$$

$$\rho \frac{Du_i}{Dt} = -p_{,i} + \tau_{ij,j} + \rho g \delta_{i3}, \qquad\qquad i, j = 1,3 \tag{36}$$

$$\frac{D\lambda}{Dt} = R(p, \tau_*)(1 - \lambda)^{-\beta} \qquad\qquad \beta > 0, \tag{37}$$

where $\dfrac{D}{Dt} = \dfrac{\partial}{\partial t} + (\vec{u} \cdot \vec{\nabla})$, τ_{ij} is the tensor of viscous stresses, $\tau_* = (\frac{1}{2}\tau_{ij}\tau_{ij})^{1/2}$ is the intensity of shear stresses. The system (35)-(37) includes the mass conservation law (35), Navier-Stokes equations (36) and kinetic equation (37). The volume compressibility will be ignored in the stress tensor τ_{ij}.

From the first law of thermodynamics for a unit mass flow, the following equation can be obtained

$$\frac{DS}{Dt} + \bar{u}\frac{D\bar{u}}{Dt} - gu_3 + \chi_{ext} + \frac{1}{\rho}tr(\vec{\tau}\cdot\nabla\bar{u}) = 0 \tag{38}$$

where $s = E + p/\rho$ is the specific enthalpy, the value χ_{ext} characterizes heat-exchange with the surrounding medium

Now the closed system of equations can be written for the undisturbed state corresponding to a steady laminar stream with the "frozen" value $\lambda = \lambda_*$ and negligibly small value $\frac{D\vec{Q}^0(\vec{x},t)}{Dt} \ll 1$. Taking the uniform field as the simplest solution of undisturbed state, the system of equations of the disturbed state is derived. Considering traveling waves along the streamline yields the dispersion equation

$$F(k,\omega,\xi) = 0 \tag{39}$$

where k is the wave number, ω is the angular frequency, ξ is a perturbation parameter. . Then the analysis is carried out in the neighborhood of critical Morse points (CMP). At the isolated nondegenerated CMP $W_m \in C^3$ the following conditions hold

$$\frac{\partial F}{\partial k}\bigg|_{W_m} = \frac{\partial F}{\partial \omega}\bigg|_{W_m} = \frac{\partial F}{\partial \xi}\bigg|_{W_m} = 0 \tag{40}$$

$$\det J(k,\omega,\xi)\big|_{W_m} \neq 0 \tag{41}$$

where J is the Gessian of F, ξ is a nonspectral parameter

$$F(k,\omega,\xi): C\times C\times C \to C$$

According to Morse' lemma the complex hypersurface in the vicinity of CMP has the standart representation of dispersion equation in the following form

$$\lambda_1 k^2 + \lambda_2 \omega^2 + \lambda_3 \xi^2 + 2\delta = 0 \tag{42}$$

Now it is possible to pass to the configuration space $k \to i\partial/\partial s$, $\omega \to i\partial/\partial \tau$, where s,τ are linear combinations of l,t.

Pre-ruptured state takes place when $(1-\lambda) \to 0$, $\lambda \in [0,1]$, and in this case the system of equations predicts solutions of jump type leading to triggering phenomena.

5. Conclusion

Brief review of tsunami excitation by earthquakes is presented showing undeterminate data to state corresponding IBV-problems.

The IBV-problem for tsunami wave generation is stated and solved on the basis of Fourier-Bessel expansions and Tikhonov regularization procedure. The results of calculations are presented demonstrating the initial development and evolution of tsunami waves. The dependence of tsunami generation on the sharpness of a source is analysed.

Evolution equations for propagation of nonlinear-dispersive water waves over an excitable bottom are derived starting from the original 3-D statement.

A new wave model for triggering mechanisms of earthquakes is presented. The problem includes the equation for the damage ratio λ whose a critical value can essentially influence the solution. The analysis is based on the critical Morse' point approach.

Acknowledgement: This investigation is supported by the Research Project INTAS (Grant 99 - 1637) and SFFR Project of Ukraine (Grant № 01.07/00079).

References

1. Alpar B., Yüksel Y., Dodan E., Gavioglu C., Cevik E. and Altmok Y. (2001) An estimate of detailed depth soundings in limit bay before and after 17 August 1999 eartquake. Turkish *J. Marine Sciences*, 7(1), 3-18.
2. Braddock R.D., van den Driesshe and Peady G.W. (1973) Tsunami generation. *J. Fluid Mech.*, 59, part 4, 817-828.
3. Brevdo L. (2000) Three-dimensional logarithmic resonances in a homogeneous elastic wave guide. *Eur. J. Mech. A/Solids*, 19, 121-137.
4. Cadek O., Yuen D.A. and Cizkova H. (1997) Mantle viscosity inferrred from geoid and seismic tomography by genetic algorithms: results for layered mantle flow. University of Minnesota Supercomputing Inst. Research. *Report UMSI 97/ 118*, August 1997, 18pp.
5. Cagniard L. (1939) *Reflexion et refraction des ondes seismiques progressive*. Gauthier-Villard, Paris, 1939; English transl., Reflection and refraction of proggresive seismic waves, translated by E.A. Flinn and C.H. Dix, McGraw-Hill, New Yourk, 1962.
6. Elsasser W.M. (1969) Convection and stress propagation in the upper mantle. *In: Appl. Modern Physics. Earth, Planets. Interior*. New York, Wiley, 223-246.
7. Kajura K. (1963) The leading wave of a tsunami. *Bull. Eartquake Res. Inst.*, 42, 535-571.
8. Ladbury R. (1998) Energy budget of deep-focus eartquakes suggests they may be slip-sliding away. *Physics Today*, April 1998, 19-21.
9. Murty T.S. (1977) *Seismic sea waves tsunami*. Ottava, Canada, 1977. Russian translated edition, 1981.
10. Nikolaevsky V.N. and Ramazanov T.K. (1985) The theory of fast tectonic waves (in Russian). *Applied Mathematics and Mechanics*, 49, N3, 462-469.
11. Papanicolaou G. (1998) Mathematical problems in geophysical wave propagation .*Proc. Int. Congress of Mathematicians. Vol.1: Plenary Lectures and Ceremonies*, Berlin, Germany, August 18-27, 403-427.
12. Pelinovsky E.N. (1996) *Hydrodynamics of tsunami waves* (in Russian). Nizhny Novgorod. Institute of Applied Physics, Russian Acad. Sci.
13. Selezov I.T. (1999) Interaction of water waves with engineering constructions and topography in coastal area. Proc. Fifth Int. Conference on Coastal and Port Engineering in Developing Countries COPEDEC 5, Cape Town, South Africa, April 19-23, 1, 1-12.
14. Selezov I. (1999) Some hyperbolic models for wave propagation. Int. Series of Numerical Mathematics. Hyperbolic Problems: Theory, Numerics, Applications. Edited by M.Fey and R.Jeltsch. Birkhauser Verlag Basel/Switzerland, 130, 833-842.
15. Selezov I.T. (2000) Some problems of surface gravity waves from M.A. Lavrentyev. Abstracts, Int. Conf. dedicated to M.A. Lavrentyev on the occasion of his birthday centenary, Kiev, Ukraine, Oct 31 Nov 3, 54-55.
16. Selezov I.T. (2000) Wave processes in fluid and solid media (in Russian, to be translated in USA). *Applied Hydromechanics*, 2(74), N 4, 99-118.
17. Selezov I.T. and Ostroverkh B.N. (1996) Modelling of seismic underwater centers of generation and transformation of tsunami waves in seismic active regions. *Marine Geophys. J.*, N 1, 66-77.
18. Selezov I. and Spirtus V. (1998) Propagation of disturbances along axial lines of tectonic flows under conditions of continuum destruction. Proc. Int. Simposium "Trends in Continuum Physics, TRECOP' 98". Poznan, Poland, 17-20 Aug., 1998. Ed. B.T.Maruszewski, W.Muchik and A.Radowicz. Singapore-New Jersey, London-Hong Kong, World Scientific, 1998, 322-333.
19. Solovyev S.L. (1980) Tsunami. *Earth and Universe (Zemlya i Vselennaya)*, 3, 12 -16.

20. Wang Y.-S. and Yu G.-L. (1999) Re-polarization of elastic waves at a frictional contact interface. 2. Incidence of a P or $_{SV}$ wave. *Int. J. Solids and Struct.*, 36, 4563-4586.

EFFECTS OF TSUNAMI AT SISSANO LAGOON, PAPUA NEW GUINEA: SUBMARINE-LANDSLIDE AND TECTONICS ORIGINS

MASAFUMI MATSUYAMA[1], HARRY YEH[2]
[1]*Central Research Institute of Electric Power Industry, Abiko, Japan*
[2]*University of Washington, Seattle, WA, USA*

Abstract

A submarine landslide has been conjectured at the outset as a source of the 1998 PNG tsunami, because the observed run-up pattern was much more localized than the predictions of numerical simulations with a co-seismic fault source. The discrepancy is however resulted from the faulty bathymetry data and insufficient grid size in the simulations. An earthquake fault source combined with the seafloor geometry taken from the published nautical chart can explain the tsunami focusing. A smaller source area of the submarine landslide results a runup distribution pattern distinct from the fault-source model.

1. Background

Compared with other recent tsunami events, the 1998 PNG tsunami has been more scrutinized by the tsunami community [1, 2, 3, 4, 5]. Part of the strong motivation is because it caused significant number of casualty: more than 2,000 people were killed. But it is more to the scientific fact that the tsunami destruction was much greater than the initial expectation made by the seismic signals. Only a narrow region centered around Sissano Lagoon was significantly hit by the tsunami (see Fig. 1), while the numerical simulations based on the seismic data indicate otherwise. We will discuss later that the bathymetry data used for the preliminary numerical simulations made immediately after the event were grossly inaccurate: the published nautical charts were not used. Because of the discrepancy, a giant submarine landslide was speculated at the outset as the tsunami generation. The submarine landslide scenario made many scientists alert, because it can be a major potential hazard that might occur elsewhere.

While submarine landslide has been the primary focus in research to explain the PNG tsunami, Matsuyama et al. [6] demonstrated numerically that the co-seismic fault-dislocation model can reproduce the tsunami concentration toward Sissano Lagoon, if the simulation were made with the accurate bathymetry data obtained by JAMSTEC [5].

In spite of the work by Matsuyama et al. [6], the hypothesis of submarine-landslide origin could not be refuted. It is mainly because tsunami arrival times along the coast appear to

A. C. Yalçıner, E. Pelinovsky, E. Okal, C. E. Synolakis (eds.),
Submarine Landslides and Tsunamis 151-162.

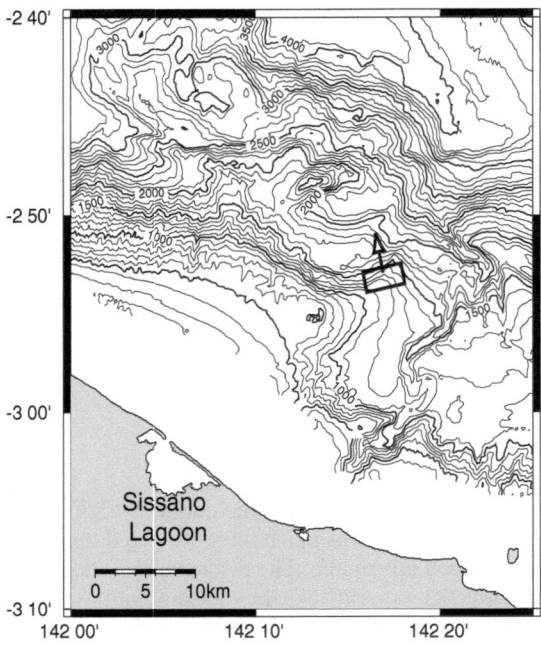

Figure 1. The location and orientation of the submarine slump. Note that the longitude is in East, and the latitude in South.

contradict with eyewitness accounts. Many eyewitness accounts indicate that the tsunami reached the coast after the first aftershock (09:09 GMT with Mw 5.9) instead of the main shock of Mw 7.1 at 08:49 GMT. The multi-channel seismic reflection profiles discovered a major landslide image [8], although the seismic reflection profiles cannot identify when the event occurred. Based on the seismic profiles, the center of the landslide is estimated at the coordinates (142°16.11''E, 2°52.06''S), and the slide direction is approximately in N22°W as illustrated in Fig. 1. The estimated landslide size is that the maximum thickness = 760 m; the length = 4.5 km; the width = 2.5 – 3.0 km; the cross-sectional area = 2.3 km^2; the total volume = 3.8 – 4.6 km^3; the movement of the center of mass along the slide plane = 980 m [8]. According to Iwabuchi [9], this slide location is believed to be the only potential "major" landslide site based on the detailed bathymetry obtained by JAMSTEC [5], although there are evidently numerous small landslides. The occurrence of submarine landslide is detected by the T-wave analysis [7]. The analysis indicates that the slide must occur at 09:02 GMT, lasting more than 45 – 50 sec. It is, however, noted that this analysis cannot provide how large the slide was. Because of these factors, Synolakis et al. [7] focused on the submarine landslide source and discounted a possibility of the seismic dislocation cause.

If indeed this landslide were triggered by the 1998 earthquake, it must be a slow and cohesive slump [10], instead of a fast moving slide with significant sediment disturbance. It is so because the slide movement is short, 980m [8], and turbidities were not detected on the seabed in any of the piston core samples taken by JAMSTEC [5].

2. Simulations

Based on the above information, we performed numerical simulation; first for the tsunami source due to co-seismic fault dislocation, then for the landslide based on the seismic profile data [8]. The numerical code is based on the fully nonlinear shallow-water-wave theory and no bottom friction is included. The shallow-water-wave theory is based on the depth-averaged equations of mass and momentum conservation:

$$\frac{\partial \eta}{\partial t} + \frac{\partial M}{\partial x} + \frac{\partial N}{\partial y} = 0 \quad , \tag{1}$$

$$\frac{\partial M}{\partial t} + \frac{\partial}{\partial x}\left[\frac{M^2}{D}\right] + \frac{\partial}{\partial y}\left[\frac{MN}{D}\right] + gD\frac{\partial \eta}{\partial x} = 0 \quad , \tag{2}$$

$$\frac{\partial N}{\partial t} + \frac{\partial}{\partial x}\left[\frac{MN}{D}\right] + \frac{\partial}{\partial y}\left[\frac{N^2}{D}\right] + gD\frac{\partial \eta}{\partial y} = 0 \quad , \tag{3}$$

where η is the water-surface elevation from its equilibrium state, M and N are the depth-averaged volumetric flux in the (x, y) horizontal directions, respectively, t is the time, D is the total water depth, g is gravitational acceleration. The equations are solved numerically for the unknowns M, N, and η, using the staggered-grid leap-frog numerical scheme, e.g. [11]. Note that this numerical scheme has been widely used and is proven to provide adequate predictions for real tsunami effects, especially those generated by co-seismic fault displacements [12].

2.1. SEISMIC FAULT SOURCE

For the present numerical simulations, the initial fault displacement is inferred from one of the two Harvard CMT solutions: seismic moment (M_0) = 3.67 x 10^{19} N-m with the strike, dip, rake of 287°, 75°, 78°, respectively (i.e., the high dip-angle reverse fault scenario). Note that this high dip-angle scenario was discounted by Synolakis et al. [7] based on their interpretations of the seismic data. The displaced area for our model is estimated from the JAMSTEC bathymetry data and aftershock locations [5], and is set to be 40 km long and 20 km wide (see Fig. 2). The resulting sea floor deformation was computed by the method developed by Mansinha and Smylie [13]. Note that both the position and the deformation of the fault are uncertain and could be in error as much as by 50% [14].

Three different grid sizes are tested for the numerical simulations using the newly acquired bathymetry data (JAMSTEC data) reported in [5]. Note that the JAMSTEC data are only available in the region deeper than approximately 200m. The bathymetry data in the region shallower than 200m are available by the existing nautical charts. The bathymetry data we used for the simulations are based on the montage of these two data sets. The computational domain is set 109.8km (E-W) by 82.8km (N-S) as shown in Fig. 2. The entire domain is divided by sub-

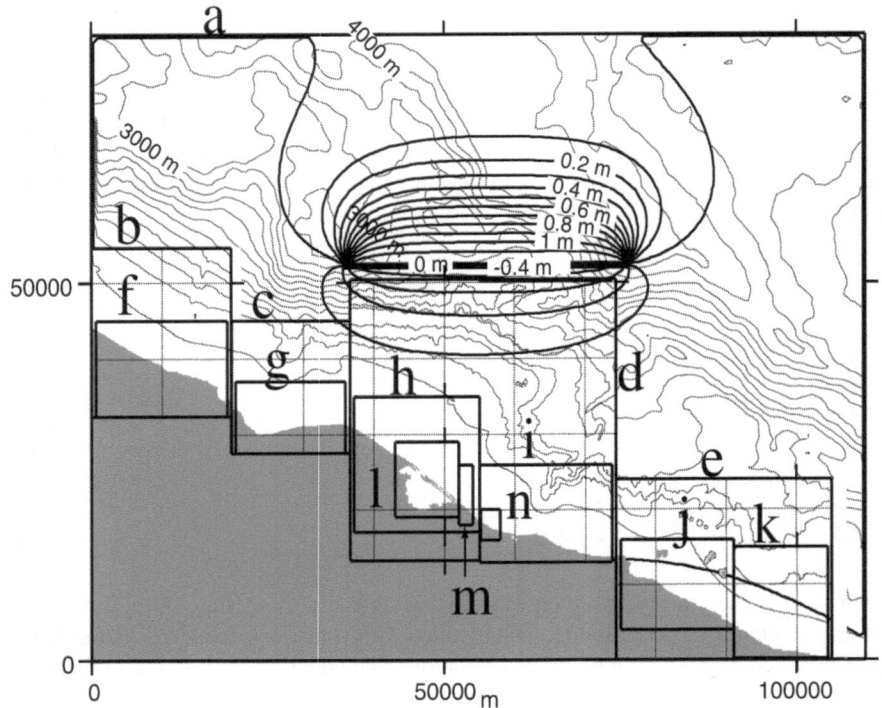

Figure 2. The domain of numerical simulation and the initial sea-surface deformation. The domain is 109.8km in E-W direction and 82.8km in N-S direction. The sub-domains are identified by alphabet letters. The N-S origin is 141°60'E and the E-W origin is 3°45'S.

domains with different grid sizes, viz. the grid size is $\Delta = 600m$ in the region a (see Fig. 2) and smaller grid sizes in other sub-regions b through n. Figure 3 shows the runup heights computed with three different grid-size combinations: in Fig. 3a, $\Delta = 200m$ in b ~ n; in Fig. 3b, $\Delta = 200m$ in b ~ e, and $\Delta = 66.6m$ in f ~ n; in Fig. 3c, $\Delta = 200m$ in b ~ e, $\Delta = 66.6m$ in f ~ k, and $\Delta = 22.2m$ in l ~ n. Comparing the results in Fig. 3, the grid-size effect is evident. The maximum runup height in Fig. 3a is 5.6 m, 7.3 m in Fig.3b, and 9.0 m in Fig. 3c. Also note that the simulation with the coarse grid size (Fig. 3a) cannot adequately model the tsunami focusing effect toward Sissano Lagoon. It is because the convex shelf formation in front of the lagoon cannot be effective for tsunami refraction unless the grid size is sufficiently small. In addition, numerical dissipation and dispersion appear to be substantial for the simulation with the coarse grid size.

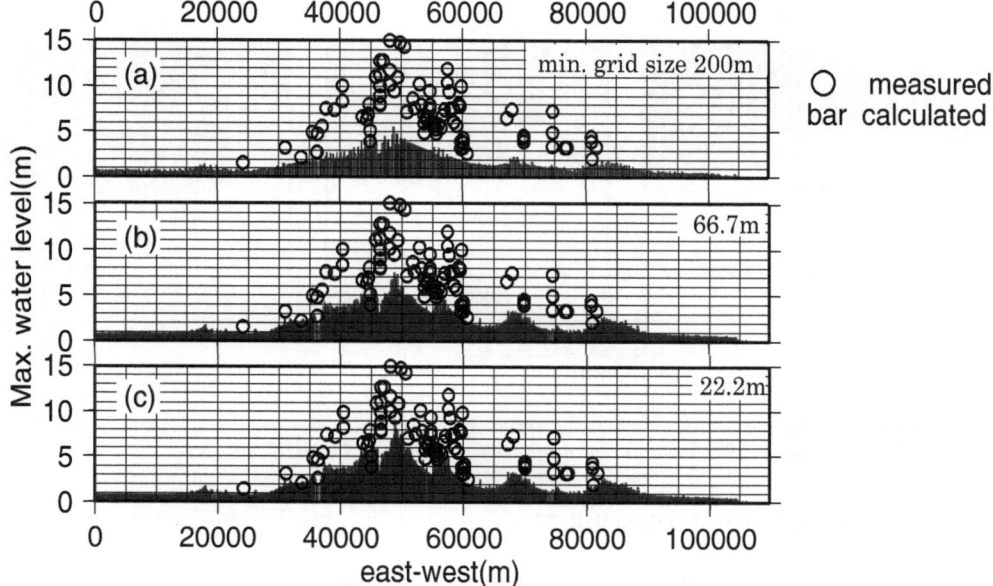

Figure 3. The maximum run-up distribution along Sissano coast; o, the surveyed run-up heights, and the vertical bars show simulated run-up heights with three different grid schemes in the sub-domains as indicated in Fig. 3. The grid sizes Δ are a) 600m in a, and 200m in b ~ n; b) 600m in a, 200m in b ~ e, and 66.6m in f ~ n; c) 600m in a, 200m in b ~ e, 66.6m in f ~ k, and 22.2m in l ~ n.

Most, if not all, of numerical simulations made prior to the first JAMSTEC cruise [5], are based on the bathymetry shown in Fig. 4a. Matsuyama et al. [8] demonstrated that it is this inaccurate bathymetry data that caused gross discrepancy in coastal tsunami effects, and this is indeed the reason why submarine landslide became initially a strong speculation for an alternate tsunami source. There is, however, another bathymetry data, as shown in Fig. 4b, which was based on the published nautical charts [15]. It is remarkable to point out that Fig. 4b captures most of the large-scale bathymetry features of the JAMSTEC data shown in Fig. 4c, whereas Fig. 4a is grossly different from Figs. 4b and c. Results from the numerical simulations using those three different bathymetry data with the finest grid size (Fig. 3c) are shown in Fig. 5. The results show that the run-up pattern similar to that with the JAMSTEC data could have been simulated with the nautical chart data, i.e. the results in Figs. 5b and c are similar.

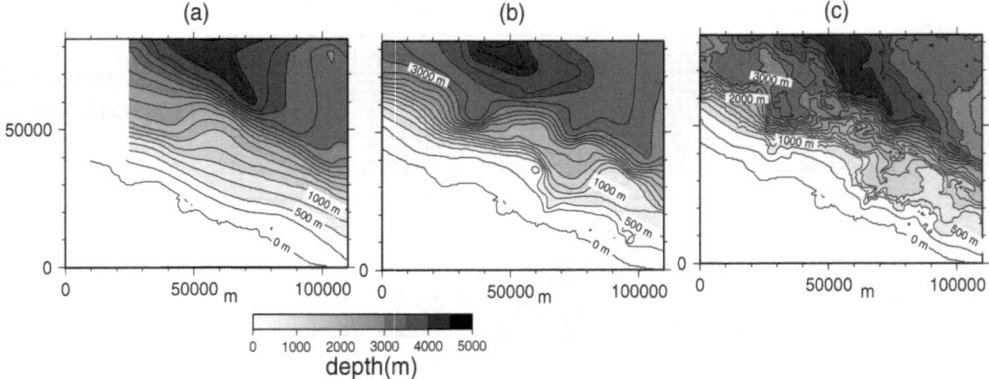

Figure 4. Three different bathymetries based on a) that used for the preliminary simulations prior to the first JAMSTEC cruise in 1999, b) the existing nautical charts, and c) the JAMSTEC data.

2.2. SUBMARINE LANDSLIDE SOURCE

To estimate the initial condition created by submarine landslide, we used a static water-surface deformation estimated by the empirical formula [16]. Their model is for wave generation by sliding a semi-ellipse-shaped rigid body down on a uniform slope. The slide speed is modeled by a hyperbolic-tangent function of time. The empirical prediction for the maximum free-surface depression η_{max} caused by a landslide was introduced as

$$\left|\eta_{max}\right| \approx 0.218T\left(\sin\theta\right)^{1.38}\left(\frac{b}{d}\right)^{1.25}$$

(4)

where T is the thickness, b is the length of the sliding rigid body, d is the initial depth at the middle of the sliding body, and θ is the seabed slope angle.

The water depth at the center of the PNG landslide is approximately 1750m on the slope of $\theta \approx 12°$ [8] — see Fig. 1 for the location and orientation of the slump. To use (4), the effective thickness of the landslide that can affect the water must be estimated. Note that most of the landslide took place well below the sea floor [8], which plays no role in generating tsunamis. Figure 6a shows the seabed profiles before and after the landslide; those are reproduced from [8]. We infer from Fig. 6a and the bathymetry, that the estimated effective thickness T is at most 350m - equivalent to the thickness of an elliptic-shaped rigid body in the model of (4) - and the equivalent initial submergence of the rigid body is $d \approx 1200$m. Also estimated from Fig. 6a is the slide distance of the effective mass, $L \approx 2,000$m, which is different from the movement of the center of the actual mass along the slide plane that was estimated 980 m [8]. The length of the effective sliding body is also estimated $b \approx 2,000$m. The rigid body slide used for the model in (4) is sketched in Fig. 6b. Using $T = 350$m, $b = 2,000$m, $d = 1,200$m, and $\theta = 12°$, the maximum free-surface depression is found from (4) to be $\eta_{max} = 16$ m.

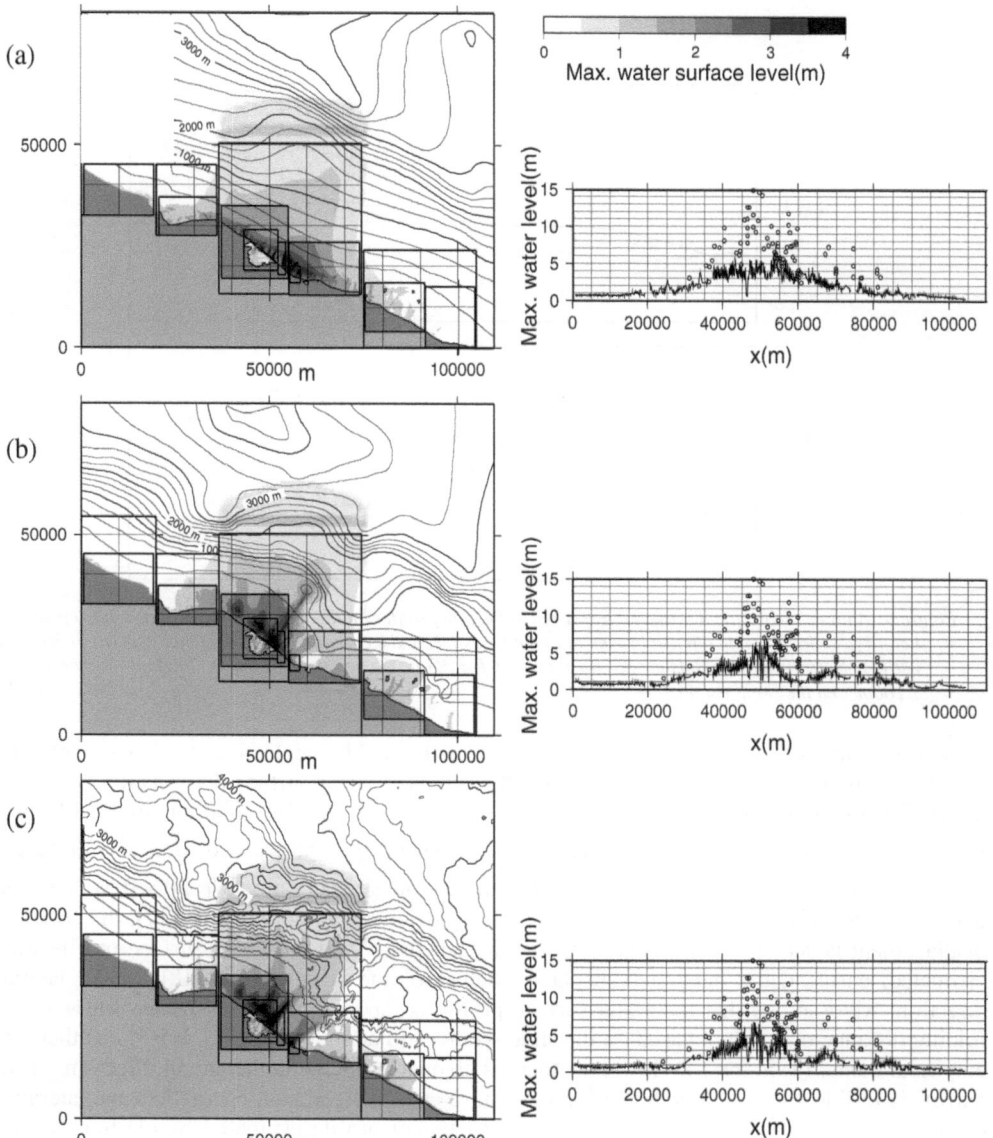

Figure 5. Based on the bathymetry shown in Fig. 4, the maximum water-surface profile of tsunami and the maximum run-up distribution along Sissano coast; o, the surveyed run-up heights; ----, the simulated run-up heights; a) with the bathymetry data shown in Fig. 4a, b) in Fig. 4b, and c) in Fig. 4c.

158

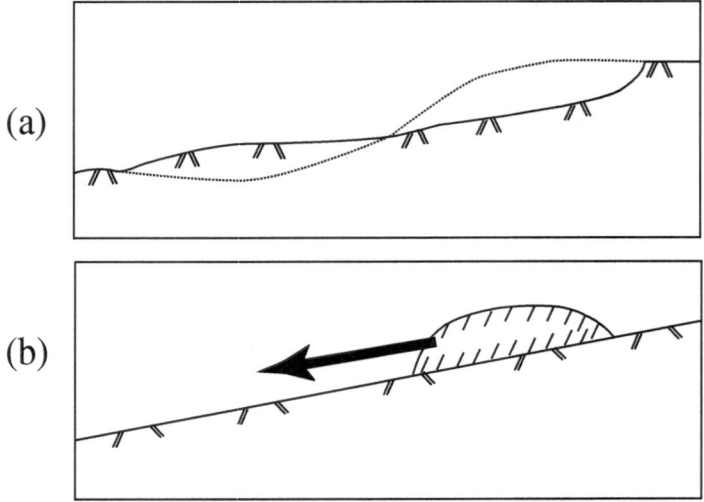

Figure 6. A landslide profile discovered by seismic reflection profiles off the Sissano coast [8]; a) seabed profiles before (a solid line) and after (a broken line) the landslide, and b) a rigid body sliding model of (a) in order to use (4). Note the model landslide has the thickness T = 350m, the length b = 2,000m, the submergence d = 1,200m, and the slope θ = 12°.

For a submarine landslide in "deep" water, errors caused by neglecting the dynamic effects of the slide motion should be insignificant. The estimated water-surface deformation is shown in Fig. 7 with the maximum water surface depression of 16m and the initial longitudinal length scale $\lambda \approx 6.5$ km. The initial wavelength is estimated by the wave speed, $\sqrt{gd} \approx 130$ m/sec (where d is the mean depth of the slide = 1750 m) and the slide duration $t_0 \approx 50$ sec, which was detected by the T-wave analysis [7]. Based on the previous laboratory results [17], the positive displacement is set to one third the negative displacement, and the water-surface profile was estimated, ensuring the integrated displaced water volume is conserved [18]. In the lateral direction, we assumed that the water-surface profile is tapered by a Gaussian profile with e-folding distance of 1,500 m based on the slide-width estimation of 3 km [8]. It is noted that the empirical model (4) is for 2-D flows, i.e. there is no energy spread in the lateral directions. The narrow landslide width of less than 3 km must have caused significant energy leakage laterally during the generation process; hence the initial tsunami profile estimated by (4) must over-estimate the initial strength of tsunamis.

With our estimated initial condition, the result of numerical simulation is shown in Fig. 8. Simulated tsunami runup heights along Sissano are approximately 1/4 of the measured values, in spite of our exaggerated initial wave amplitude. More importantly, the "distribution pattern" of the computed wave-heights along the coast is distinct from the observations, as well as the results for the co-seismic sources shown in Fig. 5.

There are several reasons why the landslide model yields the wave-height distribution different from the co-seismic fault model (and the field observations): the reasons are the small source size and its location. Even if we fictitiously increase the initial amplitude, as long as the source size and location are the same, the tsunami-height "distribution pattern" should remain the same.

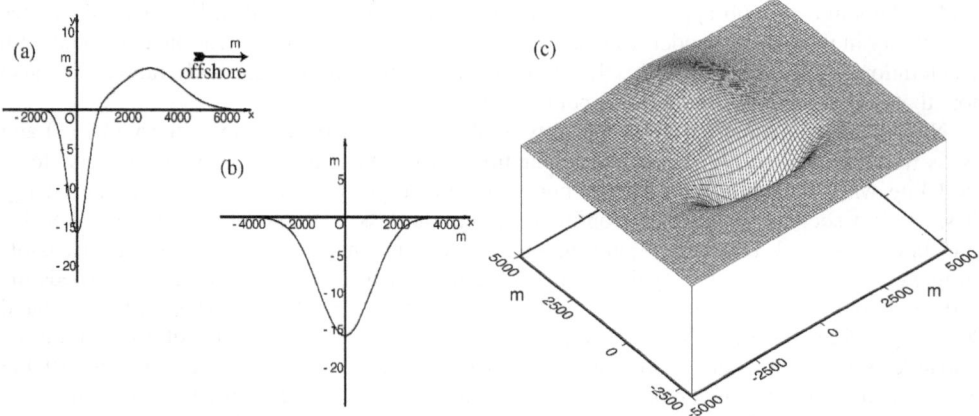

Figure 7. The initial water-surface deformation due to the submarine landslide; a) the profile in the slide direction, b) the profile in the lateral direction at the maximum depression, and c) the perspective view.

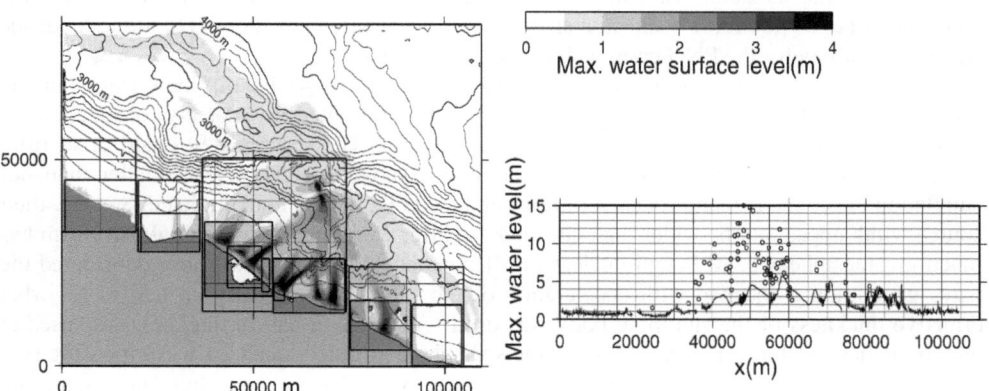

Figure 8. The maximum water-surface profile for landslide-generated tsunami and the maximum run-up distribution along Sissano coast; o, the surveyed run-up heights; ----, the simulated run-up heights.

3. Discussion

The source model (4) that we adopted for the submarine landslide is for sliding a rigid body down on a uniformly sloping seabed. The event discovered off the Sissano resembles a "slump" rather than a landslide [8, 10]. In fact, most of the slump took place under the deformed sea floor [8]. To model the slump, we estimated the equivalent values to fit the rigid-body slide model. Although Fig. 6b appears to correspond to the event presented in Fig, 6a, this creates uncertainty in the present model results. We also neglected the wave generation associated with deceleration of the slumping body: the model is based only on the rigid body runout. We also note that (4) has not been validated in publication.

Numerical simulation based on the fully nonlinear shallow-water-wave theory (1), (2) and (3) is traditionally used for tsunami computations. It is believed that the theory is capable of modeling wave-breaking, not in detail but as a flow-property discontinuity, including energy dissipation associated with the discontinuity. None the less, extreme cares must be taken for landslide generated tsunamis. Landslide-generated tsunamis tend to be shorter (tens-of-kilometer long at most) than typical tsunamis generated by fault displacement (hundreds-of-kilometer long). Even the co-seismic fault source of the present case, which was estimated 20km by 40km, is not as large as typical tsunami sources. The source of the submarine landslide is much smaller (approximately 3km by 5 km). When the horizontal scale of the source is not very large in comparison with the source depth, the sea floor deformation may not be translated effectively to the water-surface deformation. It is especially true for a submarine landslide case. The deeper the source is the smaller the water-surface deformation results, as demonstrated by (4). The reduction is anticipated much greater if we consider its small breadth dimension of the landslide. In other words, to generate significant tsunami by a submarine landslide, a large slide must occur in a shallow area. Note that the water depth at the landslide site is not shallow for the PNG case: well deeper than 1,000m.

Our numerical result for the submarine landslide appears to be different from that by Synolakis et al. [7] – contrary to our simulation; they presented no sparsely localized wave focusing and nearly perfect match to the observed data. The discrepancy could arise from several factors. First, the computational domain of Synolakis et al. [7] is smaller than our simulation hence the tsunami focus at the east end (see Fig. 8) was not simulated. Second, their source width appears to be wider than our model. Third, their longitudinal initial waveform has much greater upheaval. Although unclear in [7], Synolakis et al. [7] must have considered the wave generation associated with deceleration of the slump. It is equally unclear in [7] what effective thickness of the slumping body was used – note that a 350 m thickness was used in our simulation as shown in Fig. 6. Lastly, the sparsely distributed localized wave focusing (see Fig. 8) can be achieved by using the refined grid size ($\Delta = 22$ m) in our simulation, while no detailed model information including the grid size was given in [7]. In any case, recognizing the uncertainty and vagueness involved in the prediction of submarine-landslide-generated tsunami, it is fair to declare that neither of the simulations is reliable at the present time.

Because the small horizontal dimension of the tsunami source (especially for the landslide case), frequency dispersion effects on tsunami propagation might become important, which is not modeled in (2) and (3). Frequency dispersion effect enhances, as the pressure field becomes non-hydrostatic. The assumption of hydrostatic pressure field is deviated by square of the product of wavenumber $k = 2\pi/\lambda$ and the depth d, where λ is the wavelength. If we use our estimates of the model landslide, $\lambda = 6.5$ km, and $d = 1750$m, the value of $(k\,d)^2$ is 2.9, which is evidently not small and our hydrostatic assumption breaks down. Incidentally, the value of $(k\,d)^2$ for our co-seismic fault model is estimated approximately 0.4, which is not sufficiently small but, hopefully, adequate enough for the approximation. Another important parameter is the Ursell number U_r, which is the ratio of the nonlinear effect to frequency dispersion effect: $U_r = k\,a/(k\,d)^3$. Using the wave amplitude $a = \eta_{max} = 16$m, the Ursell number for the submarine landslide case is found to be small $U_r = 0.003$. This means that frequency dispersion effect dominates over nonlinear effect (or amplitude dispersion effect) at the source, although as the tsunami approaches the shore, the nonlinear effect must become increasingly important due to shoaling. This simple evaluation clearly demonstrates that our simulation for the landslide-generated tsunami may have serious flaws.

When frequency dispersion is important (the landslide case), the generated wave tends to disperse due to different energy-propagation speeds associated with the wave components of the initial water-surface deformation at the source. Hence tsunami energy should spread quickly, which suggests that our simulation of non-dispersive waves should provide greater tsunami amplitudes than those affected by frequency dispersion. This tendency may not be the case for tsunami behavior near the shore, however. In shallow-water region, tsunami tends to increase its amplitude and to steepen the front face due to shoaling. Our fully nonlinear shallow-water-wave model is believed capable of modeling wave breaking, not in detail but as a flow-property discontinuity. However, the simulated wave tends to break too early in the offshore region, forming the characteristic saw-tooth shape wave, and dissipating energy in the model due to the non-dispersive character of the equations [12]. Frequency dispersion effect should delay such early wave breaking and may yield higher tsunami run-up on the shore. Note that all of our simulations exhibited the occurrence of wave breaking offshore and substantial reduction in wave amplitudes toward the shore due to energy dissipation associated with the discontinuity at the wave front.

4. Conclusion

By performing numerical simulations for the co-seismic fault source, we have demonstrated that the primary cause of the discrepancy in tsunami run-up height and pattern between the observation and the initial numerical prediction is the use of faulty bathymetry data. No such a gross discrepancy could have resulted if we had used the available nautical chart data [15] for the quick preliminary computations. The grid size used in the numerical simulation is critical in order to capture the local bathymetry effects as well as to minimize the numerical dispersion and dissipation.

Many uncertainties involved in modeling the submarine landslide. The present model demonstrated that the tsunami runup is different, qualitatively and quantitatively, from that caused by the co-seismic fault dislocation. All of our numerical simulations are based on the fully nonlinear shallow-water-wave theory, which might have serious flaws, especially simulating for landslide-generated tsunamis; the initial wave form is too short to justify the use of shallow-water-wave theory.

If the eyewitness accounts are correct, then tsunami must have arrived after 09:09 GMT. This means that the tsunami was not generated by the initial main shock of Mw 7.1 at 08:49 GMT. One possible conjecture is that the tsunami was generated by the co-seismic fault motion at the second major aftershock of 09:09 GMT with Mw 5.9. This could be related to the rupture of a splay fault. Incidentally, this possibility was also discounted by Synolakis et al. [7] based on their interpretation of seismic information.

Acknowledgements

This work was supported by the PNG government, Japan Marine Science and Technology Center (JAMSTEC), South Pacific Applied Geoscience Commission (SOPAC), and the US National Science Foundation (CMS-9978399, and partially via the Univ. of Southern California).

References

1. Kawata, Y., B.C. Benson, J.C. Borrero, J.L. Borrero, H.L. Da vies, W.P. de Lange, F. Imamura, H. Letz, J. Nott, and C.E. Synolakis. (1999) Tsunami in Papua New Guinea was as intense as first thought, *EOS, Trans., AGU*, **80**, 101, 104-105.
2. Geist, E.L. (1999) http://walrus.wr.usgs.gov/tsunami/PNGhome.html
3. Davies, H.L. (1998) The Sissano tsunami 1998: Earth Talk, in *The National* (PNG newspaper), Boroko, Papua New Guinea.
4. Dengler, L. and Preuss, J. (1999) Reconnaissance report on the Papua New Guinea tsunami of July 17, 1998. *EERI News Letter*, **33-1**.
5. JAMSTEC Cruise Reports (1999) Report Nos. NT00-01, NT00-02, and NT00-03, Japan Marine Science and Technology Center, Kanagawa. No, NT00-01, 1999
6. Matsuyama, M., Walsh, J.P., and Yeh, H. (1999) The effect of bathymetry on tsunami characteristics at Sissano Lagoon, Papua New Guinea. *Geophys. Res. Lett.* **26**, 3513-3516.
7. Synolakis, C.E., Bardet, J.-P., Borrero, J.C., Davies, H.L., Okal, E.A., Silver, E.A., Sweet, S., and Tappin, D.R. (2002) The slump origin of the 1998 Papua New Guinea Tsunami. *Proc. R. Soc. Lond. A* **458**, 763-789.
8. Sweet, S. and Silver, E.A. (2001) Tectonics and slumping in the source region of the 1998 Papua New Guinea tsunami from seismic reflection images.
9. Iwabuchi, Y. (2000) Presentation at Western Pacific Geophysics Meeting, AGU, Tokyo, Japan.
10. Tappin, D.R., T. Matsumoto, P. Watts, K. Satake, G.M. McMurtry, M. Matsuyama, Y. Lafoy, Y. Tsuji, T. Kanamatsu, W. Lus, Y. Iwabuchi, H. Yeh, Y. Matsumoto, M. Nakamura, M. Mahoi, P. Hill, K. Crook, L. Anton, J.P. Walsh. (1999) Sediment slump likely caused 1998 Papua New Guinea Tsunami, *EOS*, **80**, 329, 334, 340.
11. Goto, C. and Shuto, N. (1983) Numerical simulation of tsunami propagation and run-up, in Iida and Iwasaki (eds.), *Tsunamis: Their Science and Engineering*, Terra Science Pub. Co., pp. 439-451.
12. Yeh, H., Liu, P., and Synolakis, C. (Eds) (1996) *Long-Wave Runup Models: Proceedings of the International Workshop, Friday Harbor, USA, 1995*. World Scientific Publishing Co., Singapore.
13. Mansinha, L., and Smylie, D.E. (1971) The displacement of the earthquake fault model, *Bull. Seismol. Soc. Amer.* **61**, 1433-1400.
14. Synolakis, C., Liu, P, Carrier, G., and Yeh, H. 1997. Tsunamigenic seafloor deformations. *Science*, **278**, 598-600.
15. Nautical Charts. Papua New Guinea – North Coast: Kairiru Island to Vanimo Harbour, Chart No. Aus 389, the Hydrographic Service, Australia, 1994; New Guinea – North Coast: Tanjong Narwaku to Wuvulu Island, Chart No. 3250, the Admiralty, London, 1972.
16. Grilli, S.T. and Watts, P. (1999) Modeling of waves generated by a moving submerged body. Applications to underwater landslides. *Engng Anal Boundary Elements.* **23**, 645-656.
17. Rzadkiewics S. A., Mariotti, C., and Heinrich, P. (1996) Numerical simulation of submarine landslides and their hydraulic effects. *J. Wtrwy., Port, Coast., and Oc. Engrg.*, ASCE, **123**, 149-157.
18. Matsuyama, M., Watts, P. and Yeh, H. (1999) Numerical simulation, In JAMSTEC report SOS-01, 86-105.

NATURAL GAS HYDRATES AS A CAUSE OF UNDERWATER LANDSLIDES: A REVIEW

M. PARLAKTUNA
Department of Petroleum and Natural Gas Engineering
Middle East Technical University
06531 Ankara-TURKEY

Abstract

Natural gas hydrates occur worldwide in polar regions, normally associated with onshore and offshore permafrost, and in sediment of outer continental margins. The total amount of methane in gas hydrates likely doubles the recoverable and non-recoverable fossil fuels. Three aspects of gas hydrates are important: their fossil fuel resource potential, their role as a submarine geohazard, and their effects on global climate change. Since gas hydrates represent huge amounts of methane within 2000 m of the Earth's surface, they are considered to be an unconventional, unproven source of fossil fuel. Because gas hydrates are metastable, changes of pressure and temperature affect their stability. Destabilized gas hydrates beneath the seafloor lead to geologic hazards such as submarine slumps and slides. Destabilized gas hydrates may also affect climate through the release of methane, a "greenhouse" gas, which may enhance global warming.

1. What are Hydrates?

Hydrates are the members of the class of compounds labeled "clathrates" after the Latin "clathratus" meaning, "To encage". All hydrate structures have repetitive crystal units composed of asymmetric, spherical "cages" of hydrogen-bonded water molecules. Each cage contains at most one guest (gas) molecule held within the cage by dispersion forces. There is no chemical union between the gas and water molecules. The water molecules that form the lattice are strongly hydrogen bonded with each other and the gas molecule interacts with water molecules through van der Waals type dispersion force.

There are three types of gas hydrate structures: Structure I (sI), structure II (sII), and structure H (sH). Figure 1 shows the shape of these structures. As it was mentioned before, natural gas hydrates consist of hydrogen-bonded lattices with cages in which the gas molecules are trapped. The basic cage, common for all type of gas hydrate structures, is the pentagonal dodecahedron that consists of twelve pentagons joined together as a small ball. At the cross of the pentagon there are oxygen atoms and line of the pentagon is occurred by hydrogen bonds. Five types of cavities shown in Figure 2 have been known; a) 5^{12} b) $5^{12}6^2$ c) $5^{12}6^4$ d) $4^3 5^6 6^3$, and e) $5^{12}6^8$.

A. C. Yalçıner, E. Pelinovsky, E. Okal, C. E. Synolakis (eds.),
Submarine Landslides and Tsunamis 163-169.
©2003 Kluwer Academic Publishers.

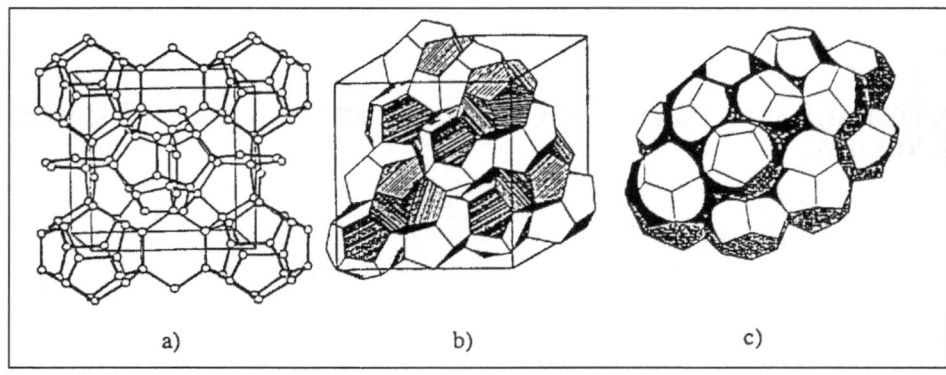

Figure 1. The unit cells of structure I (a), structure II (b) and structure H (c).

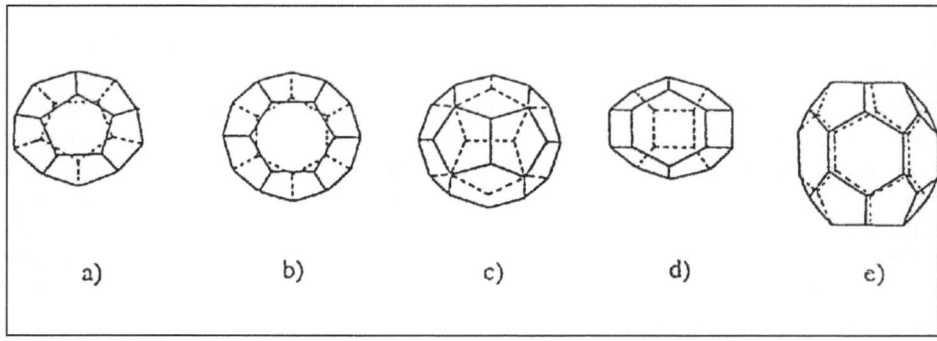

Figure 2. The hydrate cages; a) 5^{12}, b) $5^{12}6^2$, c) $5^{12}6^4$, d) $4^35^66^3$, and e) $5^{12}6^8$.

The $5^{12}6^2$ cavity consists of 12 pentagonal faces and 2 hexagonal faces and consists of 24 water molecules. The $5^{12}6^4$ cavity has 12 pentagonal, 4 hexagonal faces and consists of 28 water molecules. A $4^35^66^3$ cage consists of three fairly strained square faces, six pentagonal and three hexagonal faces. Finally the bulky $5^{12}6^8$ cavity is built of 12 pentagonal and 8 hexagonal faces.

The numbers of the different cavities in the unit cell of the different hydrate structures are tabulated in Table 1.

TABLE 1. The Number of the Cavities in the Hydrate Structures

Structure	5^{12}	$5^{12}6^2$	$5^{12}6^4$	$4^35^66^3$	$5^{12}6^8$
I	2	6	-	-	-
II	16	-	8	-	-
H	3	-	-	2	1

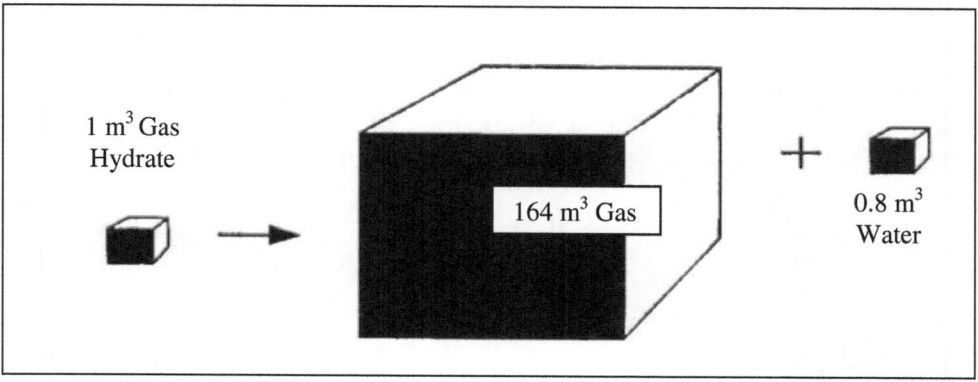

Figure 3. Yield of 1 m³ of methane hydrate at standard temperature and pressure.

One important feature of gas hydrates is the amount of gas that is stored in hydrate structure. One cubic meter of methane hydrate, if gas molecules occupy all the cavities, yields 164 m³ of gas and 0.8 m³ of water at standard temperature and pressure (Figure 3).

2. Where hydrates are found?

Hydrates are plentiful in nature, both underwater and under permafrost. More than 60 large gas hydrate fields have been revealed to date in oceanic sediments and eight on land (Figure 4).

Hydrates form only under certain temperature and pressure conditions. A phase diagram showing the boundary between free methane gas and methane hydrate for the pure water and pure methane system is given in Figure 5. The addition of NaCl to water shifts the curve to the left. Adding CO_2, H_2S, C_2H_6 and C_3H_8 to methane shifts the boundary to the right and thus increases the area of the hydrate stability field. Stable methane hydrates are found at the temperature and pressure conditions that exist near and just beneath the sea floor where water depth exceeds 300 to 500 meters. Hydrate is also stable in conjunction with permafrost at high latitudes. Hydrates can exist up to depths of about 3100 m below the ocean floor. Below that level heat tends to keep the methane free in the form of gas.

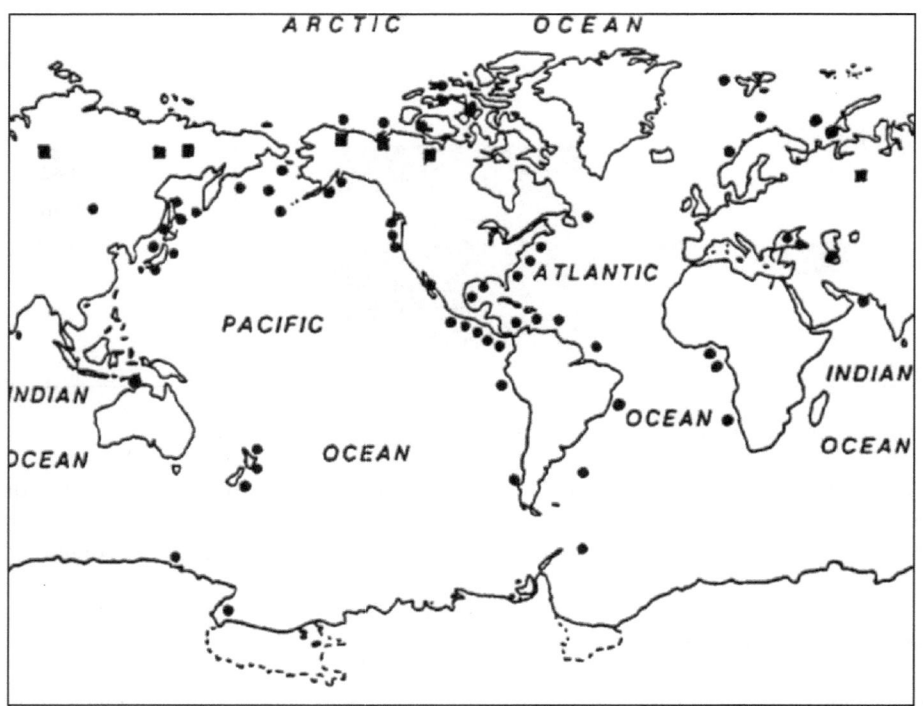

Figure 4. Map showing worldwide locations of known and inferred gas hydrates in oceanic (solid circle) and in continental regions (solid square) [1].

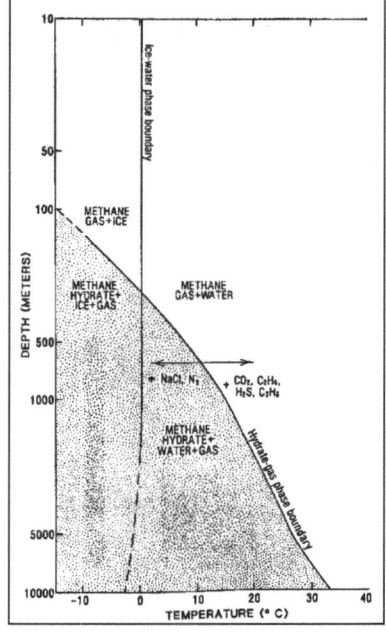

Figure 5. Phase diagram of methane hydrate [1].

There are two processes by which hydrates are formed, organic process and gas venting.

- Organic process: Most methane gas hydrate is formed from biogenic methane, excreted by bacteria that eat organic matter that has been washed into the ocean. This type of hydrate is concentrated where there is a rapid accumulation of organic detritus and also where there is a rapid accumulation of sediments.

- Venting: Hydrates also form when faults permit natural gas (or other gases) to migrate from deeper inside the Earth's crust to the surface of the seabed at places with appropriate temperature and pressure levels.

Scientists generally believe that most natural gas hydrate is formed from biogenic methane, produced by bacteria. Hydrates produced by the organic process are generally very pure; they tend to contain only water and methane. Hydrates formed from venting tend to have many gases mixed in, in addition to methane.

Hydrates may exist as outcropping or mounds on the seafloor. Hydrates may also exist in layers separated by sediments. These layers may be largely hydrate, or the layers may consist of sediments mixed in with hydrates so that the sediments are cemented or sealed. Some layers trap free methane beneath them.

3. Hydrates as Geohazard

Hydrates affect the strength of the sediments in which they are found. Areas with hydrates appear to be less stable than other areas of the seafloor. Consequently, it is important to assess their presence in the framework of the construction of underwater structures, for example in relation to military defense and to gas and oil exploration and production. Lack of stability might also be a factor in climate change.

Hydrates can cement loose sediments in the surface layer several hundred meters thick. That might lead one to believe that hydrate stabilizes the seafloor. In fact, the reverse appears to be true. When hydrates are created in loosely consolidated sedimentary rocks, the hydrate will be a cementing material. If the hydrate dissociates, the rock formation becomes unconsolidated and loses its strength. Natural gas hydrates are dangerous during the construction and operation of wells, platforms, pipelines, and other offshore engineering structures. They may even cause tsunamis.

Seafloor slopes of 5 degrees and less should be stable on the Atlantic continental margin. Yet many landslide scars are present on such gentle slopes. The top of these scars is near the top of the hydrate zone, and seismic profiles of these scars indicate that there is less hydrate in the sediment beneath slide scars.

As a result, scientists believe there is a link between hydrates and the occurrence of landslides on the continental margin. Landslides may begin when hydrates at the base of the hydrate layer break down, so that the bottom of the hydrate deposits is no longer semi-cemented but is instead full of free methane. Such a zone would be weak and would likely facilitate sliding (Figure 6).

Slides might also result from the melting of the top of a hydrate layer that is covered by sediment. As the hydrates revert to water and methane, they would likely disturb the sediment and promote shifting. This form of breakdown might occur during drops in sea level, such as occurred during glacial periods when ocean water became isolated on land in great ice sheets.

168

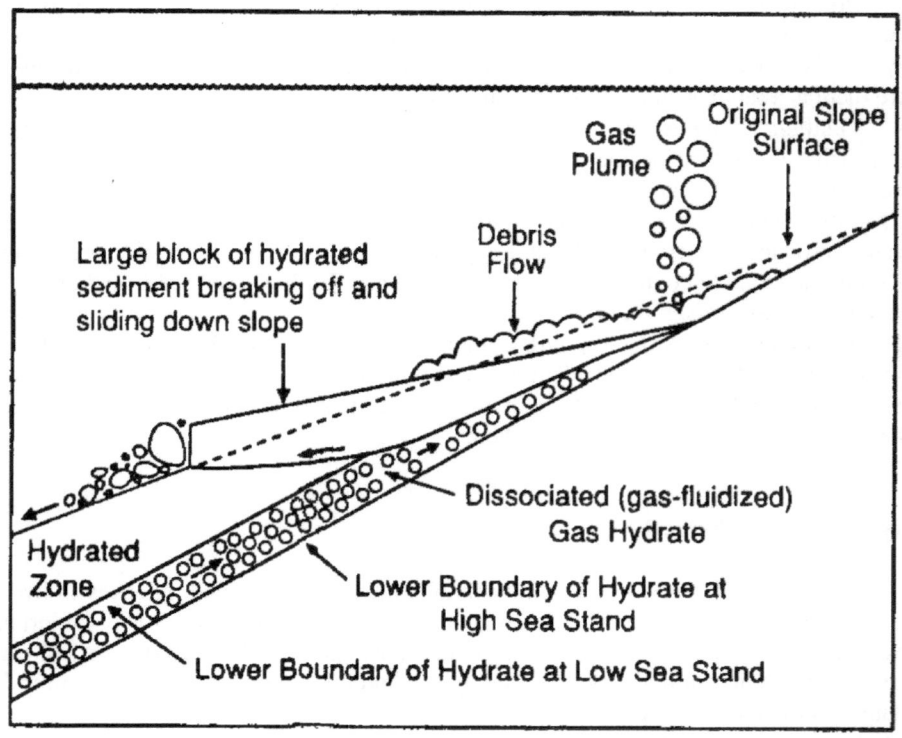

Figure 6. Diagram showing seafloor failures and gas release due to hydrate dissociation [2].

Seafloor landslides that result, for instance, from earthquakes can also cause breakdown of hydrate if, as a result, hydrate layers are repositioned so as to reduce the pressure that maintains the hydrate stability.

All of the processes above may interact. The result may be cascading slides, which could result in even further breakdown of hydrate and release of methane to the surrounding water or into the atmosphere.

In the past, hydrates have been associated with significant movement of earth in deepwater ocean environments. Examples include surficial slides and slumps on the continental slope and rise of South West Africa [3], slumps on the U.S. Atlantic continental slope [4], marine slides on the Norwegian continental margin [5], [6] and massive bedding-plane slides and rotational slumps on the Alaskan Beaufort Sea continental margin [7].

References:

1. Kvenvolden, K.A., (1988) "Methane Hydrate-A Major Reservoir of Carbon in the Shallow Geosphere?", <u>Chem. Geol.</u>, **71**, 41-51.
2. McIver, R.D. (1982) "Role of Naturally Occurring Gas Hydrates in Sediment Transport", <u>AAPG Bull.</u>, **66**, 789-792.
3. Summerhayes, C.P., Bornold, B.D., and Embley, R.W. (1979) "Surficial Slides and Slumps on the Continental Slope and Rise of South West Africa: A Reconnaissance Study", <u>Mar. Geol.</u>, **31**, 265-277.
4. Carpenter, G. (1981) "Coincident Sediment Slump/Clathrate Complexes on the U.S. Atlantic Continental Slope", <u>Geo Mar. Lett.</u>, **1**, 29-32.
5. Jansen, E., Befring, S., Bugge, T., Eidvin, T., Holtedahl, H., and Sejrup, H.P. (1987) "Large Submarine Slides on the Norwegian Continental Margin: Sediments, Transport and Timing", <u>Mar. Geol.</u>, **78**, 77-107.
6. Bugge, T.S., Befring, S., Belderson, R.H., Eidvin, T., Jansen, E., Kenyon, N.H., Holdedahl, H., and Sejrup, H.P., (1987) "A Giant Three-Stage Submarine Slide off Norway", <u>Geo Mar. Lett.</u>, **7**, 191-198.
7. Kayen, R.E., and Lee, H.J. (1991) "Pleistocene Slope Instability of Gas Hydrate –Laden Sediment on the Beaufort Sea Margin", <u>Mar. Geotechnol.</u>, **10**, 125-141.

COASTAL DEFORMATION OCCURRED DURING THE AUGUST 17, 1999 İZMİT EARTHQUAKE

A. BARKA[1], W. LETTIS[2], E. ALTUNEL[3]

[1] Deceased, Formerly ITU, Eurasian Earth Science Institute, İstanbul, Turkey
[2] William Lettis &Association, Walnut Creek, CA, USA
[3] Osmangazi University, Engineering Faculty, Department of Geology,
Eskişehir, Turkey

Abstract

Sudden elastic motions, deformation of the continental slopes such as large mass movements, and landslides are major reasons for the tsunamis. The types of nearshore deformations, resulted from the August 17, 1999 İzmit Earthquake are investigated and described under the basic headings of faulting, lateral spreading, and slumping in this study. Analysis of near shore deformation can help to prevent the future losses of both lifes and properties in similar areas.

1. Introduction

Sudden elastic motion and deformation of the continental slopes such as large mass movements are major reasons for the tsunamis. In this paper, types of near shore deformation resulted from the August 17, 1999 İzmit Earthquake are described. Basically, three main types of coastal deformation are observed in the Gulf of İzmit and Sapanca Lake; faulting, lateral spreading and slumping (Fig. 1).

2. Faulting

Coastal deformation was characterized by two main types of faulting during the August 17, 1999 İzmit earthquake; normal and strike-slip faulting or their combination. Strike-slip faulting in some cases was responsible for local slumping such as in Değirmendere. In Değirmendere, about 150 m by 100 m portion of the delta submerged about 35 m under the sea due to unstable slope cut by the reactivation of the Gölcük-Karamürsel strike-slip segment. The Gölcük delta also submerged about 2.3 m due to normal faulting along the 2 km long secondary oblique fault connecting the Gölcük-Karamürsel and Sapanca segments of the rupture zone (Fig. 1). As a result of this submerge, the sea advanced into the land more than 1 km in places (Fig. 2).

A. C. Yalçıner, E. Pelinovsky, E. Okal, C. E. Synolakis (eds.),
Submarine Landslides and Tsunamis 171-174.
©2003 Kluwer Academic Publishers.

172

3. Lateral spreading

Lateral spreadings were most common in the recent depositional environments such as deltas and beaches along the coast of Sapanca Lake and Gulf of İzmit. Lateral spreading caused extensive damages both in the north and south sides of Sapanca Lake (Fig. 1). The Sapanca Hotel, located in the south side of the lake (Fig. 3a), was tilted towards north as a result of lateral spreading (Fig. 3b). In the Gulf of İzmit, Gölcük delta, Halıdere, Değirmendere and the Derince regions were effected by lateral spreadings (Fig. 4).

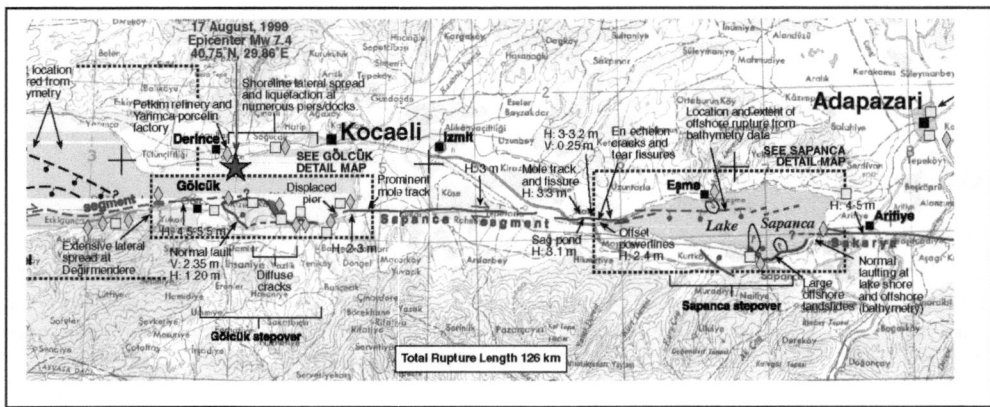

Figure 1. Coastal deformations observed in the Gulf of İzmit and Sapanca Lake

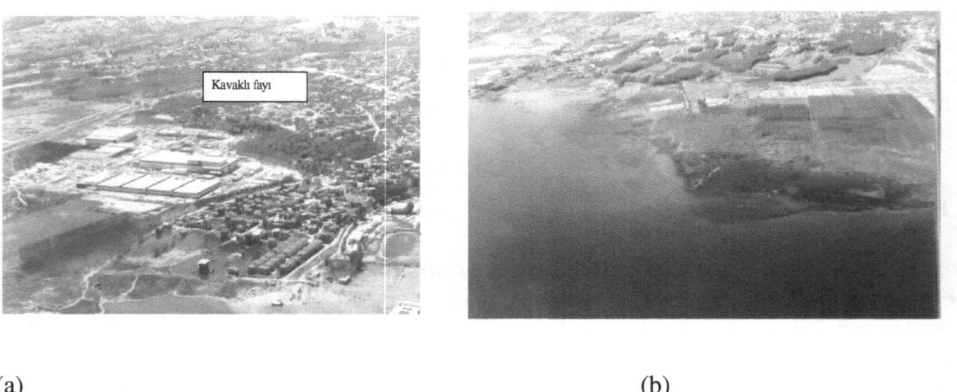

(a)   (b)

Figure 2. (a) Normal fault on the Gölcük delta,(b) pre and post earthquake shoreline on the delta.

(a)   (b)

Figure 3. Sapanca Hotel (a) before the August earthquake, (b) after the earthquake.

(a)   (b)

Figure 4. Two lateral spreading examples,in north of Sapanca Lake (a) and Gölcük delta (b)

4. Slumping

As mentioned above, the major slumping occurred in Değirmendere during the August 17, 1999 İzmit earthquake. Değirmendere is located on a delta in the south coast of İzmit Gulf (Fig. 1). Foot of the unstable delta was cut by the reactivation of the Gölcük-Karamürsel segment and a major part of the delta submerged into the see (Fig. 5). Except for Değirmendere, there are some evidence indicate that some submarine slumping might have occurred in north of Aksa and Yalova in the south coast of the Gulf.

174

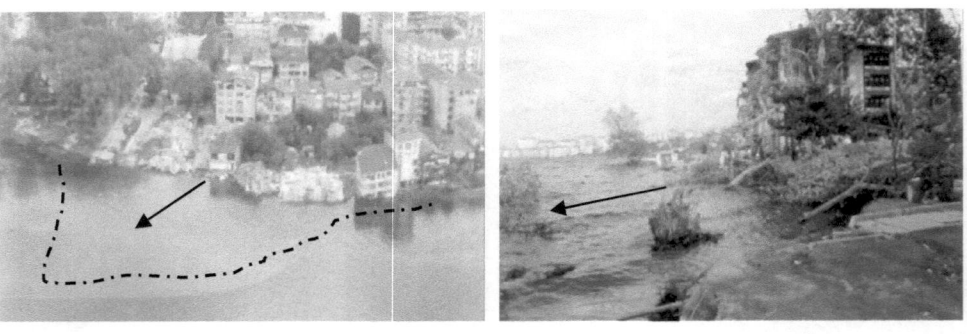

(a)   (b)

*Figure 5. Slumping in Değirmendere. Arrows show the slumping direction .(a) A general view.
Dashed red line shows approximate delta border before slumping. (b) Close-up view.*

5. Conclusion

The lesson learned from the August 17, 1999 İzmit Earthquake is applicable both to the historical and future earthquakes. After examining the coastal deformation of the August 17, 1999 İzmit earthquake, it is possible to predict the future sites of similar submarine and near coast deformation and to prevent the future losses of both lifes and properties. This information is also valuable for analysing tsunami potentials in certain regions. This experience can also help to enlarge our understanding of near shore deformation which occurred during the large historical earthquakes.

Part 4

Tsunami Propagation and Coastal Impact

A REVIEW OF SOME TSUNAMIS IN CANADA

T.S. MURTY
W.F. Baird & Associates Coastal Engineers Ltd.
1145 Hunt Club Road, Suite 500
Ottawa, Ontario K1V 0Y3, Canada

Abstract

Some past tsunamis in various regions of Canada have been reviewed. These tsunamis have occurred from a variety of sources: under-ocean earthquakes, sub-marine landslides and human-made inadvertent large chemical explosions. Resonance amplification of the tsunami has been identified as important in the Alberni Inlet on the Vancouver Island, British Columbia and also in the Burin Inlet in Newfoundland. The results of the various numerical models for tsunami generation, propagation onto the shelf and into the coastal inlets that formed the scientific basis for the British Columbia tsunami warning system have been briefly discussed.

1. Introduction

Tsunamis have occurred in Canada due to under ocean earthquakes, submarine landslides an even from large human made explosions. Geographically these events took place on the Pacific coast, the Straits of Georgia and Juan de Fuca, the St. Lawrence Estuary, Nova Scotia and in Newfoundland. In the twentieth century the following is a partial chronological list of tsunamis in Canada.

1. December 6th , 1917: Large tsunami in Halifax Harbor due to an explosion.
2. November 18th, 1929: Large tsunami in Burin Inlet due to an earthquake off the coast of Newfoundland.
3. June 23, 1946: Small tsunami in the Strait of Georgia due to an earthquake on the Vancouver Island.
4. March 28th 1964: Large tsunami on the coast of British Columbia due to an earthquake in Alaska.
5. April 27, 1975: Large tsunami in Kitimat Inlet due to a submarine landslide.

All these tsunamis have quite different characteristics as can be seen below. During the First World War, a munitions ship caught fire and exploded in Halifax harbor on December 6th, 1917. In the harbour narrows the amplitude of the tsunami that was generated was about ten meters in amplitude.

The Grand Banks earthquake of November 18th, 1929 created a tsunami of at least 12.2 metres in amplitude in the Burin Inlet. Some estimates put the maximum amplitude at

A. C. Yalçıner, E. Pelinovsky, E. Okal, C. E. Synolakis (eds.),
Submarine Landslides and Tsunamis 175-183.
©2003 Kluwer Academic Publishers.

about 35 meters, which could be an overestimate. The turbidity currents following the earthquake caused numerous cable breaks in the Atlantic Ocean. On June 23, 1946, a small tsunami occurred in the northern part of the Strait of Georgia (between Vancouver Island and the main land) following an earthquake on Vancouver Island close to the western shore of the Strait.

The May 1960 Chilean earthquake tsunami was observed at several locations of the Pacific Coast of Canada.

Following a large earthquake in Alaska on March 28th, 1964, a major Pacific wide tsunami was generated. Outside of Alaska, the largest tsunami amplitude anywhere on the west coast of North America occurred not exactly on the open coast, but inland at Port Alberni at the head of the Alberni Inlet. This can be explained as due to the amplification of the tsunami through quarter wave resonance as the tsunami traveled from the mouth to the head of the inlet.

On April 27th, 1975, following a major submarine landslide, a tsunami was generated in the Kitimat Inlet in the Douglas Channel system of the northern part of the coast of British Columbia.

2. The Halifax Harbour Explosion Tsunami

In the midst of World War I, shortly before 9 a.m. on December 6, 1917 the munitions ship Mont Blanc collided with the relief ship Imo in the Narrows of the Canadian port Halifax. The Mont Blanc was carrying the equivalent of about 2900 short tons (2000lb/ton) of TNT. A fire broke out onboard the Mont Blanc and the crew abandoned ship, which drifted into the Halifax side and grounded near one of the piers. Shortly about 9:00 a.m. the cargo exploded devastating a large section of the city. Estimates of the casualties are as high as 2000 dead and 9000 injured. Of these, many are thought to have died from the affects of the tsunami that followed. The documented damage of this, the greatest man made explosion to that date, would later be used in the Manhattan project to estimate the devastation of the first nuclear bombs. [1, 2].

There was no operational tide gauge at the time recording the changes in sea level, but there are several narrative reports of extreme high and low water and useful information might be obtained from them [3]. The explosion occurred very close to the time of low tide, about 0.5 m below mean water level.

To hindcast the tsunami, [1] ran a linearized finite element model covering the full area of Halifax Harbour from the head of the Bedford Basin to the seaward entrance to the harbour approach. The model was initialized with a setup derived using an empirical formula for waves resulting from explosions in water. Figures 1 to 2 respectively show the tsunami travel times and the tsunami amplitudes.

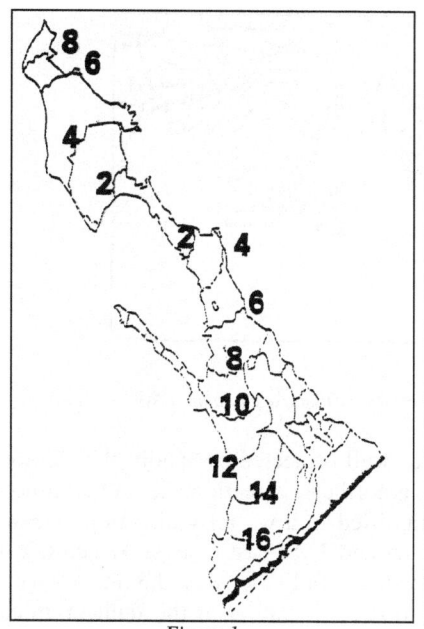

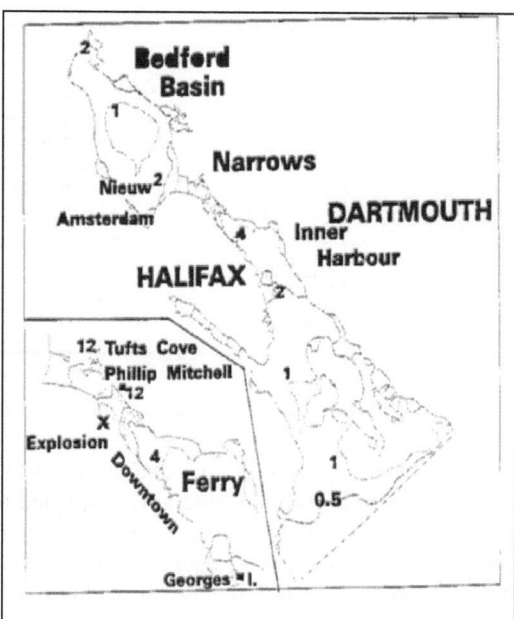

Figure 1   Figure 2

*Figure 1. The time of arrival in minutes of the
initial wave taken to be the time the sea level
first exceeded 0.2 m. From Greenberg et al,
(1994)*

*Figure 2. The maximum elevation (m) achieved during
the model run and locations (estimated) of the anecdotal
observations. The detail of the Narrows is shown in the
lower left. From Greenberg et al, (1994).*

The model results when put together with anecdotal reports give a consistent overview
of the tsunami amplitude and progression. The study by [1, 2] indicates that in the Narrows
at the explosion site, the wave was over 10 m high, but that the amplitude diminished
greatly further away. In the Bedford Basin the tsunami would not have been damaging and
in the outer reaches of the harbour, it would have been noticeable only to those looking for
it.

3. The Grand Banks Earthquake Tsunami

The so-called Grand Banks earthquake occurred on Nov. 18, 1929, in the Atlantic Ocean,
southeast of Newfoundland. The epicenter was at 44°30'N, 57°15'W and the time of
occurrence was 0.432:8 (Newfoundland Standard Time). The turbidity currents following
this earthquake caused numerous cable breaks in the Atlantic Ocean.

This earthquake generated a tsunami with amplitudes of at least 12.2 m [4] in Burin
Inlet on the south coast of Newfoundland and killed 26 people. Gregory [5] gave 30.5 m as
the maximum amplitude in Burin Inlet. McIntosh [6] gave a value of 4.6 m for the tsunami
amplitude at Lamaline. According to Johnstone, no tsunami was observed at Sable Island,

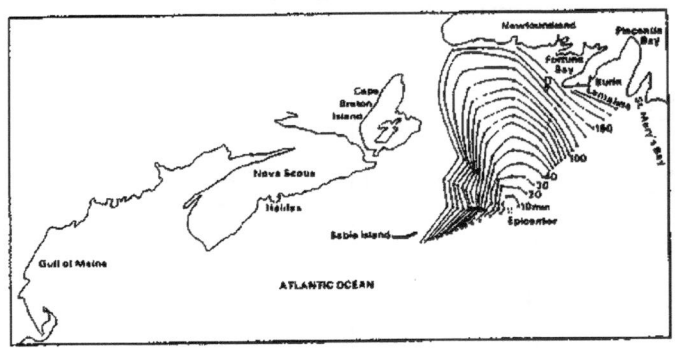

Figure 3. Travel-time curves (minutes) for the Grand Banks earthquake tsunami of Nov. 18, 1929.

Probably it's because of the fact that the island was well protected by sandbanks. Although the tsunami waves were considerably amplified in a northerly direction (i.e. in the direction of Newfoundland), they were not significantly amplified in a westerly direction as can be seen from the water- level records on the Canadian and U.S. east coasts. At Halifax, the amplitude was only 0.5 m and at Atlantic City the amplitude was 0.3 m. Murty [7] explained the nonamplification westward as due to the orientation of the fault in an east-west direction. Murty and Wigen [8] studied this tsunami in detail and showed that resonance in the V-shaped Burin Inlet accounts for the great amplification of the tsunami in that inlet. Figure 3 shows the travel-time curves for this tsunami and it can be seen that the tsunami energy traveled preferentially towards the south coast of Newfoundland.

Although turbidity currents and tsunamis are not directly related, nevertheless both could be caused by earthquakes such as the 1929 Grand Banks earthquake. Turbidity currents and tsunamis could also be generated by landslides whether or not an earthquake occurred.

Most submarine telegraph cables from North America to Europe pass south of Newfoundland. At about 2032 h GCT, Nov. 18, 1929, an earthquake of magnitude 7.2 occurred on the continental slope southeast of the Cabot Trench. This generated a tsunami that caused considerable property damage and loss of life along the shores of Placentia Bay, which was discussed above.

Another important consequence occurred during the 13 h and 17 min following the earthquake. An orderly sequence of breaks occurred in the telegraph cables of 483 km south of the epicenter. According to Heezen and Ewing [10], although all cables along the continental slope and on the floor of the ocean south of the epicenter were broken, none on the continental shelf were disturbed. The exact times and locations of the cable breaks were known, respectively, from the telegraph records and resistance measurements.

Bucher [9] hypothesized that erosion in the submarine canyons caused by the tsunami left the cables unsupported, thus leading to breakage. Heezen and Ewing [10] criticized Bucher's explanation on the grounds that the cable breaks were too regular in time and the cables were of different ages and breaking strengths.

Heezen and Ewing [10] showed rather convincingly that the successive series of breaks in the telegraph cables following the Grand Banks earthquake of 1929 were caused by the turbidity current generated.

4. Tsunami in Kitimat Inlet Due To a Landslide

A major submarine slide occurred on April 27, 1975 in Kitimat Inlet in the Douglas Channel system on the West Coast of Canada (Figure 4). Following this slide, at least two water waves were observed and it was estimated that the range (crest to trough) of the first wave could have been 8.2 m. Two simple theories have been used by Murty [11] to estimate the wave height. Considering the uncertainties both in the observed data as well as in the calculated wave height, there is reasonable agreement.

Kitimat Inlet on the West Coast of British Columbia has a history of landslides: several slides occurred during the period 1952 to 1968, and also in 1971. On October 17, 1974, following a submarine slide, a water wave of 2.8 m amplitude was generated. On April 27, 1975, following a major slide, water waves with ranges up to 8.2 m were generated. In other parts of British Columbia, also major slides occur – for example, in Howe Sound, in 1995.

Depending on the nature of movement of the sediment, a slide will have two components: horizontal and vertical. Usually, one component dominates the other. For theoretical and laboratory studies, it is convenient to distinguish between vertical and horizontal slides, depending upon which component dominates. Another type of classification of slides is: (a)

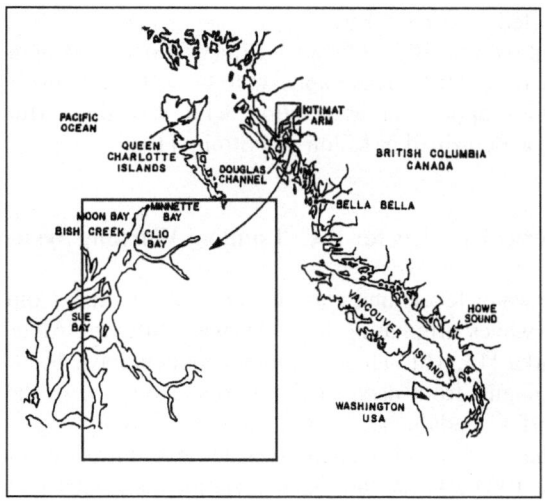

Figure 4. Geography of the Pacific Coast of Canada

land-slides, (b) submarine or under-water slides. Actually, in several instances, there may be visual manifestations in the surrounding land, even in the case of a submarine slide.

Here, we concern ourselves specifically with the submarine slide of April 27, 1975 in the Kitimat Inlet. There are visual manifestations of the slide in the Moon Bay area even now, and it is estimated that about $3 \times 10^6 m^3$ of material was involved in this bay alone. Casagrande [12] estimated the total volume of the material involved in the slide that gave rise to the water waves to be about $10^7 m^3$. Based on an examination of the hydrographic charts prepared before and after the slide, and some rough measurements of shore features, Murty [11] suggested that an upper limit for the total amount of material involved in the whole slide is about 26 x 10^6m. Actually, it is not the volume of the material involved in the slide that directly enters into the calculation of the amplitude of the water waves generated: various factors associated with the slide together determine the amplitude of the waves.

At the time of occurrence of the slide of April 27, 1975, there was no major seismic event reported in that area. Also, there was no meteorological event at that time which could have caused large water-level oscillations.

Hence, there is no doubt that water waves were generated by the submarine slide. At least two water waves (and possibly three) were generated, and propagated into the connecting bays and channels. The largest wave was estimated to be 8.2 m in range (crest to trough).

Brown [13] mentions that the first wave was observed at 10:05 a.m. (Pacific Daylight Savings Time) and, at 10:15 a.m., the bottom was visible. The whole water-level disturbance lasted about an hour. In Bish Creek and Clio Bay, which are about 8 km from the site of the slide, at least one wave was observed. Some damage occurred in Bish Creek and the range in Clio Bay was estimated to be about 6.7 m. No wave was noted in Sue Bay, probably because of the complicated path needed for the wave to travel into that bay. Although Minette Bay is at the head of the Kitimat Arm, it is an extremely shallow bay (most of the time there is no water in it) and, further, the bay bottom is very rough because of vegetation and tree trunks and it is quite conceivable that the wave could not have penetrated and traveled in Minette Bay.

The Kitimat Submarine Slide of April 27, 1975 occurred approximately 53 minutes after the occurrence of low tide. There appears to be sufficient observational data to suggest that submarine slides appear in association with low tide. Murty [11] modeled this following a technique described in Miloh and Streim [14].

5. Results of Numerical Models for B.C. Tsunami Warning System

Maximum tsunami water levels and currents along the British Columbia outer coast have been computed for waves originating from Alaska, Chile, the Aleutian Islands (Shumagin Gap), and Kamchatka [15, 16]. Three computer models have been developed to generate and propagate a tsunami from each of these source regions in the Pacific Ocean to the continental shelf off Canada's west coast, and into twenty separate inlet systems. The model predictions have been verified against water level measurements made at tide gauges after the March 28, 1964 Alaska earthquake. Simulated seabed motions giving rise to the Alaskan and Chilean tsunamis have been based on surveys of the vertical displacements made after the great earthquakes of 1964 (Alaska) and 1960 (Chile). Hypothetical bottom motions have been used for the Shumagin Gap and Kamachatka simulation. These simulations represent the largest tsuanamigenic events to be expected from these areas.

Maximum wave and current amplitudes have been tabulated for each simulated tsunami at 185 key locations along the British Columbia coast. On the north coast of British Columbia, the Alaska tsunami generated the largest amplitudes. In all other regions of the west coast, the largest amplitudes were generated by the Shumagin Gap simulations. Wave amplitudes in excess of 9 m were predicted at several locations along the coast and current speeds of 3 to 4 m/s were produced. The most vulnerable regions are the outer coast of Vancouver Island, the west coast of Graham Island, and the central coast of the mainland.

Some areas, such as the north central coast, are sheltered enough to limit expected maximum water levels to less than 3 m.

We note that the effects of dry land flooding have not been included owing to the large amount of additional topographic data that is required and the special demands created for high-resolution, local-area models in many locations. This does not pose serious problems inmost areas. However, near the heads of inlets (where these is often an extensive area of flatlands associated with a river mouth) modeled wave heights may be appreciably overestimated. The four source areas shown in Figure 5 have been identified as likely sites for generation of tsunamis that could threaten Canada's coastline. This is based on the occurrence of previous tsunamigenic earthquakes and on estimates of the likelihood of future great earthquakes in each area.

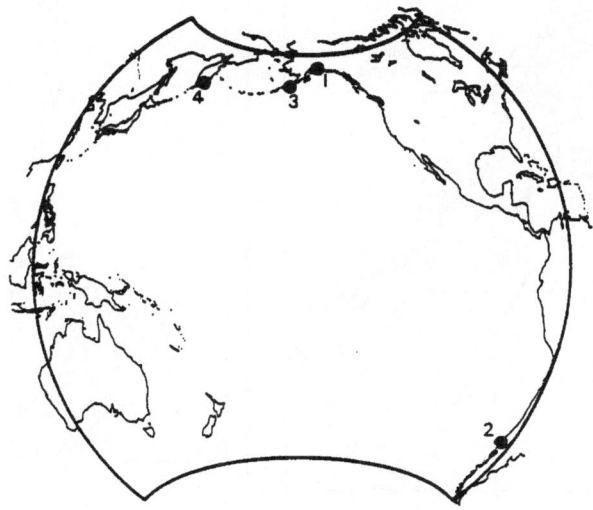

Figure 5. Epicenters of earthquakes used in tsunami simulations. The bold line is the boundary of the deep ocean model (DOM). (1: Alaska, 2: Chile, 3: Shumagin Gap, 4: Kamchatka). [16]

Tsunamis arriving at British Columbia's outer coast propagate into all exposed inlet systems. Twenty of the more exposed systems have been identified and incorporated into this study. Each has a corresponding numerical model that uses time-varying water elevations at one or more connections to the continental shelf to calculate the water surface response and currents within the system. Construction of each model required detailed extraction of bathymetry and dimensional data (cross-sectional and surface areas) and calibration.

Seismic activity in the Pacific Ocean is confined primarily to zones adjacent to the continental margins where subduction of the oceanic plates under the continental landmasses episodically releases bursts of energy in the form of an earthquake. Other areas, such as the Hawaiian Islands, are also sites of large earthquakes but these do not pose a threat of producing a destructive tsunami in British Columbia waters.

If an earthquake results in vertical (dip-slip) motion of the oceanic crust, then it is tsunamigenic, that is, it will result in the deformation of the water surface and subsequent propagation of the disturbance outward form the source as a seismic sea wave (tsunami).

182

Tsunamis from even moderately small earthquakes may result in significant damage within a short distance of the source. If the earthquake is sufficiently large (with a Richter scale magnitude greater than 7.5 to 8.0), however, then a large tsunami may be generated and damage will result at great distances from the source. This latter situation is most relevant to the west coast of Canada.

Three distinct numerical models were used to simulate tsunami generation and propagation to the inlets of British Columbia. A deep ocean model (DOM) with 0.5° resolution (Figure 5) has been used to simulate bottom motions that give rise to tsunamis and to propagate the resulting waves to the continental shelf (Figure 6) off Canada's west coast. There, a 5-km

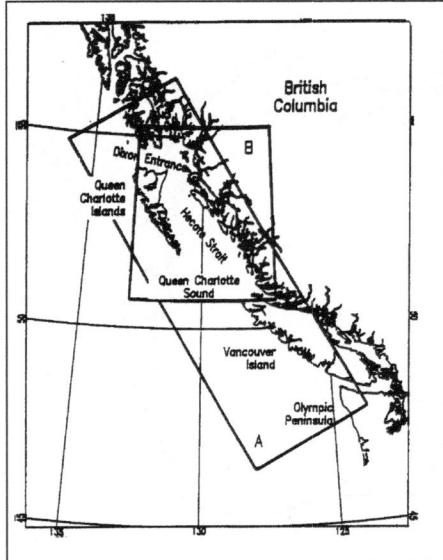
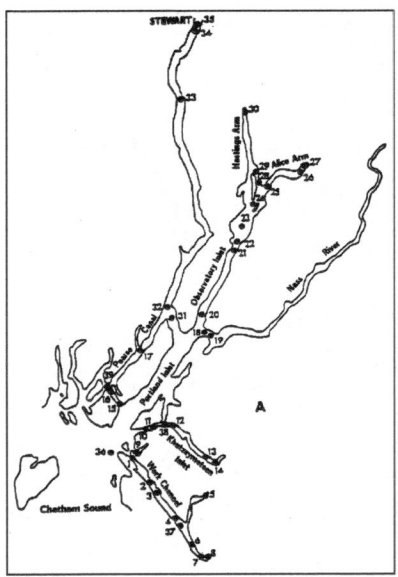

Figure 6. Shelf model grid outline (A) and outline of region used for field plots of northern British Columbia (B). From Dunbar et al, 1989.

Figure 7. Map showing water level and current locations in some inlets. From Dunbar et al, 1989

resolution model (C2D) covering the shelf propagates the waves to the entrances of the inlets. Finally, 2-km resolution models (FJORDID) determine water levels and current velocities in the inlet (Figure 7). The shelf and inlet models were run simultaneously as a coupled system. Wave amplitudes on the edge of the shelf were specified from a preceding run of the deep ocean model reported in [16].

After confirming the correctness of all model components using historical tsunami data, the three models were run for a set of simulated tsunamigenic earthquakes. These included simulations for measured bottom displacements at the source regions of the 1960 Chilean and 1964 Alaskan tsuanmigenic earthquakes, and hypothetical earthquakes at the Shumagin Gap and Kamchatka Peninsula sites. In addition, the Alaskan simulation was repeated for a case where bottom motions were amplified by 25%. The simulations provided wave amplitudes and currents.

References

1. Greenberg, D.A., Murty, T.S. and Ruffman, A. (1993). A Numerical Model for the Tsunami in Halifax Harbour due to the Explosion in December 1971. Marine☐ odesy, 16:153-167.
2. Greenberg, D.A., Murty, T.S. and Ruffman, A. (1994). Modeling the Tsunami From the 1917 Halifax Harbor Explosion, Science of Tsunami Hazards, Vol. 11, No. 2, 67-80.
3. Ruffman, A, Greenberg, D.A. and Murty, T.S. (1995). The Tsunami from the Explosion in Halifax Harbour, Ground Zero, pp. 327-344, Edited by A. Ruffman And C.D. Howell, Nimbus Publishing Ltd. Halifax.
4. Johnstone, J.H. L. 1930. The Acadian-Newfoundland earthquake of November 18, 1929. Proc. Trans. N.S. Inst. Sci. Halifax, N.S. 17:223-237.
5. Gregory, J.W. 1929. The earthquake south of Newfoundland and submarine canyons. Nature 124: 945-946.
6. McIntosh, D.S. 1930. The Acadian-Newfoundland earthquake. Proc. Trans. N.S. Inst. Sci. 17:213:222.
7. Murty, T.S. (1977). Seismic Sea Waves-Tsunamis, Bulletin No. 198, Fisheries Research Board of Canada, Ottawa, 337p.
8. Murty, T.S., and Wigen, S.O. 1976. Tsunami behavior on the Atlantic coast of Canada and some similarities to the Peru coast. Proc. IUGG Symp. tsunamis and tsunami Res. Jan. 29-Feb. 1, 1974. Wellington, N.Z., R. Soc. N.Z. Bull. 15:51-60.
9. Bucher, W.H. 1940. Submarine Valleys and Related Geologic Problems of the North Atlantic. Geol. Soc. Am. Bull. 51: 489-512.
10. Heezen, B.C., and Ewing, M. 1952. Turbidity currents and submarine slumps and the 1929 Grand Banks earthquake. Am. J. Sci. 250: 849-873.
11. Murty, T.S. (1979). Submarine Slide-Generated Water Waves in Kitimat Inlet, J. of British Columbia, Phys. Res., Vol. 84, No. C12, 7777-7779.
12. Casagrande, L. (1977). Kitimat Arm B.C. Slide of April 27, 1975. Unpubl. manuscript. Casagrande Consultants, 40 Massachusetts Avenue, Arlington, Mass. 30 pp. and Appendices.
13. Brown, W.E. (1975). Underwater Subsidence at Kitimat: Sunday 27 April, 1975 (Unpubl. man., Institute of Ocean Sciences, Patricia Bay, Fisheries and Environment Canada, Sidney, B.C. 8 pp.
14. Miloh, T and H.L. Strein (1978). Tsunami effects at coastal sites due to offshore faulting, tectonophysics, vol. 46, 347-356.
15. Dunbar, D.S., LeBlond, P.H. and Murty, T.S. (1989a). Evaluation of Tsunami Amplitudes for Pacific Coast of Canada, Progress in Oceanography, Vol. 26, 115-177.
16. Dunbar, D.S., LeBlond, P.H. and Murty, T.S. (1989b). Maximum Tsunami Amplitudes and Associated Currents on the Coast of British Columbia, Science of Tsunami Hazards, Vol. 7, No. 1, 3-44.

References

1. (illegible reference text)
2. (illegible reference text)
3. (illegible reference text)
4. (illegible reference text)
5. (illegible reference text)
6. (illegible reference text)
7. (illegible reference text)
8. (illegible reference text)
9. (illegible reference text)
10. (illegible reference text)
11. (illegible reference text)
12. (illegible reference text)
13. (illegible reference text)
14. (illegible reference text)
15. (illegible reference text)
16. (illegible reference text)

SYNTHETIC TSUNAMI SIMULATIONS FOR THE FRENCH COASTS

M. FRANCIUS [1], E. PELINOVSKY [2, 3], I. RIABOV [3)], C. KHARIF [4]

[1]*Laboratoire de Mecanique et d'Acoustique, 31 Chemin Joseph Aiguier, 13402 Marseille, Cedex 20, France*
[2]*Institute of Applied Physics of the Russian Academy of Sciences 46 Ulyanov Street, 603600 Nizhniy Novgorod, Russia*
[3]*Applied Mathematics Departmen, Nizhny Novgorod State Technical University, 24 Minin Street, 603000, Nizhny Novgorod, Russia*
[4]*Institut de Recherche sur les Phenomenes Hors Equilibre, BP 146 49 rue Frederic Joliot-Curie, 13384 Marseille cedex 13, France*

Abstract

We analyzed the manifestation of tsunamis for the French coasts of the Mediterranean. Historical data of past tsunamis are analyzed to provide a representative set of potential "seismic" tsunami sources for the Ligurian Sea. Nonlinear shallow water model was used to simulate the propagation from initial displacements in the tsunami sources that were located at the places of historical tsunamis. The synthetic method was applied to study the characteristics of possible earthquake-generated tsunamis in the Ligurian Sea. It is found that tsunami manifestation on the French coasts has a very local character and that the Western French Mediterranean coasts are protected from tsunamis generated in the Ligurian Sea.

1. Introduction

The tsunami hazard has been less studied for the coasts of the French metropolis than for the coasts of French overseas territories. In fact, this problem is rather difficult because of the lack of historical information and measurements regarding tsunami inundation. Thus a direct statistical analysis is not possible. To provide a preliminary study of the tsunami hazard for the French coast, the synthetic method was used, as for instance did Curtis & Pelinovsky [1]. Large number of numerical simulations of past real events and of future probable tsunamis can be analysed to compare the characteristics of tsunami propagation for different coastal areas, and to make prognostic modelling of the propagation. This method was used for the Ligurian Sea in order to give a preliminary analysis of the tsunami hazard for the French Mediterranean coasts. This study focuses on tsunami induced by earthquakes.

2. Analysis and modelling tsunami propagation

To analyse the tsunami manifestation in the Ligurian Sea, we considered four groups of experiments (see fig. 1). Those groups were chosen according to the analysis of the

A. C. Yalçıner, E. Pelinovsky, E. Okal, C. E. Synolakis (eds.),
Submarine Landslides and Tsunamis 185-190.
©2003 Kluwer Academic Publishers.

186

catalogues of Soloviev et al. [2] and of Tinti & Maramai [3]. The first group concerns

Figure 1. Epicentres of earthquakes induced tsunami on the French coast of the Mediterranean

tsunamis induced by earthquakes with epicentres in the vicinity of Nice and Imperia. In this very shallow zone, tsunamigenic earthquakes have 10-20 km focal depth and can induce significant waves in spite of their relative small magnitude (6.2 – 6.5). The location of epicentres of computed tsunamis was the same as for historical tsunamis 1818, 1831, 1854 and 1887. The second group considers the tsunami propagation from northern Italy to the French coast, which seems to have a very local character. Indeed the weak earthquake of July 2, 1703 in Genoa (magnitude 3.2) generated a local tsunami. In particular the sea level in the *Harbour of Genoa* decreased by1.5-2 m. The third group considers the tsunami propagation from the southern Italy. Such tsunamis were never registered in France, but taking into account the large seismicity of southern Italy, this area can be considered like a probable place for generation of strong tsunamis. Finally, we considered the tsunami propagation from the Algerian coast, where a strong earthquake occurred on September 9, 1954 (magnitude 6.7).

The tsunami propagation was considered in the framework of the non-linear shallow-water theory in the form of Saint-Venant equations. For rough estimates of the tsunami source parameters, the simplified piston model was used with a two-parameter tsunami source model (earthquake magnitude and focal depth). The initial data is the fault length in the earthquake source l_0 related with the earthquake magnitude M_0. For prognostic modelling we used for all events an earthquake magnitude $M_0 = 6.8$, an earthquake focal depth of 20 km, a north-east orientation of the tsunami source, and a roughness coefficient $m = 0.0012$ which is a general hydraulic parameter which accounts for bottom friction. This value of the earthquake magnitude corresponds to the mean value for tsunami-genetic earthquakes in the whole Mediterranean basin. The chosen source orientation is typical for earthquakes of the northern part of the Ligurian Sea. The parameters have been chosen identical for all simulations to analyse the general features of the tsunami propagation in the Ligurian Sea. We could also compare the global characteristics of tsunami waves along the coasts, namely the relation between the tsunami heights in different coastal points. In fact, this simplified model can not provide accurate calculations of tsunami heights but the relative distribution of tsunami heights along the coast should be more realistic since it depends mainly on the coastal topography and on the very rough characteristics of the tsunami source (in particular, the source orientation).

Each "computer" tide-gauge is located in the "last" sea point $\eta_w(t)$ (depth of about 20 m). In this point the assumption of a vertical wall is used and the oscillations of the water level on the shoreline are calculated according to the formula of Kaistrenko et al. [4]. With the assumptions that the beach is plane and that the waves come on almost onshore direction, the run-up height may be calculated with the formula

$$R(t) = \int_{0}^{t-T} \sqrt{(t-\tau)^2 - T^2} \, \frac{d^2 \eta_w}{d\tau^2} \, d\tau$$

where T is a travel time from last sea point to shore, and $t > T$. This combination of 2D model with 1D model has been tested, and it generally improves the agreement between observed and computed data, see Choi et al. [5].

3. Results and discussion

Our model is tested with data of the 1887 tsunami. Characteristic amplitude and period of the spectral maximum are in good agreement with observed tide-gauge records and calculations of Eva & Rabinovich [6]. The western part of the Mediterranean coast is protected from these tsunamis by the southern part of the French Riviera. Tsunamis generated in southern Italy affect more Corsica which "screens" partly the French coasts. Tsunamis generated in the vicinity of the African continent are weak on French coasts. In general long duration of tsunami record is characteristic of the French Riviera coasts and this should be accounted in tsunami warning and forecasting. The average total duration of the tsunami record is of about 10 – 20 hrs (see fig. 2) and this should be accounted by tsunami warning. Characteristic period of oscillations varies from one point to another point (20 – 30 min), but it is less for the northern part of Ligurian Sea than for the western part of the Mediterranean (30 – 45 min), see fig. 3. The travel time of tsunamis depends on the location of the earthquake and it is several hrs for Italian and African earthquakes. If the tsunami is generated in the vicinity of Nice – Imperia, the travel time to Nice is a few minutes for Nice and Cannes, and 1-2 hrs – for the western part of the Mediterranean. Wave heights on the tsunami records have an order of a few cm for the points far from earthquakes, and a few ten cm for the nearest points. The run-up process is characterised by an average amplification factor of 2.5.

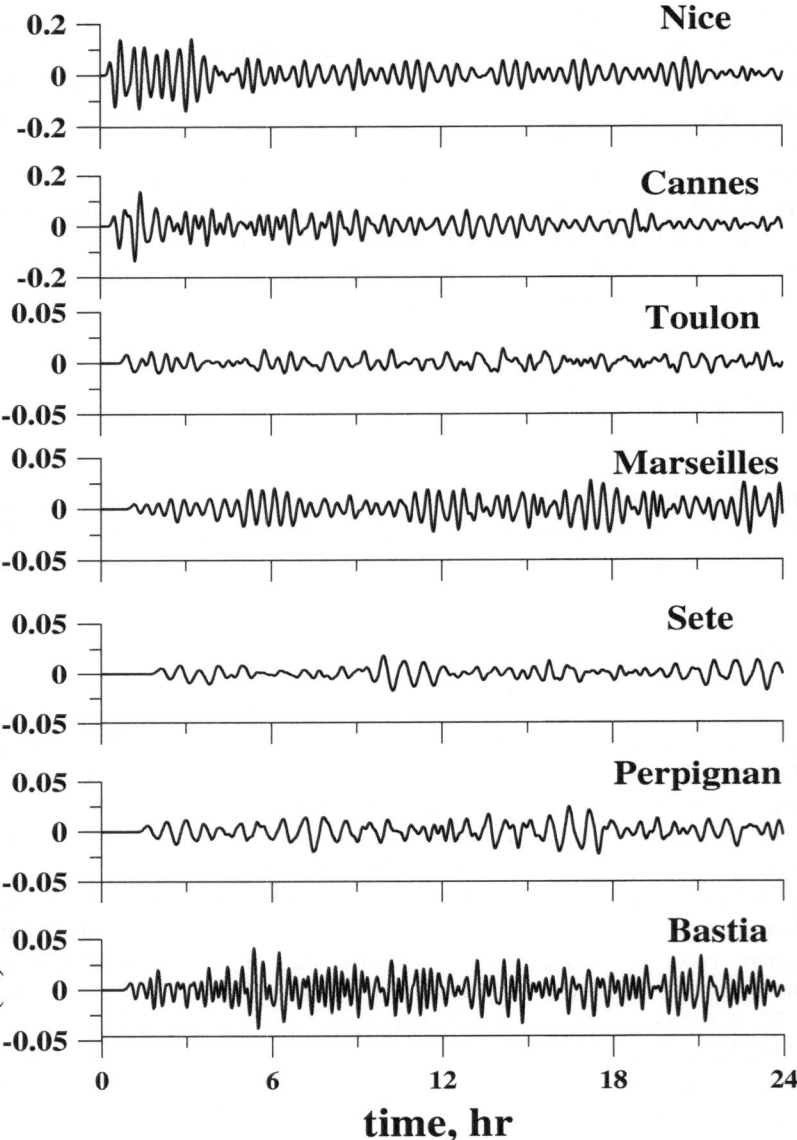

Figure 2. Computed tsunami records for the 1887 event.

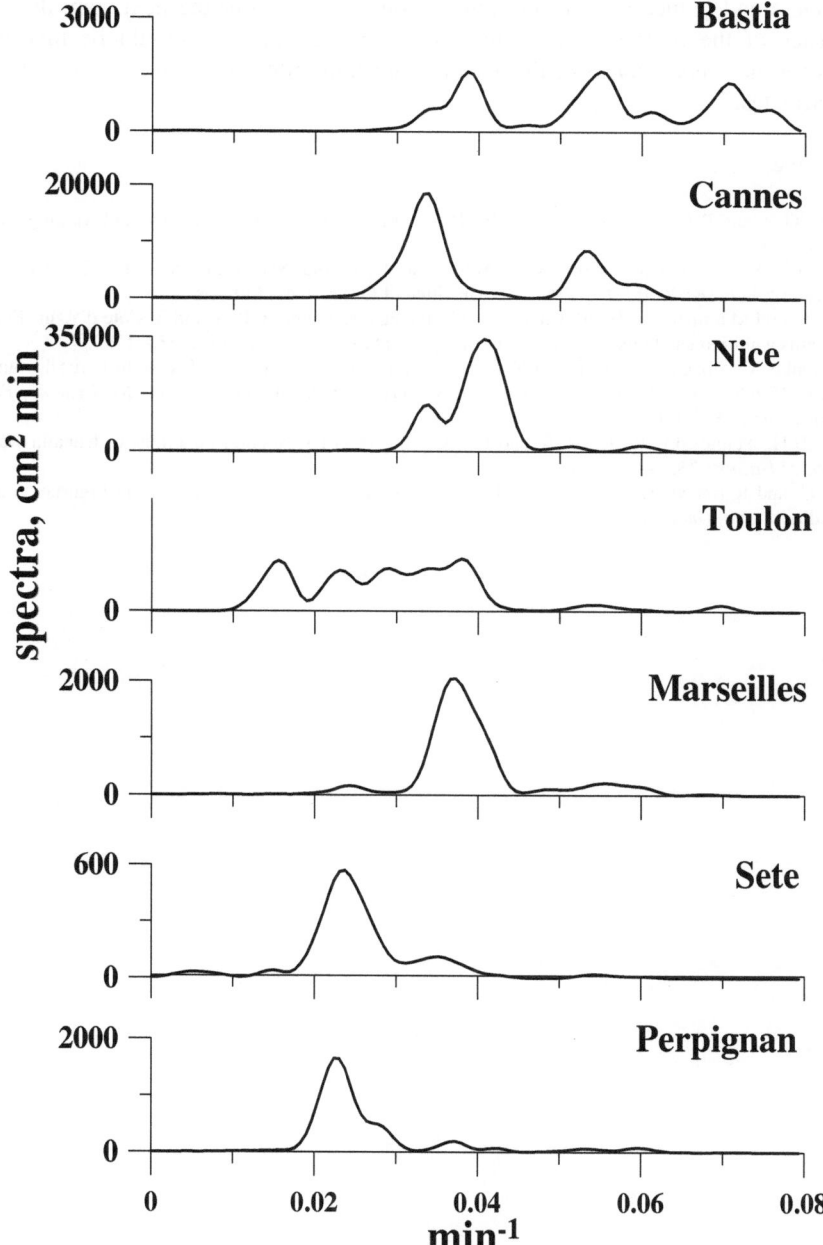

Figure 3. Computed spectra of the 1887 event

Nevertheless, it is important to notice that information on tsunami tide-gauge records and observed wave heights on the beach for the 1887 Ligurian and 1979 Nice tsunamis, the amplification factor can reach 10 at selected points. It means that the nearest earthquakes in the vicinity of Cannes – Imperia can generate tsunami waves with maximum run-up of a

1-2 meters and induce significant damages on the coast. For the next step, the resonance properties of the bottom topography and the run-up process should be investigated in details for the French coasts of the Mediterranean in order to obtain more accurate tsunami risk estimates.

References

1. Curtis, G.D. and Pelinovsky, E.N. (1999) Evaluation of tsunami risk for mitigation and warning, *Sci. Tsunami Hazards* **17,** n° 3, 187-192.
2. Soloviev, S.L., Go, C.N., Kim, K.S., Solovieva, O.N. and Shetnikov, N.A. (1997), *Tsunamis in the Mediterranean sea 2000 B.C. – 1991 A.D.*, Institute of Oceanology, Moscow.
3. Tinti, S. and Maramai, A. (1996), Catalogue of tsunamis generated in Italy and in Cote d'Azur, France: a step towards a unified catalogue of tsunamis in Europe, *Annali di Geofisica*, **39**, 1253 – 1299.
4. Kaistrenko, V., Go, C.N. and Chung, J.Y. (1999) A simple method for tsunami wave form recalculation through the shelf, *IOC-IUGG Joint International Workshop on Tsunami Warning Beyond 2000 Theory, Practice and Plans, Korea, Seoul*, 15.
5. Choi, B.H., Pelinovsky, E., Hong, S. and Ryabov, I. (2002) Distribution Functions of tsunami wave heights, *Natural Hazards* **25,** 1-21.
6. Eva, C. and Rabinovich, A. (1997) The February 23, 1887 tsunami recorded on the Ligurian coast, western Mediterranean, *Geophys. Research Letters* **24**, n° 17, 2211–2214.

INUNDATION MODELING OF THE 1964 TSUNAMI IN KODIAK ISLAND, ALASKA

ELENA N. SULEIMANI[1] and ROGER A. HANSEN[1],
ZYGMUNT KOWALIK[2]
[1]*Geophysical Institute*
University of Alaska Fairbanks, USA
[2]*Institute of Marine Science*
University of Alaska Fairbanks, USA

Abstract

In this work a numerical modeling method is used to study tsunami waves generated by the Great Alaska earthquake of 1964 and their impact on Kodiak Island communities. The numerical model is based on the nonlinear shallow water equations of motion and continuity which are solved by a finite-difference method. We compare two different source models for the 1964 deformation. It is shown that the results of the near-field inundation modeling strongly depend on the slip distribution within the rupture area. These results are used for evaluation of the tsunami hazard for the Kodiak Island communities.

1. Introduction

Seismic events that occur within the Alaska-Aleutian subduction zone have a high potential for generating both local and Pacific-wide tsunamis. Seismic water waves originating in Alaska can travel across the Pacific and destroy coastal towns hours after they are generated. However, they are considered to be a near-field hazard for Alaska, and can reach Alaskan coastal communities within minutes after the earthquake. Therefore, community preparadeness plays a key role in saving lives and property. Evacuation areas and routes have to be planned in advance, which makes it essential to have an estimate for the potential flooding area of coastal zones in the case of a local tsunami. To help mitigate the large risk these earthquakes and tsunamis pose to Alaskan coastal towns, the authors participate in the National Tsunami Hazard Mitigation Program through evaluating and mapping the inundation of Alaska coastlines using numerical modeling of tsunami wave dynamics.

The communities for inundation modeling are selected in coordination with the Alaska Division of the Emergency Services with consideration to location, infrastructure, availability of bathymetric and topographic data, and willingness for a community to incorporate the results in a comprehensive mitigation plan. Kodiak Island was identified as a high-priority region for Alaska inundation mapping. There is a number of communities with relatively large populations and significant commercial resources. They are in need of tsunami evacuation maps which show the extent of inundation with respect to human and

A. C. Yalçıner, E. Pelinovsky, E. Okal, C. E. Synolakis (eds.),
Submarine Landslides and Tsunamis 191-201.
©2003 Kluwer Academic Publishers.

cultural features, and evacuation routes. The purpose of this case study is to model tsunami waves generated by the 1964 earthquake and to compare the extent of the computed inundation zone with the observations. This experiment is set to test the model that will be applied to produce tsunami inundation maps for Alaskan coasts. The maps will summarize information on historical tsunamis in the region and the results of model runs for different source scenarios.

2. Numerical Model

The numerical model used in this study is based on the vertically integrated nonlinear shallow water equations of motion and continuity with friction and Coriolis force. Written in a spherical coordinate system, they are [1];[2]:

$$\frac{\partial U}{\partial t} + \frac{U}{R\cos\varphi}\frac{\partial U}{\partial \lambda} + \frac{V}{R}\frac{\partial U}{\partial \varphi} - fV = -\frac{g}{R\cos\varphi}\frac{\partial \xi}{\partial \lambda} - \frac{rUW}{D} \tag{1}$$

$$\frac{\partial V}{\partial t} + \frac{U}{R\cos\varphi}\frac{\partial V}{\partial \lambda} + \frac{V}{R}\frac{\partial V}{\partial \varphi} + fU = -\frac{g}{R}\frac{\partial \xi}{\partial \varphi} - \frac{rVW}{D} \tag{2}$$

$$\frac{\partial \xi}{\partial t} = \frac{\partial \eta}{\partial t} - \frac{1}{R\cos\varphi}\left[\frac{\partial(DU)}{\partial \lambda} + \cos\varphi\frac{\partial(DV)}{\partial \varphi}\right], \tag{3}$$

where λ is longitude, φ is latitude, t is time, U and V are horizontal velocity components along longitude and latitude, $W = \sqrt{U^2 + V^2}$, ξ is variation of sea level from equilibrium, η is the bottom displacement, g is the gravity acceleration, R is radius of the Earth, f is the Coriolis parameter, $D = (H + \xi - \eta)$ is the total water depth, and r is the bottom friction coefficient.

Various approaches to deriving a numerical solution of the above system of equations were outlined in Imamura (1996) [3] and Titov and Synolakis (1998) [4]. In this study, we apply a space-staggered grid which requires either sea level or velocity as a boundary condition. The first order scheme is applied in time and the second order scheme is applied in space. Integration is performed along the north-south and west-east directions separately in a way that is described in Kowalik and Murty (1993) [5]. To apply this procedure, equations (1) and (2) are split in time into two subsets. First, these equations are solved along the longitudinal direction,

$$\frac{U^{m+1} - U^m}{T} + \left(\frac{U}{R\cos\varphi}\frac{\partial U}{\partial \lambda}\right)^m + \left(\frac{V}{R}\frac{\partial U}{\partial \varphi}\right)^m - (fV)^m =$$

$$= -\frac{g}{R\cos\varphi}\left(\frac{\partial\xi}{\partial\lambda}\right)^m - \left(\frac{rUW}{D}\right)^m \tag{4}$$

$$\frac{1}{2}\frac{\xi^* - \xi^m}{0.5T} = \left(\frac{\partial\eta}{\partial t}\right)^m - \frac{1}{R\cos\varphi}\frac{\partial(D^m U^{m+1})}{\partial\lambda}, \tag{5}$$

and next along the latitudinal direction

$$\frac{V^{m+1} - V^m}{T} + \left(\frac{U}{R\cos\varphi}\frac{\partial V}{\partial\lambda}\right)^m + \left(\frac{V}{R}\frac{\partial V}{\partial\varphi}\right)^m + (fU)^{m+1} =$$

$$= -\frac{g}{R}\left(\frac{\partial\xi}{\partial\varphi}\right)^m - \left(\frac{rVW}{D}\right)^m \tag{6}$$

$$\tag{7}$$

$$\frac{1}{2}\frac{\xi^{m+1} - \xi^*}{0.5T} = -\frac{1}{R}\frac{\partial(D^m V^{m+1})}{\partial\varphi}.$$

The calculation of sea level starts from time step m, and the intermediate value of sea level ξ^* is obtained after integration along the first direction. Afterwards, this value is carried over to the other direction to derive the sea level at the $(m+1)$ time step.

In order to propagate the wave from a source to various coastal locations we use embedded grids, placing a coarse grid in a deep water region and coupling it with finer grids in shallow water areas. We use an interactive grid splicing, therefore the equations are solved on all grids at each time step, and the values along the grid boundaries are interpolated at the end of every time step [6]. The radiation condition is applied at the open (ocean) boundaries [7]. At the water-land boundary, the moving boundary condition is used in those grids that cover areas selected for inundation mapping [8]. At all other land boundaries, the velocity component normal to the coastline is assumed to be zero.

3. The Source Model for the 1964 Tsunami

We have started this project with the modeling of the Alaska 1964 tsunami, because this event is most probably the worst case scenario of a tsunami for the Kodiak Island communities. The 1964 Prince William Sound earthquake generated one of the most destructive tsunamis observed in Alaska and the west coast of the US and Canada. The major tectonic tsunami was generated in the trench area and affected all the communities in Kodiak and the nearby islands. There were also about 20 local submarine and subaerial landslide-genereated tsunamis in Alaska that account for 75% of all tsunami fatalities in the region. In Kodiak, the tsunami caused 6 fatalities and about $30 million in damage. This tsunami was studied in depth by a number of authors [9]; [10]. The observed inundation patterns for several locations on Kodiak Island are available for calibration of the model. Tsunami propagation models use output of the submarine seismic source models as an initial condition for ocean surface displacement which then propagates away from the source. The amplitude of this initial disturbance is one of the major factors that affect the resulting runup amplitudes along the shore line. One of the source models used in the tsunami generation problem is a double-couple model of an earthquake source. Okada (1985) [11] developed an algorithm to calculate the distribution of coseismic uplift and subsidence resulting from the motion of the buried fault. The fault parameters that are required to compute the deformation of the ocean bottom are location of the epicenter, area of the fault, dip, rake, strike and amount of slip on the fault. The rupture area of the 1964 earthquake was too large to be adequately described by a simple one-fault model. It was demonstrated by Christensen and Beck [12] that there were two areas of high moment release, representing the two major asperities of the 1964 rupture zone: the Prince William Sound asperity and the Kodiak Island asperity. The segmentation of the source area suggests that the single-fault model with uniform slip is not adequate to describe the slip distribution of this earthquake. A detailed analysis of the 1964 rupture zone was presented by Johnson et. al. (1996) [13] through joint inversion of the tsunami waveforms and geodetic data. Authors derived a detailed slip distribution for the 1964 earthquake, which is shown in Figure 1.

To construct a source function for the 1964 event, we used their fault model that has 8 subfaults representing the Kodiak asperity of the 1964 rupture zone, and 9 subfaults in the Prince William Sound asperity. The authors didn't include the Patton Bay fault on Montague Island in the source mosaic, because the contribution of this fault to the tsunami waveforms was negligible. However, they removed the effect of this fault by subtracting the deformation due to it from all geodetic observations. We used the equations of Okada (1985) [11] to calculate the distribution of coseismic uplift and subsidence resulting from the given slip distribution (Figure 2). Then, the derived surface deformation pattern was used as the initial condition for tsunami propagation.

4. Modeling of the 1964 Tsunami in Kodiak

Initially, the 1964 tsunami wave was modeled using the above described source function consisting of 17 subfaults, each having its own parameters. To verify the accuracy of the far-field calculations, we compared numerical results with the observations at the Sitka and Yakutat tide gauges. Tsunami amplitudes were computed at the grid points closest to the tide gauges. Figure 3a,b show the observed and calculated time-dependent sea level at the two locations. The plots indicate that the time of arrival of the fist wave and its amplitude are in good agreement with the observations. Our goal was also to estimate the importance of the detailed slip distribution of the rupture zone for the near- field inundation modeling. To accomplish that, we compared two source models of the 1964 event: one, consisting of a single fault with uniform slip distribution and the other of 17 subfaults. The amount of slip on the single fault was calculated in a way that preserves the seismic moment. The resulting surface deformation was computed using the Okada (1985) [11] algorithm and used in the tsunami model as the initial condition.

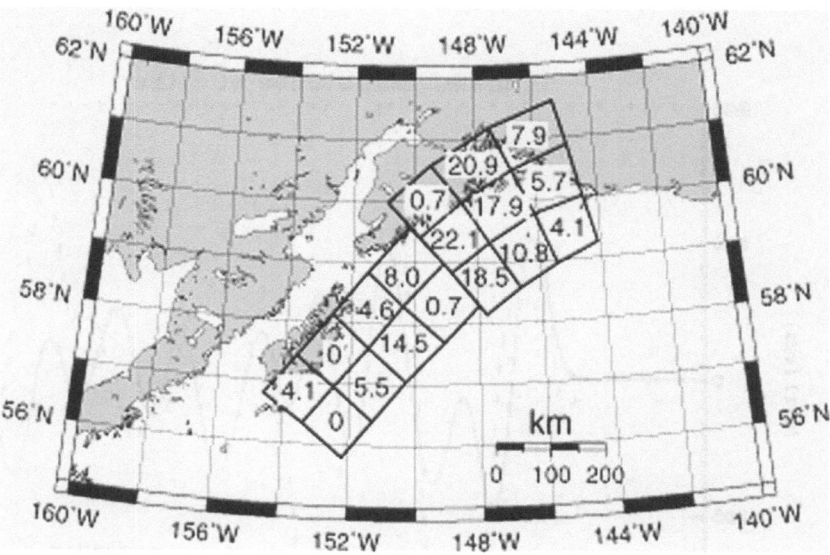

Figure 1. Slip distribution of the 1964 earthquake, from [13]. Numbers represent slip in meters on every subfault.

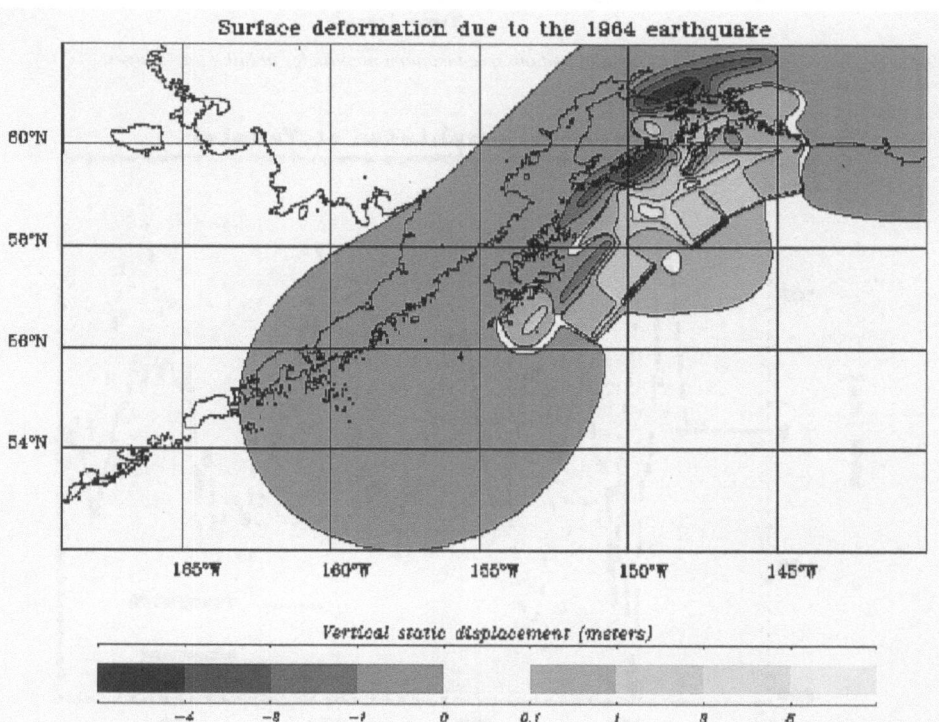

Figure 2. Surface deformation of the 1964 earthquake, calculated from the slip distribution given in Figure 1.

196

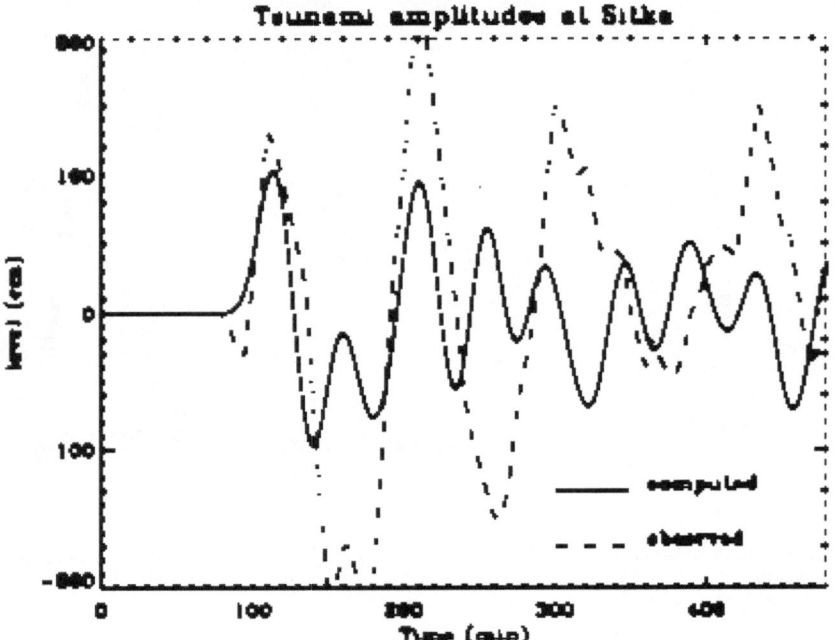

Figure 3a. Computed and observed tsunami amplitudes at Sitka tide gauges.

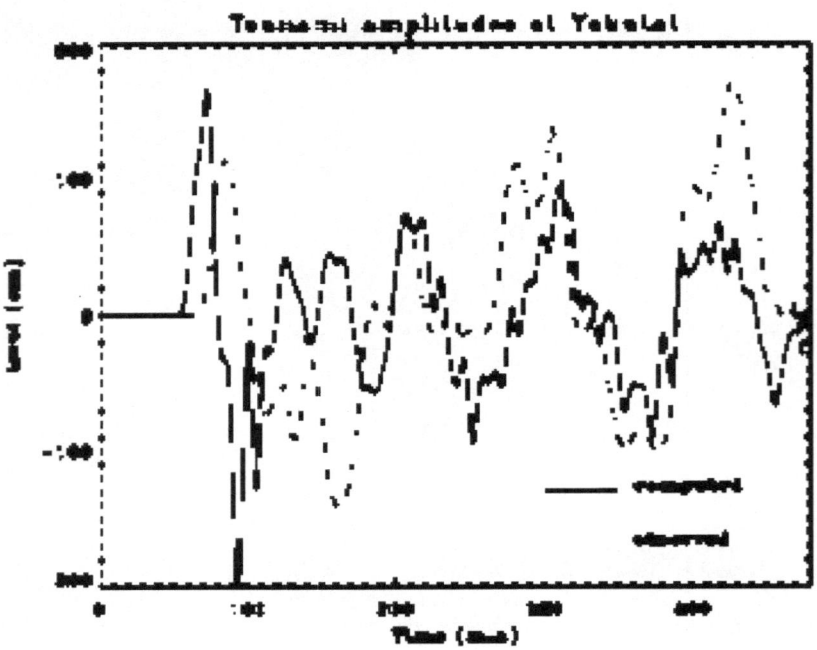

Figure 3b. Computed and observed tsunami amplitudes at Yakutat tide gauges.

The region examined in this study is shown in Figure 1. This area is covered by the largest grid of 2 minute resolution. We use four embedded grids in order to increase resolution from 2 minutes (2 km x 3.7 km) in the Gulf of Alaska to 21.8 m x 27.5 m in the three grids that cover communities selected for inundation modeling. The embedded grids are shown in Figure 4. The first one covers the lower part of Cook Inlet and waters around Kodiak Island. This grid is of 24 second resolution. The north-east segment of the island is covered by the 8 second grid, and the Chiniak Bay is covered by the 3 second grid. There are 3 more fine resolution grids covering regions of Kodiak Island where runup calculations are performed. They are shown as three rectangles in the Chiniak Bay grid. In these grids, the combined bathymteric and topographic data allow for application of the moving boundary condition as well as calculating the runup heights and extent of the inundation. We assume that the initial displacement of the ocean surface from the equilibrium position is equal to the vertical component of the ocean floor deformation due to the earthquake rupture process. This ocean surface displacement is the initial condition for the computation of tsunami wave field in the region.

For the communities of Kodiak City and the Kodiak Naval Station (Figure 4), the observed inundation was documented in July of 1964 by Kachadoorian and Plafker (1967) [10]. Kodiak City, the largest community on the island, suffered the greatest damage from the tsunami. Figures 5-a and 5-b show computed and observed inundation lines for the City of Kodiak and the U.S. Coast Guard Base (formerly the Kodiak Naval Station), respectively. The blue line delineates the area inundated in 1964 following data collected after the event by the U. S. Army Corps of Engineers, U. S. Navy stations and other authorities. The solid red line shows the inundated area computed using the complex source function of 17 subfaults. The dashed red line is for computed inundation zone using the simple one-fault source model. The observed area of maximum inundation at the Kodiak Naval Station is taken from Kachadoorian and Plafker (1967) [10]. The results show that the complex source model with detailed slip distribution describes the inundation zone much better than the simple one-fault model. The one-fault model greatly underestimates the extent of flooding caused by the 1964 tsunami wave.

The computed wave history of the tsunami waves at the US Coast Guard base (formerly the Kodiak Naval Station) is represented in Figure 6. The zero time corresponds to the epicenter origin time. The arrows show observed arrivals of the first three waves at the Naval Station [9]. There is agreement with the observations only for the first wave. The arrival times for the second and the third waves are less than those observed. Using vizualization methods, we found that the first wave arrives from the fault in the Kodiak asperity with the amount of slip of 14.5 meters. The second and the third arrivals result from the interaction of the waves coming from the faults in the Prince William Sound asperity (22.1 meters and 18.5 meters of slip) and the waves already refracted from the Kodiak shores. We have assumed that the discrepancy in computed arrival times is related to the uncertainty in the source function. Numerical experiments have shown that modifying the amount of slip on individual subfaults can cause changes in the maximum amplitude arrivals at particular locations. These experiments show that the near-field inudation modeling results are very sensitive to the fine structure of the tsunami source, and it is much easier to indroduce errors in the modeling process when the complexity of the source function is combined with the proximity of the coastal zone.

198

5. Discussion

Locally-generated tsunamis pose a significant hazard for coastal Alaska, and better understanding of the tsunami source mechamism is crucial for hazard mitigation. Numerical analysis has shown that the detailed distribution of the ocean bottom uplift and subsidence due to an earthquake is a very important factor in the near-field inundation modeling. When the tsunami is generated in the vicinity of the coast, the direction of the incoming waves, their amplitudes and times of arrival are determined by the initial displacements of the ocean surface in the source area, because the distance to the shore is too small for the waves to   disperse. Comparison between the two source models for the 1964 tsunami indicates that using the source model of 17 subfaults with detailed slip distribution within the rupture area produces the inundation line closest to that observed in 1964. The results show the need for   detailed studies of the source mechanism of tsunamigenic earthquakes. Computed and observed sea level amplitude at the distant stations in Sitka and Yakutat depict good agreement. Computations in the region located close to the tsunami source (Kodiak Island)

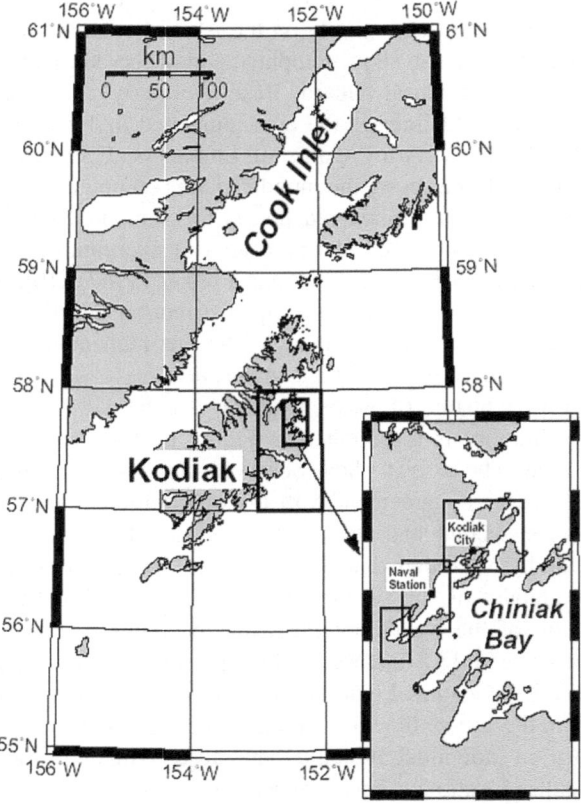

Figure 4. The Kodiak Island grid of 24 second resolution. The two rectangles delineate the 8 second and the 3 second resolution grids. The zoom is on the 3 second grid that includes finer resolution grids for the Kodiak Island communities of Kodiak City, US Coast Guard Base and Womens Bay.

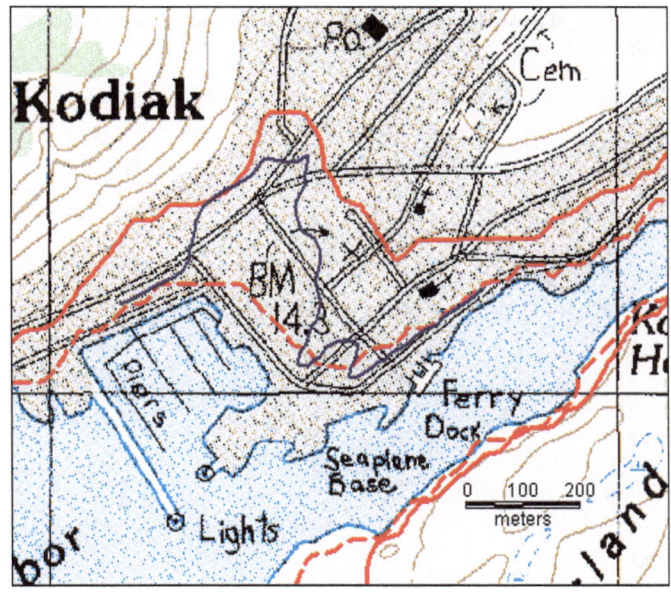

Figure 5-a. Observed (dark blue) and computed (red) inundation linesfor the Kodiak downtown area. Solid red line delineates inundation calculated using the 17-fault model, dashed red line represents the inundaiton calculated using the single fault model.

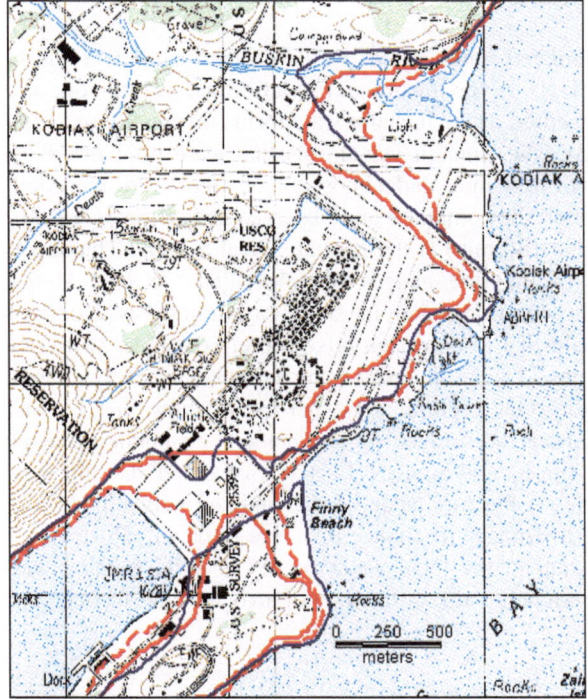

Figure 5-b. Observed (dark blue) and computed (red) inundation lines for the US Coast Guard Base, formerly the Kodiak Naval Station . Solid red line delineates inundation calculated using the 17-fault model, dashed red line represents the inundation calculated using the single fault model.

200

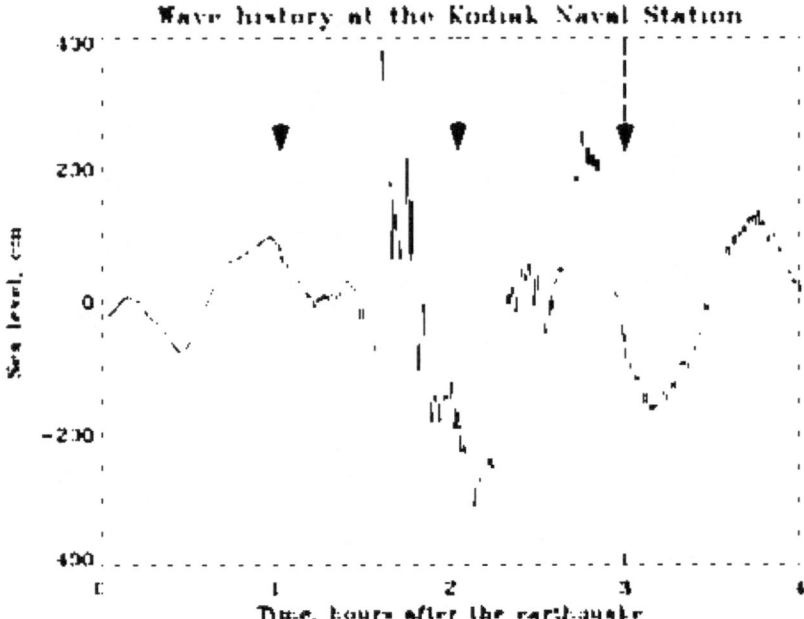

*Figure 6. Computed wave history at the Kodiak Naval Station. The arrows indicate the observed arrivals of
the first three waves.*

show that the calculated runup agrees well with the visual data but the timing of observed
and calculated arrivals of the maximum tsunami wave is often out of phase. This error may
be related to the method of the tsunami waveforms inversion used in the construction of the
source model [13]. The tsunami waveforms that were used in the joint inversion had an
average duration of 100 min, and almost all of them were recorded at distant locations.
Here we try to match the computed and observed arrival times for the second and the third
waves (120 min and 180 min) at the Kodiak Naval Station. This area is much closer to the
source than any of the tide gauges that provided the records for the joint inversion
algorithm. Also, observations taken in small bays around the Kodiak Island require further
studying. We would like to approach this problem through: a) investigation of the natural
mode of oscillations in semi-enclosed water bodies, and b) by changing the parameters of
the source function. Our preliminary experiments show that a water body whose own
periods of oscillations are close to an incident tsunami wave period responds by
generating higher run-up and higher reflected waves, and also the resonance response of a
small bay depends on the time span of the incident tsunami wave train. Longer tsunami
wave trains cause a stronger resonance response to occur. This can be used to demonstrate
that the strong runup in the local water bodies is quite possible not only at the time of
arrival of the first wave but also at the time of arrival of the second or third waves due to the
pumping action.

Acknowledgments

The authors wish to thank Dr. Fumihiko Imamura for the Fortran code of the Okada algorithm he kindly provided. This study was supported in part by the National Tsunami Hazard Mitigation Program through NOAA grant NA67RJ0147, by the Alaska Department of Military and Veteran Affairs (grant RSA-0991007) and by the Alaska Science and Technology Foundation (Grant ASTF-971002).

References

1. T.S. Murty. *Storm Surges - Meteorological Ocean Tides*, Bull. 212, Fisheries Res. Board Canada, Ottawa, 1984.
2. E.N. Pelinovsky. *Tsunami Wave Hydrodynamics*, Russian Academy of Sciences, Nizhni Novgorod, 275pp.,1996.
3. F. Imamura. Review of tsunami simulation with a finite difference method. *Long-Wave Runup Models*, H.Yeh, P. Liu and C. Synolakis, Eds., pages 25 - 42, World Scientific, 1996.
4. V.V. Titov and C.E. Synolakis. Numerical modeling of tidal wave runup. J. *Waterway, Port, Coastal and Ocean Eng.*, 124(4):157 - 171, 1998.
5. Z. Kowalik and T.S. Murty. Numerical simulation of two-dimensional tsunami runup. *Marine Geodesy*, 16:87 - 100, 1993.
6. E.N. Troshina. *Tsunami waves generated by Mt. St. Augustine volcano, Alaska*. MS thesis, University of Alaska Fairbanks, 1996.
7. R.O. Reid and B.R. Bodine. Numerical model for storm surges in Galveston Bay. *J. Waterway Harbor Div.*, 94(WWI):33 - 57, 1968.
8. Z. Kowalik and T.S. Murty. *Numerical Modeling of Ocean Dynamics*, World Scientific, 481 pp., 1993a.
9. B.W. Wilson and A. Tørum. *The Tsunami of the Alaskan Earthquake, 1964: Engineering Evaluation*, U.S. Army Corps of Engineers, Technical Memorandum No. 25, 1968.
10. R. Kachadoorian and G. Plafker. Effects of the earthquake of March 27, 1964 on the communities of Kodiak and nearby islands. *Prof. Paper 542-F, U.S. Geological Survey*, Washington, D.C., 1967.
11. Y. Okada. Surface deformation due to shear and tensile faults in a half-space. *Bull. Seismol. Soc. Am.*, 75:1135 - 1154, 1985.
12. D.H. Christensen and S.L. Beck. The rupture process and tectonic implications of the Great 1964 Prince William Sound earthquake. *Pageoph*, 142(1):29 - 53, 1994.
13. J.M. Johnson, K. Satake, S.R. Holdahl and J. Sauber. The 1964 Prince William Sound earthquake: Joint inversion of tsunami and geodetic data. *J. Geophys. Res.*, 101(B1):523 – 532, 1996.

A METHOD FOR MATHEMATICAL MODELLING OF TSUNAMI RUNUP ON A SHORE

L.B.CHUBAROV, Z.I.FEDOTOVA
Institute of Computational Technologies Lavrentyev ave. 6
630090, Novosibirsk, Russia

Abstract.

The paper is devoted to construction of numerical algorithms for modelling tsunami runup on beaches. Comparing numerical accounts both with analytical solutions and with experimental data was executed.

1. Introduction

Recent progress in numerical modelling of tsunami is due to the increasing demand for protection of disaster prone objects. A new generation of national tsunami warning systems is based on the local networks tuned to capture the geographical, morphological and social factors peculiar to the protected coast. Local networks provide an alternative to the larger regional systems developed before. They allow for better response times and higher precision predictions, and not only reduce the overall damage from natural hazards and assure protection of potentially dangerous objects as nuclear power plants situated in the coastal zone or nuclear power driven vessels, but also decrease the probability of a false alert.

With the development of such systems new efficient methods for modelling of tsunami wave's transformation in the coastal zone and runup on the shore are required.

The generally accepted mathematical model of runup is the system of equations describing the propagation of a long wave over a flat bottom towards the coast with a constant slope. Different versions of long wave equations such as the Boussinesq equation or non-dispersive shallow water equations had been proposed. Studies in one-dimensional modelling of wave runup on a slopping beach distinguish three ranges of the slope angle, α. These ranges correspond to gentle ($\tan \alpha \ll 1$) moderate ($\tan \alpha \sim 1$) and steep ($\tan \alpha \gg 1$) slopes.

The case of a *gentle* slope secures the choice of nonlinear shallow water theory equations for long waves, as the vertical projections of the acceleration of fluid particles are relatively small. Most waves were observed to break approaching the shore. In the framework of nonlinear model the characteristic nonlinearity distance is introduced. If the nonlinearity distance is close to the length of the slope, then hyperbolic nonlinear equations of shallow water theory in the form of conservation laws can be applied.

Waves interacting with a *moderate* slope typically reflect from the shore without breaking. Nonlinear shallow water equations are widely used for the runup problem, but a detailed description of the fluid motion near the point of splash requires a proper account if

A. C. Yalçıner, E. Pelinovsky, E. Okal, C. E. Synolakis (eds.),
Submarine Landslides and Tsunamis 203-216.
©2003 *Kluwer Academic Publishers.*

the vertical acceleration of the particles. Nevertheless the nonlinear model provides for an accurate estimation of the maximal value of the inundation.

The shallow water model is not applicable for *steep* slopes, as the vertical acceleration of the particles is significant. Good approximation can be achieved by substituting the shore slope by an impenetrable vertical wall with the corresponding modification of the boundary conditions. Again the maximal runup height can be estimated from the equations of shallow water theory.

The main contribution of the work is the description of a numerical model for tsunami wave propagation and runup in a bay with irregular topography of the bottom. The solution is difficult to estimate *a priori* because of the strong effect of the bottom shape on the wave transformations near the shore. Description of wave propagation over a bottom of complex shape and an estimation of the runup height is based on the shallow water theory equations. A finite-difference method of *straight-through calculation*, which captures discontinuities of the solution, had been developed for a fixed grid over a rectangular domain. The computational domain represents the water bay with the adjacent shore. The scheme allows for the computation of the shoreline at every moment $t = n\Delta t$.

2. One-Dimensional Numerical Algorithms

The mathematical model is based on nonlinear non-dispersive equations. Let

$$\mathbf{W} = \begin{pmatrix} H \\ Hu \\ Hv \end{pmatrix}, \mathbf{F} = \begin{pmatrix} Hu \\ Hu^2 + gH^2/2 \\ Huv \end{pmatrix}, \mathbf{G} = \begin{pmatrix} Hv \\ Huv \\ Hv^2 + gH^2/2 \end{pmatrix}, \mathbf{Q} = \begin{pmatrix} 0 \\ gHh_x \\ gHh_y \end{pmatrix}.$$

The matrix form of the equations over the components $u = u(x, y, t)$ and $v = v(x, y, t)$ of the depth average velocity is:

$$\mathbf{W}_t + \mathbf{F}_x + \mathbf{G}_y = \mathbf{Q}. \tag{1}$$

Here $h = h(x, y)$ is the depth measured from the still water level; $H(x, y, t) = h(x, y) + \eta(x, y, t)$ is the total water depth varying in time; and $\eta = \eta(x, y, t)$ is the free surface elevation; and g is the gravitational acceleration.

Consider one-dimensional shallow water equations in a domain with a moving boundary. A number of finite-difference schemes were compared in order to choose a suitable numerical method for the tsunami runup problem [1].

Among the schemes frequently used to approximate nonlinear shallow water equations are the MacCormack scheme, the Aracawa scheme, the Ousher scheme and the Lax scheme. The former two have the second order of approximation in space and time, while the latter are of the first order.

The choice of a scheme had been based on the comparison of the numerical solution with the exact solution or a solution obtained by more accurate models, such as nonlinear dispersive models. One of the exact solutions of this kind is represented by a solitary wave

$$\eta(x,t) = \eta_0 \operatorname{sech}^2 \left\{ \left(\frac{3\eta_0}{4h_0^2 (h_0 + \eta_0)} \right)^{\frac{1}{2}} (x - x_0 - ct) \right\}, \quad u = \frac{c\eta_0}{h_0 + \eta_0},$$

(2)

where $c = \sqrt{g(h_0 + \eta_0)}$. Experimental study of the convergence properties of the four schemes mentioned above was carried out on solutions (2), which is an analytical solution of the more accurate long-wave model (see [1]). A basin was specified by a constant depth $h_0 = 1$ and had a length five times the length of the wave. A sequence of uniform grids with increasing resolution was considered. The MacCormack scheme was shown to achieve good accuracy on a coarser grid [1].

We proceed with the description of the implemented algorithms. When the computational domain is fixed and coincides with the original domain an algorithm consists of a finite-difference scheme and a correction procedure on the nodes close to the shoreline.

Five algorithms were studied in the one-dimensional case.

Algorithm 1. A finite-difference scheme in a fixed line segment represents the simplest method for runup modelling. The numerical domain includes "underwater" and "dry" points. The method is defined by the grid resolution, time step and the minimal depth, h_0, such that a layer of water of depth less than h_0 is considered dry. The total depth and velocity over the dry regions are set to zero. Calculations in the entire domain including water and the shore were performed using several schemes as the MacCormack scheme on a non-staggered fixed grid and the second-order scheme with central differences on a staggered fixed grid. The value of h_0 can be chosen by gradually decreasing this parameter in a series of computations until the shoreline becomes stable. In practice this value corresponds to the characteristic size of the irregularities of the bottom. For a monotone scheme and a gentle bottom slope it can be set to zero.

Algorithm 2. It is possible to consider another algorithm based on the MacCormack scheme in a fixed domain. In this method dry regions of the bottom are represented as covered by a thin layer of water of constant depth [1].

Algorithm 3. Computations in the points of the mesh near the shoreline depend on the direction of its movement. In the course of modelling of the runup stage new points are involved into the computation. In the run-down stage the exclusion condition is checked for each point of the shoreline. A special procedure is used to specify the initial values at the points freshly included.

Algorithm 4. When the oscillations of the coastal line are small the points on the shoreline marked as dry can be checked on the next time step according to the following procedure. Initial approximation of the total depth H is extrapolated on each step from inside the domain of fluid, and then corrected to satisfy the continuity equation.

Algorithm 5. Computations in a domain with a moving boundary require complicated program flow control. They can be avoided by mapping a variable size domain into a fixed domain. Earlier a second-order scheme with the approximation of spatial derivatives by central differences had been used in the fixed domain. We have used the MacCormack scheme to approximate the equations after the transformation.

Algorithms providing for more accurate tracking of the shoreline are known, but they are considerably more difficult in implementation and in computational complexity, especially in the two-dimensional case.

206

Five algorithms described above were compared in the one-dimensional case on a set of problems of modelling a wave specified by a continuous initial elevation (solitary and harmonic wave) or with a single discontinuity (a bore). The physical domain represented a part of a basin with a constant depth area merged with a flat rising slope. The results were published in [1].

Another test set was based on a problem with known exact solution [2]. Let's consider initial perturbation defined by the following equation:

$$\eta = \varepsilon \left[1 - \frac{5a^3}{2\left(a^2 + \sigma^2\right)^{3/2}} + \frac{3a^5}{2\left(a^2 + \sigma^2\right)^{5/2}} \right], \quad x = -\frac{\sigma^2}{16} + \varepsilon \left[1 - \frac{5a^3}{2\left(a^2 + \sigma^2\right)^{3/2}} + \frac{3a^5}{2\left(a^2 + \sigma^2\right)^{5/2}} \right].$$

The results are shown on Figure 1. Here 500 grid nodes were used. The analytical and numerical solutions differ only near the boundary at the final time (dash lines). T contains the runup values obtained by the methods listed above.

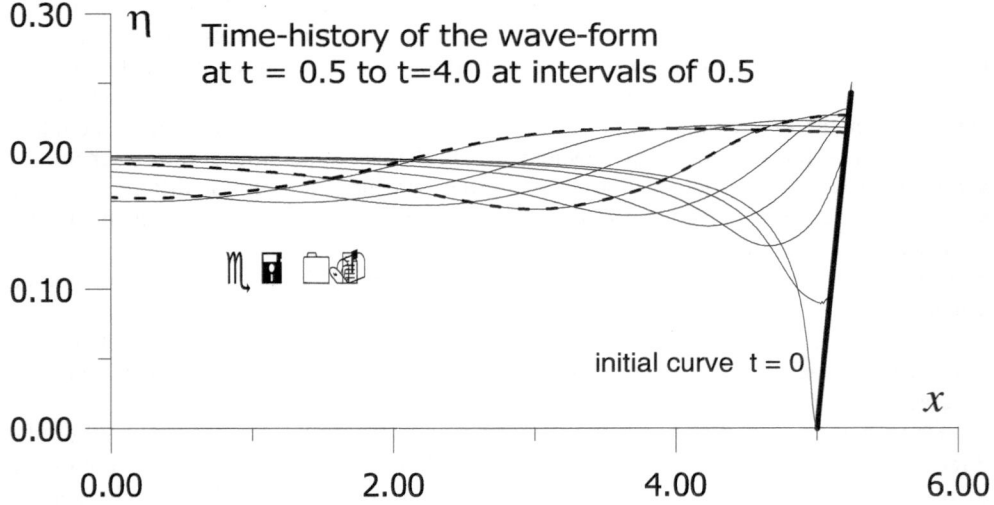

Figure 1. A comparison of analytical and numerical solutions.

TABLE 1. A comparison of runup values calculated by the methods

	Method of calculation	Runup value R
1	Analytical solution	R=0.2314
2	MacCormack scheme (moving grid tracking shoreline)	R=0.2321
3	Staggered explicit "box" scheme (moving grid)	R=0.2360
4	MacCormack scheme (author's variant, stationary grid)	R=0.2350
5	The same but with smoothing procedure of the second order of accuracy	R=0.2250
6	The same but smoothing procedure is of the first order	R<0.1750

The proposed algorithm was observed to be efficient regarding the utilisation of the computational resources, and to produce the values of maximal runup height in good agreement with the results obtained by more complicated methods. This algorithm was generalised for two-dimensional problems, and in the next sections we describe the generalised algorithm. Note that not all the other methods can be modified for the two-dimensional case.

3. Numerical Algorithm for a Basin with a Complex Shape

Finite-difference methods are frequently applied for two-dimensional problems of the long wave runup. Some of them use moving curvilinear grids while other carry computations on stationary rectangular grids. Each approach has its advantages and drawbacks. In the paper a finite-difference method on a rectangular grid in a fixed domain is consider. Having the advantage of simpler logical structure it can be used to handle arbitrary bottom topography and curvilinear coastal line. Being suitable for a wide variety of problems it looses in accuracy. The accuracy can be increased if the scheme has additional convergence properties. Theoretically an arbitrary accuracy level can be achieved at the expense of higher grid resolution.

The domain of a straight-through calculation method for tsunami wave runup problems usually contains both the water basin and the adjacent shore. The numerical domain is covered with a regular grid. The function reflecting the depth or elevation over the still water level is defined at each node. Construction of a numerical model requires special handling of the nodes corresponding to the shore. Several approaches to solve this problem had been proposed in the literature.

Method 1. Velocity and total depth are set to zero over the shore.

Method 2. A thin water layer kept from flowing down the slope by friction covers the shore.

Method 3. An artificial thin water layer of constant depth is maintained on the shore.

Despite the clear physical intuition behind and comprehensive coverage in the literature, the second method lacks mathematical justification. The third approach had been studied in one-dimensional case rather well. In the rest of the article the first method will mainly be considered. The advantage of this method is a clear physical interpretation and the mathematical motivation from the point of view of conservation laws.

Consider the system of conservation laws in the integral form, corresponding to the differential equation (1)

$$\int_{\delta\omega_t} \mathbf{W}dxdy + \mathbf{F}dtdy + \mathbf{G}dtdx + \int_{\omega_t} \mathbf{Q}dtdxdy = 0$$

$$(3)$$

in a rectangular domain Ω on the Oxy plane, which corresponds to an image of the basin with the adjacent shore under orthogonal projection on the horizontal plane. The domain Ω does not change in time and the process is restricted to the volume $Q_t = \Omega \times [0, t]$, however a part of Ω covered by water is variable and is divided from the shore by the moving shoreline.

The total water depth $H = h + \eta$ can be used to distinguish between the water area and the shore. The function $H=H(x,y,t)$ is nonnegative and the equation $H(x, y, t) = 0$ defines the moving shoreline. The function can be extended over the whole domain Ω by setting it to zero on the shore. The same symbol will be used to denote a function thus extended. The decomposition of the domain into the water area $\Omega_t^{(1)} = \{(x,y) : H(x,y,t) > 0\}$ and the shore area $\Omega_t^{(2)} = \{(x,y) : H(x,y,t) = 0\}$ can be expressed as $\Omega = \Omega_t^{(1)} \bigcup \Omega_t^{(2)}$. Velocity components u and v can also be set to zero on the shore. Finally define a nonpositive function $\tilde{h}(x,y)$ in the region $\Omega_t^{(2)}$ as the height of the surface over the still water level taken with the opposite sign. This function can be considered as a continuation of the still water depth function so $h(x,y)$ can also be considered defined in the entire domain. We assume the curves $h(x,y) = 0$ and $\tilde{h}(x,y) = 0$ to coincide.

Now all the functions appearing in the integral expression of (3) are defined in the entire domain Ω. The equation holds in the cylinder Q_t: it trivially holds in $\Omega_t^{(2)}$ as the expression under the integral sign is zero.

Thus the problem stated in a variable domain $\Omega_t^{(1)}$ is reduced to a rectangular domain Ω. Numerical solution can be obtained by a finite-difference method over the entire numerical domain, so that the shoreline will be computed automatically.

To construct the numerical algorithm we approximate the integral (3) in the domain $\Omega_0 \times [n\Delta t, (n+1)\Delta t]$ over the unit rectangle Ω_0 with sides $2 \times \Delta x$ and $2 \times \Delta y$ by means of the finite-difference scheme of the MacCormack type:

$$\left(\tilde{\mathbf{W}}_{mk}^{n+1} - \mathbf{W}_{mk}^{n}\right)\Delta x \Delta y + \left(\mathbf{F}_{mk}^{n} - \mathbf{F}_{m-1k}^{n}\right)\Delta t \Delta y + \left(\mathbf{G}_{mk}^{n} - \mathbf{G}_{mk-1}^{n}\right)\Delta t \Delta x = \mathbf{Q}_{mk}^{n}\Delta t \Delta x \Delta y,$$

$$\left(\mathbf{W}_{mk}^{n+1} - \frac{1}{2}\left(\tilde{\mathbf{W}}_{mk}^{n+1} + \mathbf{W}_{mk}^{n}\right)\right)\Delta x \Delta y + \frac{1}{2}\left(\left(\tilde{\mathbf{F}}_{m+1k}^{n+1} - \tilde{\mathbf{F}}_{mk}^{n+1}\right)\Delta t \Delta y + \left(\tilde{\mathbf{G}}_{mk+1}^{n+1} - \tilde{\mathbf{G}}_{mk}^{n+1}\right)\Delta t \Delta x\right) \tag{4}$$

$$-\frac{1}{2}\tilde{\mathbf{Q}}_{mk}^{n+1}\Delta t \Delta x \Delta y = 0,$$

$$\mathbf{Q}_{mk}^{n} = \begin{pmatrix} 0 \\ g\dfrac{H_{mk}^{n} + H_{m-1k}^{n}}{2} \cdot \dfrac{h_{mk} - h_{m-1k}}{\Delta x} \\ g\dfrac{H_{mk}^{n} + H_{mk-1}^{n}}{2} \cdot \dfrac{h_{mk} - h_{mk-1}}{\Delta y} \end{pmatrix}, \quad \tilde{\mathbf{Q}}_{mk}^{n+1} = \begin{pmatrix} 0 \\ g\dfrac{\tilde{H}_{m+1k}^{n+1} + \tilde{H}_{mk}^{n+1}}{2} \cdot \dfrac{h_{m+1k} - h_{mk}}{\Delta x} \\ g\dfrac{\tilde{H}_{mk+1}^{n+1} + \tilde{H}_{mk}^{n+1}}{2} \cdot \dfrac{h_{m+1k} - h_{mk}}{\Delta y} \end{pmatrix},$$

$\tilde{\mathbf{F}}_{mk}^{n+1} = \mathbf{F}\left(\tilde{\mathbf{W}}_{mk}^{n+1}\right)$, $\tilde{\mathbf{G}}_{mk}^{n+1} = \mathbf{G}\left(\tilde{\mathbf{W}}_{mk}^{n+1}\right)$. Here $\Delta t, \Delta x, \Delta y$ are the grid steps. As the values H_{mk}^{n}, $(Hu)_{mk}^{n}$, $(Hv)_{mk}^{n}$ are determined using the difference scheme above, velocities calculated at small values of H tend to be inaccurate. We pick a relatively small value H_{min} of the same order of magnitude as the irregularities of the bottom. The regularization procedure is applied so that a node (m, k) with $H_{mk}^{n} < H_{min}$ is considered dry, so that the constraints $H_{mk}^{n} = 0$, $u_{mk}^{n} = 0$, $v_{mk}^{n} = 0$ hold. The error arising from this correction is estimated by comparing the results of computations for a decreasing sequence of H_{min}: $H_{min} = 10^{-3}, 10^{-4}$…(in the dimensionless formulation).

The algorithm consists of the following steps:

1. In all internal nodes of the grid over the domain Ω, the values of H^n_{mk}, $(Hu)^n_{mk}$, $(Hv)^n_{mk}$ are calculated by the finite-difference scheme.

2. The condition $H^{n+1}_{mk} > H_{min}$ is being checked in each node; and if satisfied, the velocities are updated $u^{n+1}_{mk} = (Hu)^{n+1}_{mk} / H^{n+1}_{mk}$, $v^{n+1}_{mk} = (Hv)^{n+1}_{mk} / H^{n+1}_{mk}$, otherwise $H^{n+1}_{mk} = 0, u^{n+1}_{mk} = 0, v^{n+1}_{mk} = 0$.

3. The shoreline is determined as the boundary of the set of all nodes with $H^{n+1}_{mk} > 0$.

The algorithm is easy to implement. There are questions on both the accuracy of this method in general (with respect to the convergence to the solution of the approximated differential problem) and on the precision of the estimation of particular values, such as the values of maximal runup and average velocity of water approaching the shoreline. A number of test problems were considered for a solitary wave approaching the shore. The convergence of the algorithm was studied by considering a refining sequence of grids.

In order to determine the value of the maximal runup the highest position of the wave front on the beach should be obtained. In the current notation it is equivalent to computation of the boundary of a domain of all points which were covered by water at a moment $t > 0$, so that the full depth H was strictly positive.

Numerical experiments and theoretical analysis of the difference scheme show that when the grid refinement is applied the algorithm attains good estimation of the maximal runup for the whole range of shore slopes. However in practical computations the necessary number of points over the length of the slope is not always available if the slope is steep. A special extrapolation procedure had been developed to correct a position of a shoreline where the bottom surface has sharp gradients. The proposed technique is easy to implement and the question of its applicability should be decided depending on the required accuracy.

4. Testing the Numerical Algorithm

One of the most typical tasks of wave hydrodynamics is the problem of simulating a wave regime, which results from the interaction of an incoming wave with an island. This is to describe the transformation of the wave in the coastal zone, compute the maximal runup. To test the program against intended to solve such tasks, it is necessary to find a rather complete set of experimental data together with the detailed description of test basin, the initial data, and the results of observations.

Proceedings of International workshop on runup models [3] are an established source of empirical information for wave hydrodynamics. The book contains the four test problems that were presented to the scientific community for comparing and verification of mathematical models of long ocean waves. The first demonstration of the algorithm and its implementation we describe, is the solution of a test problem of a solitary wave running up a conical island [3].

PROBLEM 1

The wave tank of the Waterways Experiment Station of the U.S. Army Corps of Engineers is a rectangular basin with a flat bottom of width $L_y = 30$ m and of length $L_x = 25$m with a conical island in the center.

Let the Oy axis be parallel to the surface of the wave generator, and let the Ox axis be directed to the island at the still water level. The origin of the coordinate system coincides with the edge of the wave generator (*Figure 2*).

The center of the island is located at the point $x_0 = 13$m, $y_0 = 15$m. We will consider a planar solitary wave with the amplitude to the water depth ratio being equal to 0.1.

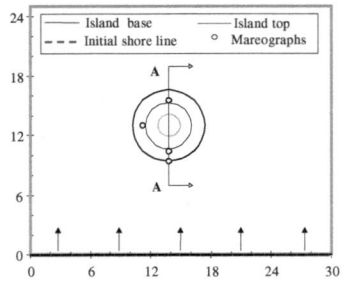 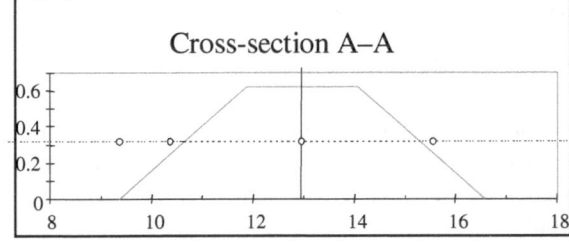

Figure 2. Scheme of water area: island contours and gages location (top view)

Figure 3. Scheme of island contours, shore line and gages locations ( ) (side view).

The experimental data and the results of our computations are presented as water level charts recorded by gages at four points inside the basin (

Figure *3*). The first gage was set in front of the island ($x_1=9.4$, $y_1=15.0$, local depth $D_1=0.32$m). The second gage was set near the still water line on the island ($x_2=10.4$, $y_2=15.0$, $D_2=0.082$m). The third point was on the left side of the island ($x_3=13.0$, $y_3=12.42$, $D_3=0.079$m). The last chart is from a gage on the rear side of the island ($x_4=15.6$, $y_4=15.0$, $D_4=0.083$m). The maximal runup height along the island was also recorded.

The external boundaries of the computational domain are the rectangle sides of the basin. On one of the sides a non-stationary boundary condition is specified to simulate the incoming wave. Three other boundaries are simulated as "transparent boundaries", transmitting the solution outwards without reflection. The purpose of the numerical experiment is to describe the evolution of the shoreline and to estimate the maximal runup height over the island boundary and at the rear side of the island in particular.

Figure 4 presents the plane solitary wave propagates over a flat bottom and the process of an interaction with the island. From $t=1.2$s to $t=5$s the wave had passed a distance equal to an effective wavelength. The front becoming steeper is characteristic to the equations of hyperbolic type.

Results of the numerical solution are shown on the figures below. The most typical features of the wave evolution in the process of its interaction with the island can be seen. Figure 4 presents the snapshots of free-surface profiles at different moments of time. The solitary wave passes along the model island with an increase in amplitude in the nearest vicinity, and forms a system of circular waves behind. The circular waves then leave the domain through the transparent border.

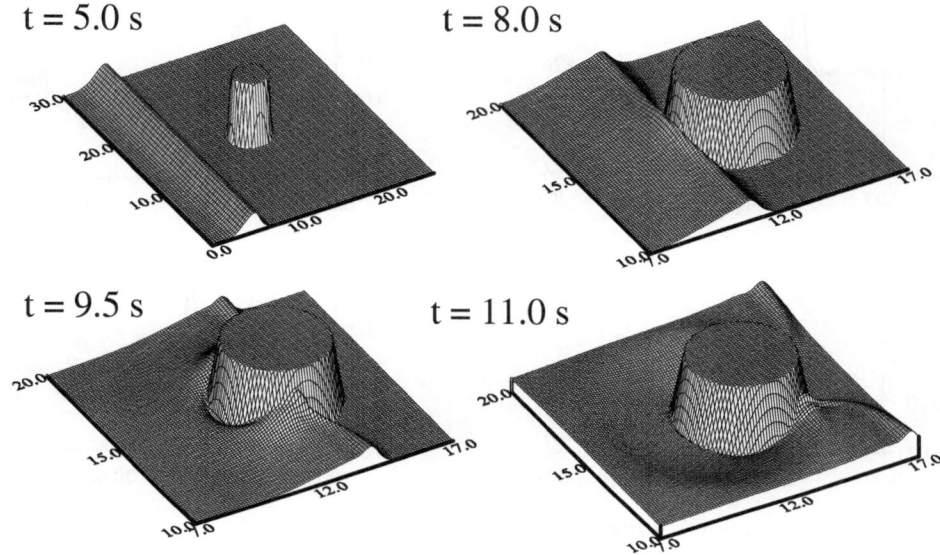

Figure 4. Solitary wave runup on a conical island. Wave profiles at different time.

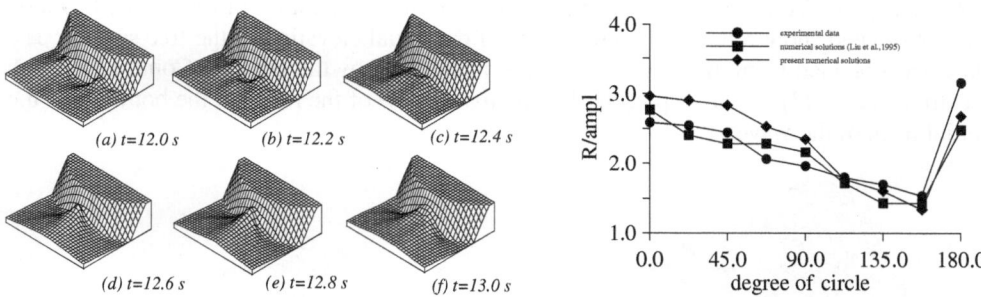

Figure 5. The wave patterns at the lee side of the island.

Figure 6. The runup heights around the island.

The wave pattern at the lee part of the island is of particular interest.

Figure 5 shows the waves passing the island to interact and to form a splash with the runup height exceeding the one at the front. Numerical estimation of the runup height behind the island is less than the experimental values, which could be explained by the effects of artificial dissipation introduced on the numerical stage (see Figure 6).

Comparison of the time charts (see Figure 7) shows the numerical solution to capture both the qualitative and quantitative features of the experimental data with acceptable accuracy. The transparent boundary conditions on the sides of a basin work almost perfectly without introducing any significant disturbance in the wave picture, while the absorbing boundaries of the experimental basin show a noticeable reflection.

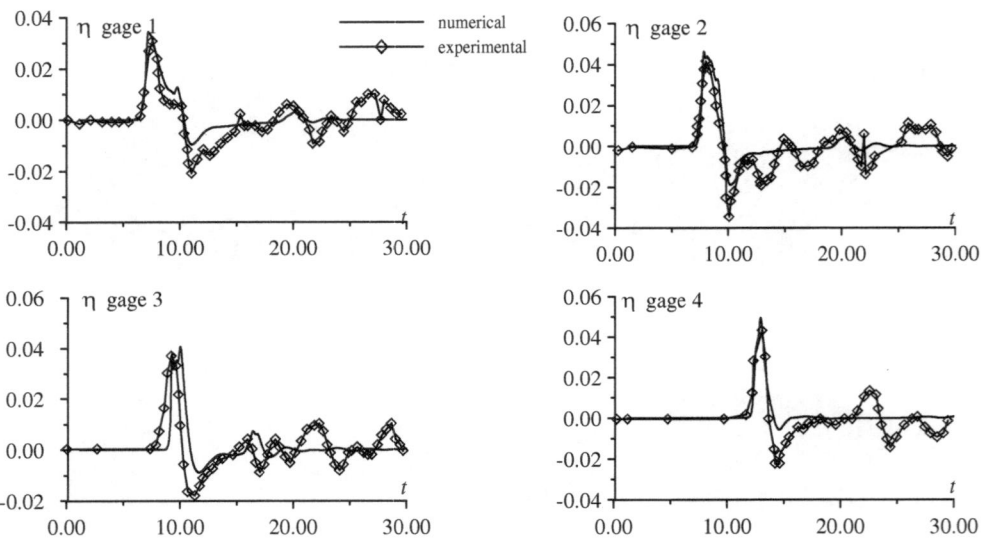

Figure 7. Water levels at the fixed points (see Figure 2).

PROBLEM 2

The second test problem is based on the real tsunami, which had occurred on July 12, 1993 in Japan. The problem had been chosen due to the availability of the detailed data on the bottom and shore relief, and on the shape of the initial elevation of the free surface. The data was obtained from the Hokusai database maintained by the Disaster Control Research Centre of Japan [3]. Figure 8 present the schematic view of the relief of the bottom and the initial form of the wave.

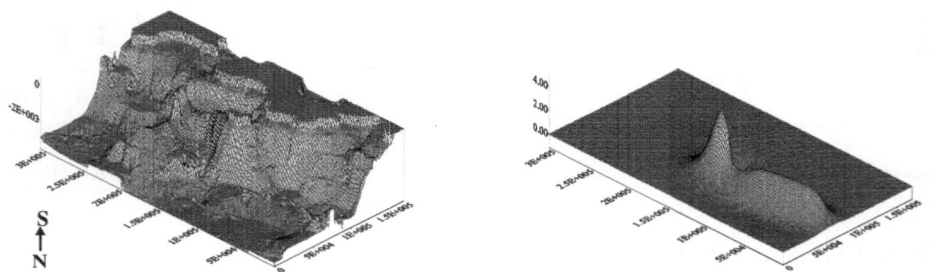

Figure 8. The bottom relief and the initial surface distribution.

The tsunami occurred near Hokkaido after a strong Nansei-oki earthquake with the magnitude of approximately 7.8 on July 12 at 22:17 NTP. At about 22:22, after the observatory in Sapporo had issued a tsunami warning message, the first wave approached the shore of the Okushiri Island. In five minutes the wave had reached the cities of Setana and Taisei on the eastern coast of Hokkaido.

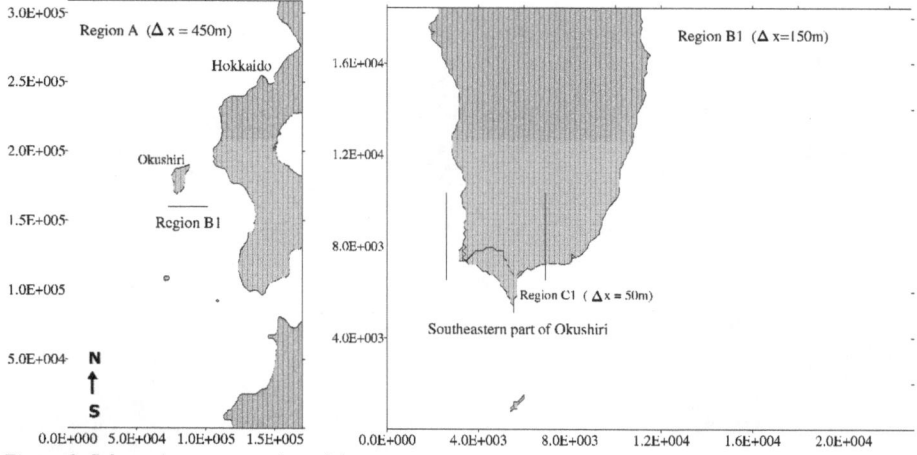

Figure 9. Schematic representation of the water area.

A sequence of three computational grids was used. A rough grid covered the entire domain (region A on the Figure 9) of the size 168.75 × 307.8km with a step corresponding to a distance of 450m. The grid was stored in an array of 379 × 687 points. Two other grids were constructed to capture the details of the interaction of the wave with the island. They cover the regions B1 and C1 on the Figure 9 with the steps corresponding to 150m and 50m distances respectively.

The Figure *10* presents the process of the wave propagation in the entire domain as computed over the coarse grid. On this level of detail the smaller islands around Okushiri can be safely treated as conical.

On the Figure11 the picture of flooding of the "tip" of the Cape Aonae (the southern coast of Okushiri) is shown. A high splash can be seen in the neighborhood of Hamatsumae located on the lee side of the island. Such degree of detail could only be obtained with the use of a grid with the smallest step of 50m.

The runup height distribution over the entire coast of the Okushiri Island as shown on the

Figure *12*, had been computed over the coarse grid (the domain A). The runup values on the coasts of Aonae and in the region of Hamatsumae obtained from the computations on the finest grid are shown on the Figure 13. The flood zone is depicted on the same picture.

214

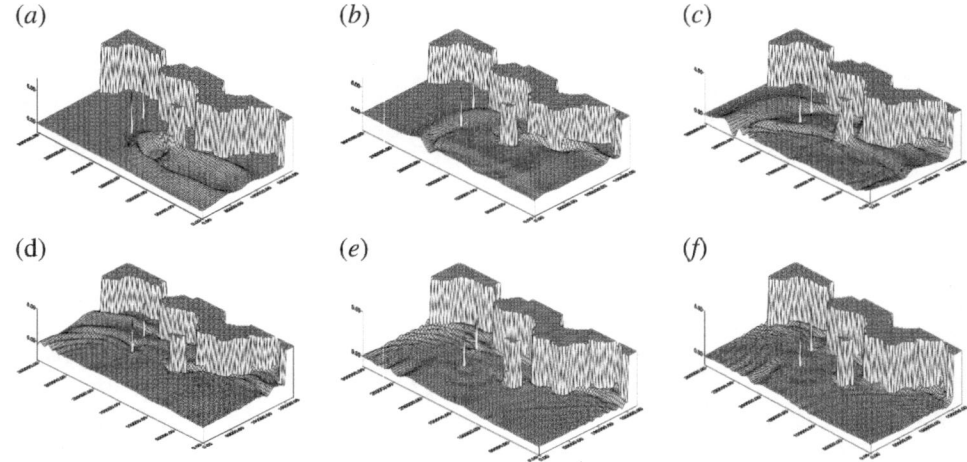

Figure 10. Wave pictures in the domain A at different moments of time.

Again the comparison of numerical results with the experimental data from the Hokusai database shows good correspondence between the two.

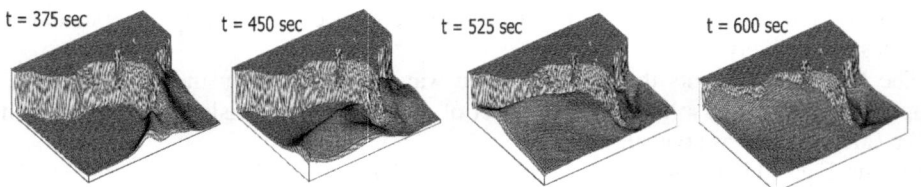

t = 375 sec   t = 450 sec   t = 525 sec   t = 600 sec

Figure11. Flooding of the Cape Aonae (the region C1). The last image shows the flooding of Hamatsumae district.

Comparing results of numerical calculations with data obtained on learning of "tracks" of real manifestation of a tsunami and of observations of the eyewitnesses (see [3]), we see rather good correlation of values of maximum runup. It is necessary to remark however, that the correct application of used here computing algorithms needs a finer grid.

5.Conclusion

We remark, that the algorithm has good potential for application to the problem of tsunami. The effect on the coastal structures can be predicted with good accuracy from estimated values of the runup height. The implementation can be extended incorporating linear dispersive and nonlinear dispersive models in addition to the nonlinear model used so far. Combining them in a single framework provides for the most accurate simulation of the natural phenomena.

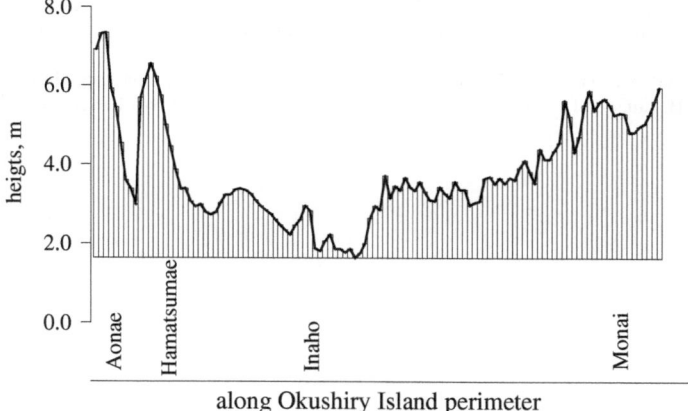

Figure 12. Maximum runup heights along the shore of Okushiri

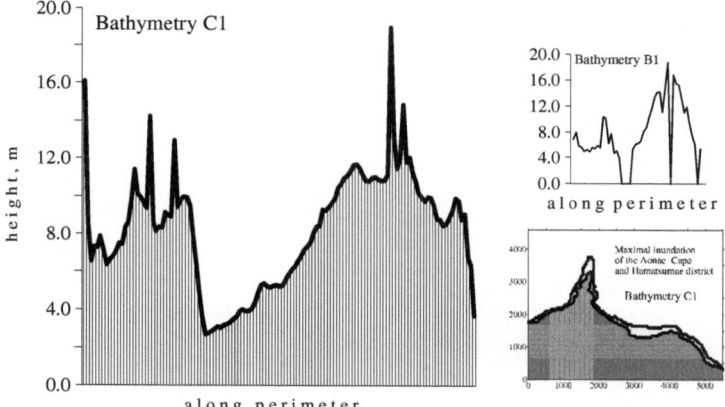

Figure 13. Maximum runup heights along the Aonae Cape.

The study has been supported by the Russian Basic Research Foundation (Grants N 00-01-00899, N 00-15-96172), Integration Program SB RAS (Grant N 1) and Federal Integration Program for Science and Education (Grant N 274).

Acknowledgement

The authors would like to express their gratitude to Dmitrii Chubarov for valuable assistance.

References

1. Kuzmicheva, T.V., Novikov, V.A., and Fedotova, Z.I. (1991) Comparison of several numerical methods for straight-through calculation of long wave runup modelling, in Yu.I. Shokin (ed.) *Proc. of the National Workshop on Numerical Methods in Wave Fluid Dynamics,* Krasnoyarsk, pp. 9-14 (in Russian)

2. Carrier, G.F. and Greenspan, H.P. (1958) Water waves of finite amplitude on a sloping beach, *J. Fluid Mech.* 4, 97-109.

3. Yeh H., Liu Ph., and Synolakis C. (*eds.*) (1996) *Long Wave Runup Models*, World Scientific Publishing, Singapore.

4. Briggs, M.J., Synolakis, C.E., Harkins, G.S., and Green, D.R. (1996) Runup of solitary waves on a circular island, in H. Yeh, P. H. Liu, and C. Synolakis (eds.), *Long Wave Runup Models*, World Scientific, Singapore, pp. 363-374.
5. Takahashi, T. (1996) Benchmark Problem 4. The 1993 Okushiri tsunami – data, conditions and phenomena, in H. Yeh, P. H. Liu, and C. Synolakis (eds.), *Long Wave Runup Models*, World Scientific, Singapore, pp. 384-403.

EVALUATION OF TSUNAMI HAZARD FOR THE SOUTHERN KAMCHATKA COAST USING HISTORICAL AND PALEOTSUNAMI DATA

V.M. KAISTRENKO[1] , T.K.PINEGINA[2] , and M.A.KLYACHKO [3]
[1] Institute of Marine Geology and Geophysics,
Nauki St., Yuzhno-Sakhalinsk, 693022 Russia.
[2] Institute of Geology and Geochemistry,
Petropavlovsk-Kamchatsky, Russia
[3] Center on Earthquake and Natural Disaster Reduction,
St.Petersburg - Petropavlovsk-Kamchatsky, Russia

Abstract. An example of solution of the problem of quantitative evaluation of the tsunami hazard is considered using tsunami data for the Kamchatka coast. This approach is based on the probability model of the Poisson type for tsunami process, with parameters calculated using geophysical data. The quality of the model used for tsunami risk estimation depends on the tsunami incidence frequency f and characteristic tsunami height H^*. The parameters f and H^* are not stationary and we refer them to their evolution as dispersion.

Three models using different tsunami data sets were considered: a homogeneous model based on a 50-year data set, a second model using the same set with additional 15-meter run-up data on Khalaktyrka coast in eastern Kamchatka caused by the 1841 tsunami, and a third model using the additional paleotsunami data from Khalaktyrka lake.

The comparison between them allows to conclude, that using the paleotsunami data for an elected point combined with the historical data makes the dispersion of the tsunami process parameters f and H^* much less.

1. Tsunamis on the Kamchatka coast

Tsunami on the Kamchatka coast are dangerous events. The catastrophic 1952 tsunami destroyed almost fully the Severo-Kurilsk city on the Paramushir Island and several settlements on the Southern Kamchatka coast. Since then, the run-up of two tsunamis was more than 7 meters, while the run-up of the 1969 tsunami reached 15 meter [1-6]. Hence quantitative evaluation of the tsunami risk on the Kamchatka coast is an important problem. Unfortunately, all the catalogues and articles [3-6] contain data only for period 250 years while only data sets related to 6 strong and moderate tsunamis occurring during the last 50 years are of sufficient quality. There are single data for other tsunamis. Tsunami data sets for the last 50 years are given in the Table.1 and incidence is shown in Figure 1.

A. C. Yalçıner, E. Pelinovsky, E. Okal ,C. E. Synolakis (eds.),
Submarine Landslides and Tsunamis 217-228

218

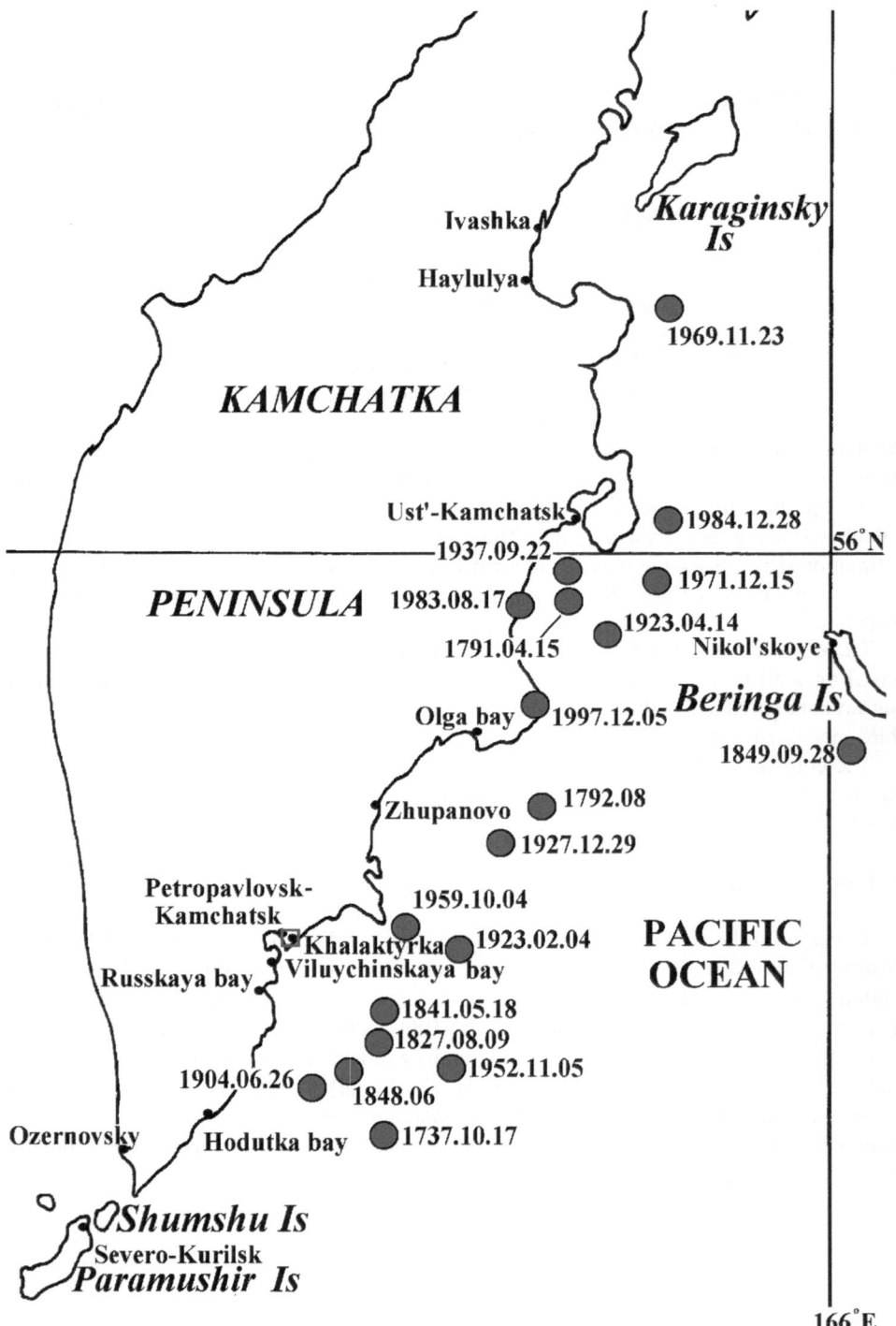

Figure 1. Tsunami sources distribution around Kamchatka

TABLE 1. Maximal tsunami run-up's for Kamchatka peninsula coast and the nearest Islands for the time period since 1952

Point	Tsunami data (year, month, day)												
	1952 11.05	1959 05.04	1960 05.04	1963 10.13	1964 03.27	1965 02.04	1969 11.23	1971 12.15	1973 02.28	1983 08.13	1984 12.28	1985 03.05	1997 12.05
Lavrova bay							2.0						
Ivashka							7.0						
Haylulya							7.0						
Ozernoy cape							5.0						
Olhovaya riv.							15.0						
Ozernaya bay							3.0						
Ozernoy riv.							7.0						
Ust'Kamchatsk (town)	0.1		0.55				0.2	0.45		0.02	0.02	0.03	
Ust'Kamch.(sea coast)	0.5		4.0					2.0			0.02		
Olga bay	13.0		4.0										1.0
Zhupanovo	5.0		4.0										0.5
Morzhovaya bay	8.0	2.0	7.0										
Shipunsky cape	12.0		3.0										
Nalychevo	7.0												
Khalaktyrka	5.0		2.0										
Bezymyanny cape	5.0												
Rokovaya bay	3.0												
Petropavlovsk.Kamch.	1.2	0.13	0.6	0.07	0.07	0.09	0.04	0.06				0.02	0.02
Tarya bay	3.0		1.0										
Mayachny bay	5.7		2.0										
Vilyuchinskaya bay	8.0		5.0										
Sarannaya bay	7.0												
Zhirovaya bay	8.0												
Russkaya bay	7.0		7.0										
Povorotny bay	10.0												
Asacha bay	7.0												
Hodutka bay			3.0										
Utashud Is.	8.6												
Lopatka cape (East)	9.5		2.0										
Lopatka cape (West)	5.0												
Ozernovsky	5.0												
SeveroKurilsk	15.0		4.7		0.8	0.1		0.1	0.76				
Medny Is.			1.0										
Nikolskoye	2.0		3.5				2.5						

220

2. Paleotsunami data at Khalaktyrka.

Because of bad situation with historical tsunami data geological deposits related to paleotsunami were investigated near Khalaktyrka beach. Two profiles in this region were investigated (Fig.2). 11 sections 309-320 were made along the profile PR-1 and 8 sections 301-308 were made along the profile PR-2. The sandy layers found in the sections were considered as tsunami traces [7-13]. Thickness of tsunami layers was 0.5 – 20 cm. and decreased along along the profile for big distance from the coast. Special tephrastratigraphy sections 2001 – 2007 were made near Khalaktyrka lake far from the coastal area.

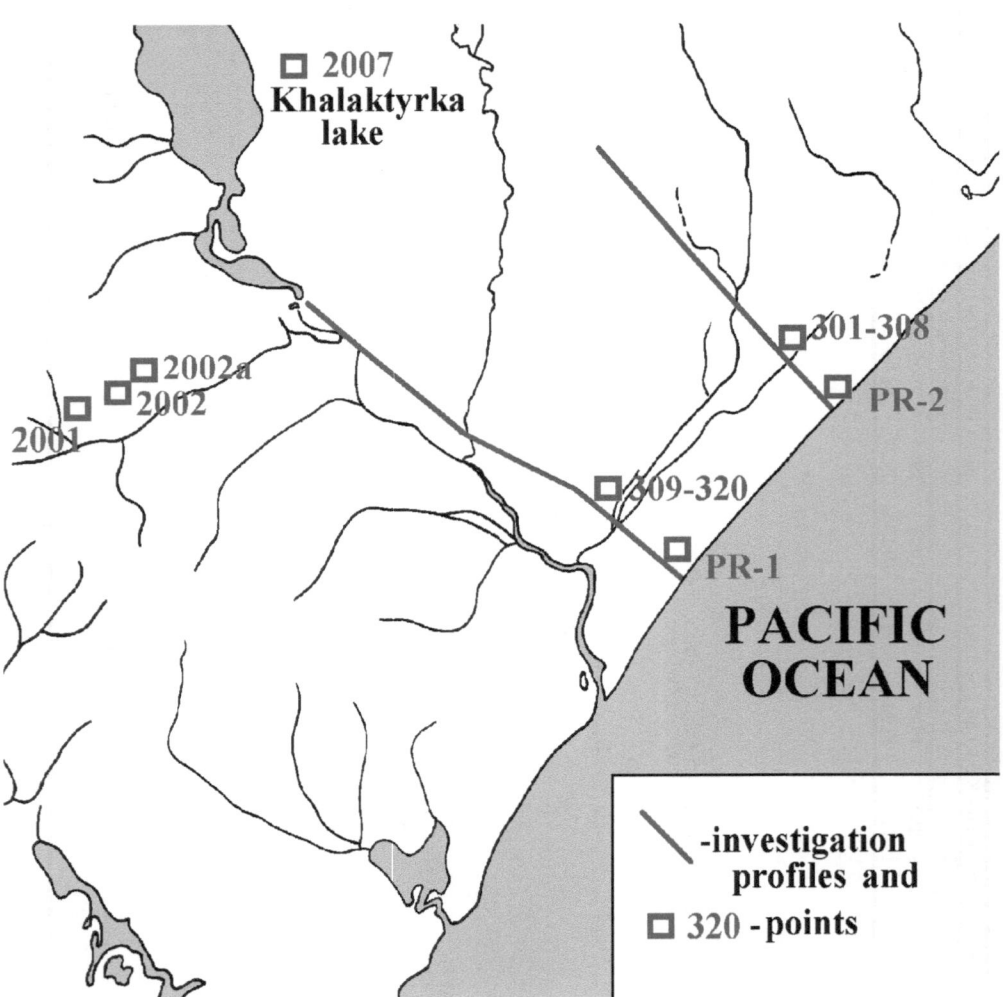

Figure 2. Investigation area near Khalaktyrka beach.

221

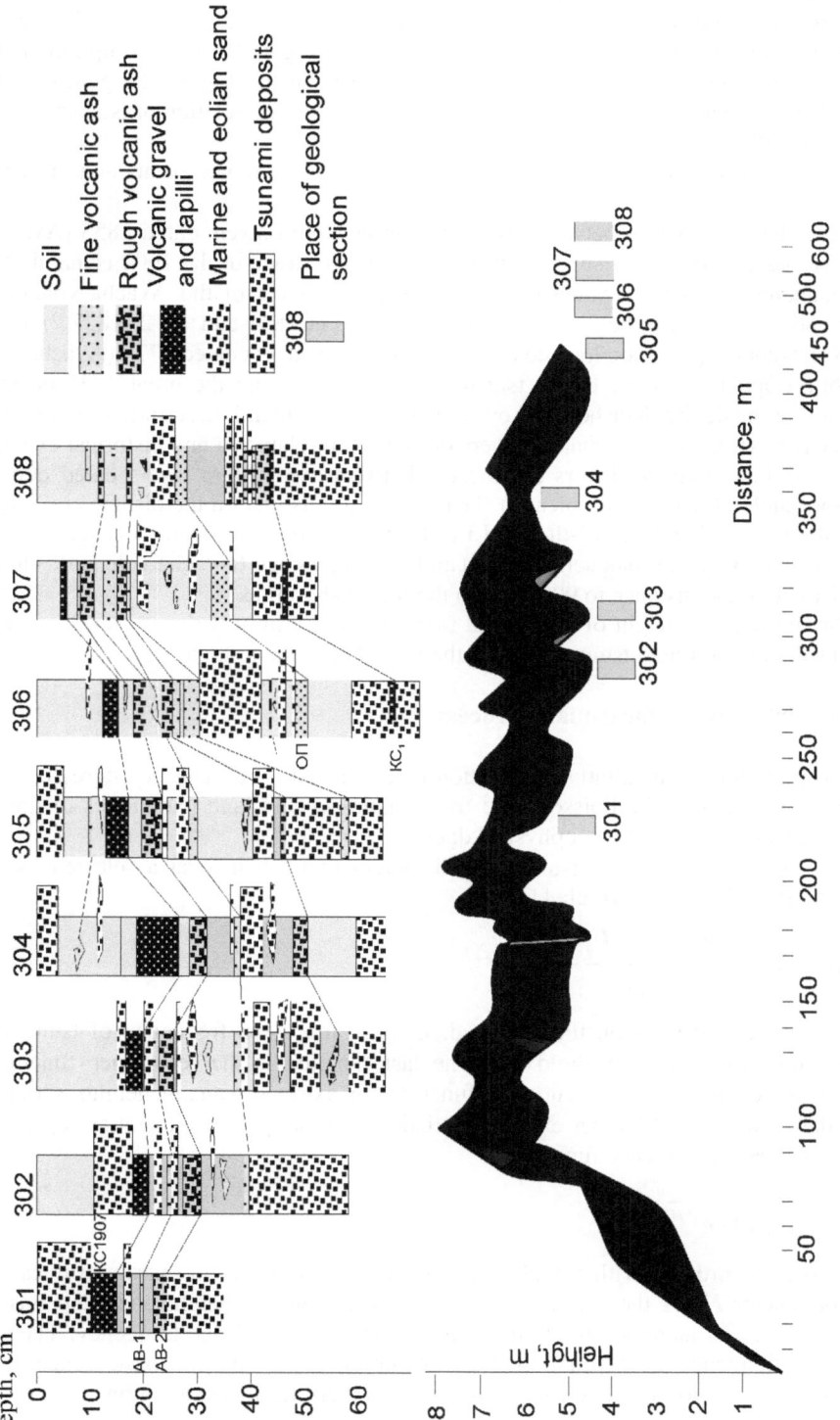

Figure 3. The section structures along the profile N 2

Several layers contained organic woody material used for ^{14}C dating. The age of the most tsunami deposit layers was estimated by tephrastratigraphy and tephrachronology methods using the tephra layers related to good investigated volcano eruptions [14]. The section structures along the profile PR-2 are shown on the Fig.3. The general synthesis related 13 tsunamis using all the geological sections after correlation of tsunami and tephra layers, as follows.

The two upper tsunami layers are related to the good known events at 1952 and 1959 (1960 ?).

According to position between two key-marker tephra layers dated 1855 (Avacha) and 1907 (Ksudach), the third tsunami layer should be the trace of the 1904 tsunami. 4-th and 5-th tsunami layers are located between two tephra layers of the Avacha volcano dated 1855 and 1779 and can be related to tsunami events at 1841 and 1792 (1827 ?). 6-th, 7-th and 8-th tsunami layers are located between two tephra layers dated 1779 (Avacha volcano) and 606 (Opala volcano). Upper tsunami trace can be from the great 1737 tsunami. Its description made by Krasheninnikov was the first tsunami description in the Russian history [6]. All the others tsunami layers do not seem related to any historical events. 9-th, 10-th and 11-th tsunami layers are located between two tephra layers dated 606 (Opala volcano) and 236 (Ksudach volcano). That layers are distinct on the distance to 700-750 m from the recent shoreline. 12-th and 13-th tsunami layers are located between two tephra layers dated 236 year (Ksudach volcano) and 3500 years ago (Avacha volcano). That layers are distinct on the distance to 900 m from the recent shoreline.

The most recent height of the coastal ridge is about 8 m over the sea level and we have to consider all the found tsunami having the wave height more than 8 m.

3. Probability model for tsunami process

The problem of quantitative evaluation of the tsunami risk can be addressed with the probability model of the Poisson type for tsunami process, and parameters of this model can be calculated using the geophysical data.

So, probability to have n tsunamis with height more than h_0 at a selected location is given by the following formula [15]:

$$P_n(h \geq h_0) = \frac{[\varphi(h_0) \cdot T]^n}{n!} \cdot e^{-\varphi(h_0) \cdot T}$$

(1)

where T is the observation time interval, $\varphi(h_0)$ is the mean frequency of tsunamis with height more then the "threshold" h_0. The last function is the recurrence function. The asymptotes of the tsunami recurrence function for extreme tsunami heights should be in accordance with good known extreme statistics suggesting on exponential approximation for an empirical recurrence function,

$$\varphi(h) = f(x) \cdot e^{-\frac{h}{H^*(x)}}$$

(2)

which is in accordance with a double negative exponential law for extreme values [16-17]. The parameter H^* is the calibrated (characteristic) tsunami height and depends on the coastal point x of tsunami observation, and f is tsunami incidence frequency. The tsunami incidence frequency is regional one varying very slowly along the coast. Hence it can be considered as a constant value for all points of Southern Kamchatka region.

In practice the functions $\varphi(h)$ are inferred from the tsunami catalogues, and they do not often contain sufficient data. The weighted least square method [3] allows the estimation of the tsunami activity parameters H^* and f. The average tsunami incidence frequencies related to the ordered tsunami heights $h_1 > h_2 > h_3 >.... h_k$ and their dispersions are presented by formulae in [15], as follows

$$\overline{\ln \varphi(h_k)} = \sum_{s=1}^{k-1} \frac{1}{s} - 0.577 - \ln T \tag{3}$$

$$\sigma(\ln \varphi(h_k)) = \sqrt{\frac{\pi^2}{6} - \sum_{s=1}^{k-1} \frac{1}{s^2}} \tag{4}$$

Clearly, Eq. 4 shows that the dispersion of logarithm of tsunami recurrence function is decreasing function of the number k of ordered tsunami run-up and incidence related to the maximal tsunami height h_1. The weighted least square method can be used to estimate the parameters H^* (for many points) and f (a single value for all region) of the empirical recurrence function with its dispersions (a priori errors) [18].

Only 6 strong and moderate tsunamis occurred during the last 50 years have been well described in catalogues (see Table 1). There are single data for other tsunamis.

The frequency f of strong tsunamis should be a stable regional parameter, and its a priori error can be considered as a general indicator of data quality. The standard deviation σ of the parameter $\ln(f)$ can be considered with a relative a priori error $\sigma(\ln(f)) \approx \frac{\sigma(f)}{f}$ for the tsunami incidence frequency.

4. Historical and paleotsunami data analysis.

The homogeneous probability model (1) for the Southern Kamchatka coast based on the 50-year data sets (see Table 1) gives the following parameters: $f=0.07$ 1/y, $\sigma(\ln(f))=0.2$ (regional) and calibrated tsunami heights H^* and its standard deviations $H^*=0.7$ m, $\sigma(1/H^*)H^*=0.6$ for Petropavlovsk-Kamchatsky and $H^*=2.8$ m, $\sigma(1/H^*)H^*=0.6$, for Khalaktyrka, where σ is standard deviation.

The product $\sigma(1/H^*) \cdot H^* \approx \frac{\sigma(H^*)}{H^*}$ can be considered as a relative a priori error of parameters H^* and its values (see Table.2) shows that having historical data do not allow calculation of needed parameters with good accuracy.

The second model uses the same set but adds the 15-meter run-up measurement on the Khalaktyrka coast caused by the 1841 tsunami. The second model gives following parameters: $f=0.07$ 1/y, $\sigma(\ln(f))=0.2$ and $H^*=0.8$ m, $\sigma(1/H^*)H^*=0.6$ (Petropavlovsk) and $H^*=4.5$ m, $\sigma(1/H^*)H^*=0.4$ (Khalaktyrka). So, using a single addition of historical measurement for Khalaktyrka changes the H^* for this point in the frames ($\pm\sigma$) of the model 1 and makes the relative error less for the same point only, while it does not change the parameters for all the others points.

To improve the model, in Autumn 2000, the traces of paleotsunami on Khalaktyrka coast were investigated. During this expedition, 13 tsunami deposits were found on the 8-meter terrace.

The model 3 uses the initial 50-year data sets and the additional 1841 tsunami date and paleotsunami data set for Khalaktyrka, results into the following parameters: $f=0.07$ 1/y,

224

$\sigma(\ln(f))=0.2$ and $H^*=0.8$ m, $\sigma(1/H^*)H^*=0.6$ (Petropavlovsk) and $H^* = 4.2$ m, $\sigma(1/H^*)H^*=0.15$ (Khalaktyrka). Using a good additional set of paleotsunami data for Khalaktyrka changes the H^* for this point from in the Model two and makes the relative error much less for the same point only while not changing the other parameters. Corresponding results are shown on the Figure 4 and Table 2.

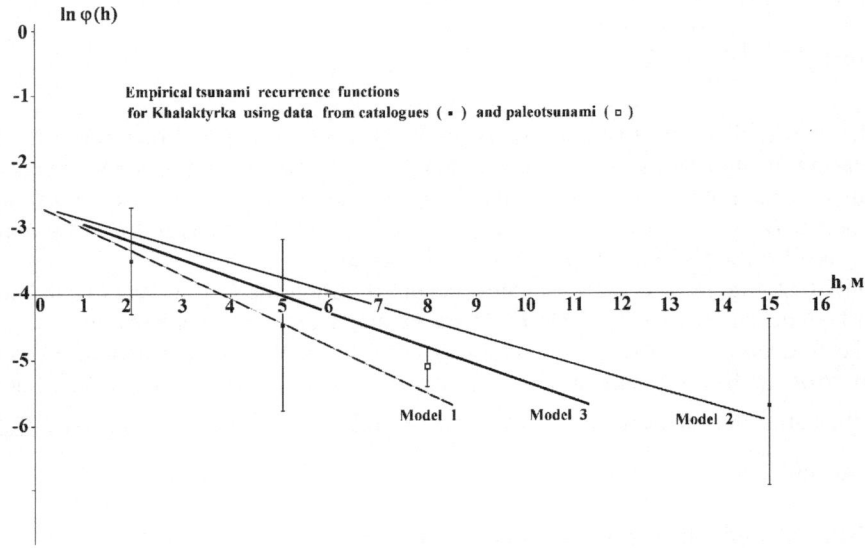

Figure 4. Empirical tsunami recurrence function for Khalaktyrka for three different models.

The most essential details on this picture are standard deviations related to different kinds of tsunami data. Tsunami heights 5 m (1952) and 15 m are (1841) are maximal values for corresponding periods and have maximal standard deviations. This fact explains the advantage of paleotsunami data. Tsunamis with a great run-up are rare events but many tsunami traces can be found for the long pre-historic period.

TABLE 2. Tsunami activity parameters f and H^* of three probability models for the Kamchatka coast and the nearest Islands.

Points	Model 1 Based on historical data for period 1952-2000 $\ln(f)=-2.65, f=0.07$ 1/year, $\sigma(\ln(f))=0.2$		Model 2 Based on historical data for period 1952-2000 with 15-meter run-up for Khalaktyrka (1841) $\ln(f)=-2.68, f=0.07$ 1/year, $\sigma(\ln(f))=0.2$		Model 3 Based on historical data for period 1952-2000 with 15-meter run-up for Khalaktyrka (1841) and paleotsunami data $\ln(f)=-2.68, f=0.07$ 1/year, $\sigma(\ln(f))=0.2$	
	H^*, m	$\sigma (H^*)/H^*$	H^*, m	$\sigma (H^*)/H^*$	H^*, m	$\sigma (H^*)/H^*$
Ust'Kamchatsk (town)	0.5	0.7	0.52	0.73	0.53	0.73
Ust'Kamch. (sea coast)	2.35	0.61	2.41	0.62	2.42	0.63
Olga bay	6.67	0.61	6.83	0.62	6.85	0.63
Zhupanovo	3.86	0.64	3.96	0.65	3.98	0.66
Morzhovaya bay	6.79	0.65	7.00	0.67	7.03	0.67
Shipunsky cape	6.14	0.63	6.26	0.64	6.28	0.64
Nalychevo	3.95	0.73	4.01	0.74	4.02	0.75
Khalaktyrka	**2.75**	**0.62**	**4.52**	**0.37**	**4.17**	**0.16**
Bezymyanny cape	2.82	0.73	2.86	0.74	2.87	0.74
Rokovaya bay	1.69	0.73	1.71	0.74	1.72	0.74
Petropavlovsk-Kamch.	0.72	0.63	0.75	0.65	0.75	0.65
Tarya bay	1.58	0.62	1.61	0.63	1.62	0.63
Mayachny bay	3.04	0.62	3.10	0.63	3.11	0.63
Vilyuchinskaya bay	5.32	0.63	5.44	0.64	5.46	0.64
Sarannaya bay	3.95	0.73	4.01	0.74	4.02	0.75
Zhirovaya bay	4.51	0.73	4.58	0.74	4.59	0.74
Russkaya bay	6.64	0.67	6.81	0.69	6.83	0.69
Povorotny cape	5.64	0.73	5.73	0.74	5.74	0.75
Asacha bay	3.95	0.73	4.01	0.74	4.02	0.75
Hodutka bay	3.87	1.07	4.01	1.11	4.03	1.11
Utashud Is.	4.85	0.73	4.93	0.74	4.94	0.75
Lopatka cape (East)	4.84	0.64	4.93	0.65	4.94	0.65
Lopatka cape (West)	2.82	0.73	2.86	0.74	2.87	0.74
Ozernovsky	2.82	0.73	2.86	0.74	2.87	0.74
Severo-Kurilsk	7.87	0.63	8.07	0.64	8.11	0.64
Medny Is.	1.3	1.07	1.34	1.11	1.35	1.12
Nikolskoye	3.35	0.7	3.47	0.72	3.49	0.73

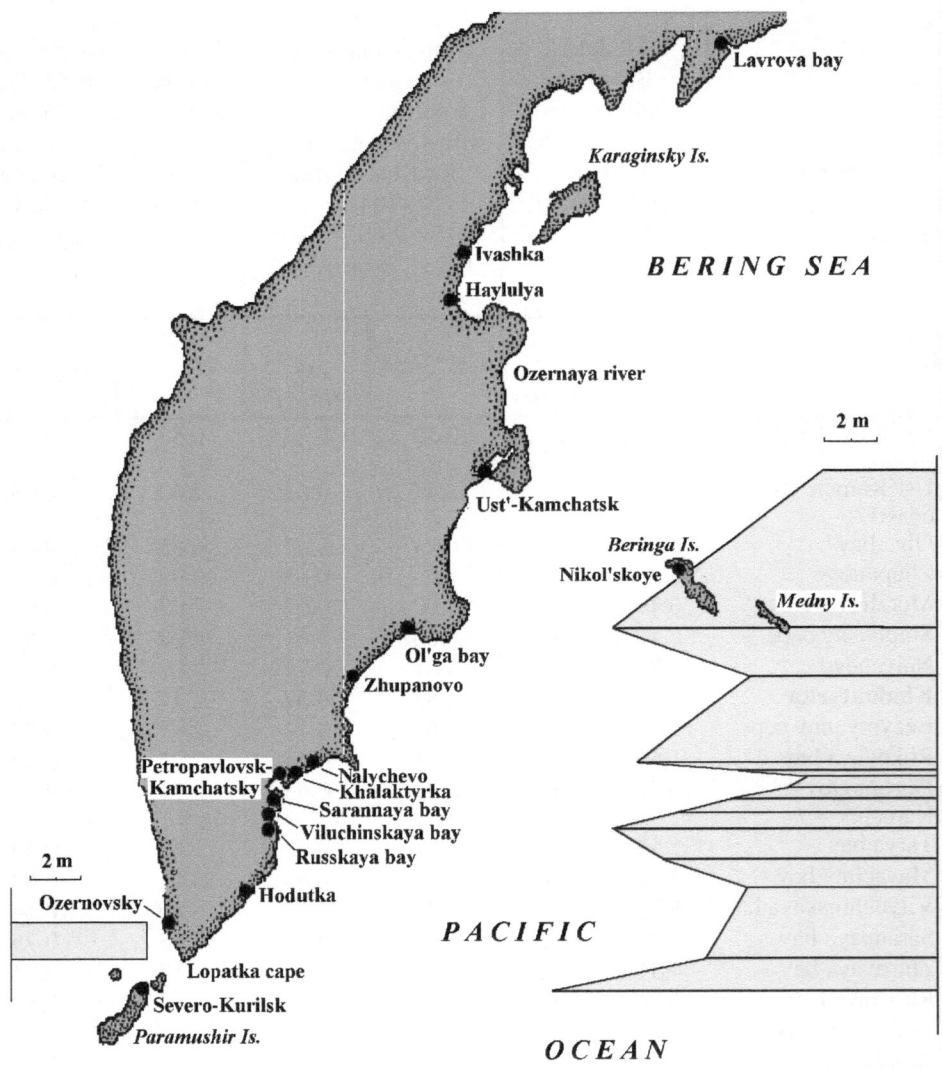

Figure 5. The observing tsunami h_{100}-zoning scheme for the Southern Kamchatka coast and the nearest Islands

Having the tsunami activity parameters f and H^*, we can estimate the tsunami hazard for any selected locale. For example the average maximal tsunami height with recurrence period t can be calculated by formula [15]:

$$h_t = H^* \cdot \ln(f \cdot t)$$

(5).

Observed tsunami zoning scheme for the Southern Kamchatka coast can be created using Model three as given in the Table 3. The results are shown in the Figure 5 for average maximal tsunami height with recurrence period of 100 years.

5. Conclusions.

Historical and paleotsunami data for the Kamchatka coast were analysed and applied to estimating tsunami hazards using a probabilistic model.

The quality of probability model used for tsunami risk estimation depends on the quality and quantity of geophysical data used for its creation. Using the good set of paleotsunami data for any selected point makes the relative error of the tsunami activity parameters less compared to calculations using just data available in catalogue

Acknowledgement

The work was supported by the Complex Federal Program 057 "Safety for Kamchatka".

References.

1. Kaistrenko V.M. and Sedaeva V.M. (2001) 1952 North Kuril tsunami: new data from archives, in G.T.Hebenstreit (ed.), *Tsunami Research at the End of a Critical Decade*, Kluwer Academic Publishers, Netherlands, pp.91-102.

2. Soloviev S.L., Go Ch.N. (1974) *Catalogue of tsunami on the Western Pacific coast.* Moscow, NAUKA, (in Russian)

3. Soloviev S.L. (1978) General description of tsunamis on the Pacific coast of USSR in 1737-1976, in E.F.Savarensky and S.L. Soloviev (eds.), *Tsunami investigation in the open sea*, Moscow, NAUKA, (in Russian).

4. Soloviev S.L., Go Ch.N., Kim Kh.S. (1986) *Catalogue of tsunami in the Pacific Ocean. 1969-1982.* Moscow, Soviet Geophysical Committee (in Russian).

5. Zayakin Yu.A. and Luchinina A.A. (1987) *Catalogue of tsunamis on Kamchatka –* Obninsk: VNIIGMI MTsD, (in Russian).

6. Zayakin Yu.A. and Pinegina T.K.(1998) Tsunami 5 December 1997 on Kamchatka, In book *Kronotskoye Earthquake 5 December 1997 on Kamchatka – precursors, peculiarities, consequences*, Petropavlovsk-Kamchatsky, (in Russian).

7   Minoura K., Nakaya S. (1983) Traces of tsunamis in marsh deposits of the Sendai Plain. North East Japan *Proc. Fourth Congr. Mar. Sci. Technology PACON.*- Tokyo; Japan, **1**, 141-144.

8. Minoura K., Nakaya S. (1991) Traces of tsunamis preserved in inter-tidal lacustrine and march deposits: some examples from northeast Japan, *Journal of Geology.* **99**, 265-287.

9. Bulgakov R.F., Ivanov I.I., Khramushin V.N., Pevzner M.M., Sulerzhitsky L.D. (1995) Paleotsunami investigations used for tsunami zoning, *Physics of the Earth*,.- **2**, 18-27.

10. Pinegina T.K., Melekestsev I.V., Braitseva O.A., Storcheus A.V. (1997) Paleotsunami traces an the Eastern Kamchatka coast, *Nature*, **4**, 102-107.

11. Pinegina T.K., Bourgeus J. (1998) Tsunami deposits and paleo-tsunami history on Peninsula Kamchatskiy (56°-57°N), Kamchatka region (Bering Sea), Russia: Preliminary report //AGU

12. Bourgeus J., Titov V., Pinegina T.K. (1999) New data about 1969 Tsunami on Bering Sea, Russia //AGU

13. Pinegina T.K., Bazanova L.I., Melekestsev I.V., Braitseva O.A., Storcheus A.V., Gusyakov V.K. (2000) Pre-historic tsunami on the Kronotsky bay coast, Volcanology and Seismology, **2**.

14. Braitseva O.A, Ponomareva V.V., Sulerzitsky L.D., Melekestsev I.V., Bailey J. (1997) Holocene Key-Marker Tephra Layers in Kamchatka, Russia, *Quaternary research* **47**, 125-139.

15. Kaistrenko V.M.(1989) Probability model for tsunami run-up... *Proceedings of the International Tsunami Symposium.* Novosibirsk, July 31 -August 10, 1989, 249-253.

16. Gumbel E. (1962) *Statistics of Extremes*, Columbia University Press, New York.

17. Galambos J. (1978) *The asymptotic theory of extreme order statistics.* John Wiley and Sons, New-York-Chichester-Brisbane-Toronto.

18. Hudson D.J. (1964) *Statistics.* Genova.

TSUNAMI HAZARDS ASSOCIATED WITH EXPLOSION-COLLAPSE PROCESSES OF A DOME COMPLEX ON MINOAN THERA

K. MINOURA[1], F. IMAMURA[2], U. KURAN[3], T. NAKAMURA[4],
G.A. PAPADOPOULOS[5], T. TAKAHASHI[6], and A.C. YALÇINER[7]

[1]*Institute of Geology and Paleontology, Faculty of Science, Tohoku University, Sendai 980-8578, Japan*

[2]*Disaster Control Research Center, Faculty of Technology, Tohoku University, Sendai 980-8579, Japan*

[3]*Disaster Affairs, Earthquake Research Department, Lodumlu, Ankara, Turkey*

[4]*Dating and Material Research Center, Nagoya University, Nagoya 464-0814, Japan*

[5]*Institute of Geodynamics, National Observatory of Athens, P.O. Box 20048, Athens, Greece*

[6]*Research Center for Disaster Reduction System, Kyoto University, Uji 611- 0011, Japan*

[7]*Coastal and Harbor Engineering Research Center, Civil Engineering Department, Middle East Technical University, 06531 Ankara, Turkey*

Abstract

The Minoan Thera eruption of the Bronze Age is the most significant Aegean explosive volcanism. The eruption resulted in caldera collapse. We studied the Minoan volcano-tectonic event from the viewpoint of sedimentology and hydraulics. On the Aegean Sea coast of Turkey and Crete we found traces of tsunamis. The tsunami layers are overlain by felsic tephra. The mineralogical characteristics of volcanic tephra and AMS radiocarbon dating on foraminiferal tests show that the tsunami layers are correlative to the Minoan event. The results of our numerical simulation on the tsunami show that a destructive train of waves caused by the caldera collapse reached the Aegean Sea shore within 2.5 hours and the great surf set up by arrival of waves washed the coast. We conclude that the Minoan tsunami hazards damaged the maritime economy in the coastal areas of Aegean islands and Turkey.

1. Introduction

The Hellenic arc is a terrane of extensive Quaternary volcanism, and one of the main centers of explosive eruptions is located on Thera (Santorini). Of all the historical eruptions of Thera, the paroxysmal volcanic explosion of late Minoan time (1600 ~ 1300 BC) was most violent and caused widespread damage. It is widely held among volcanologists that the Minoan Thera eruption, characterized by a sequence of four distinctive volcanic phases, started with strong Plinian activity and ended with collapse of the volcanic cone [1]. In the Plinian phase, the explosive eruption ejected huge amounts of volcanic aerosol and tephra into the atmosphere [2]. Westerly to northwesterly winds spread the ejecta over the Eastern Mediterranean region [3]. The phreatomagmatic eruption of the last phase led to the collapse of the Stronghyle (pre-eruption) volcano leading to formation of the present shape of the Santorini caldera Figure 1). In recent papers, it has been reported that the airborne tephra accumulated over Minoan settlements on Rhodes, Kos, and Crete [4]. [5] Archaeological arguments that the decline of the Minoan civilization was related to agricultural and/or social disruption caused by this volcanic event are, however, doubtful because there was no significant break in the life of Minoan Rhodes [6].

A. C. Yalçıner, E. Pelinovsky, E. Okal, C. E. Synolakis (eds.),
Submarine Landslides and Tsunamis 229-236.

230

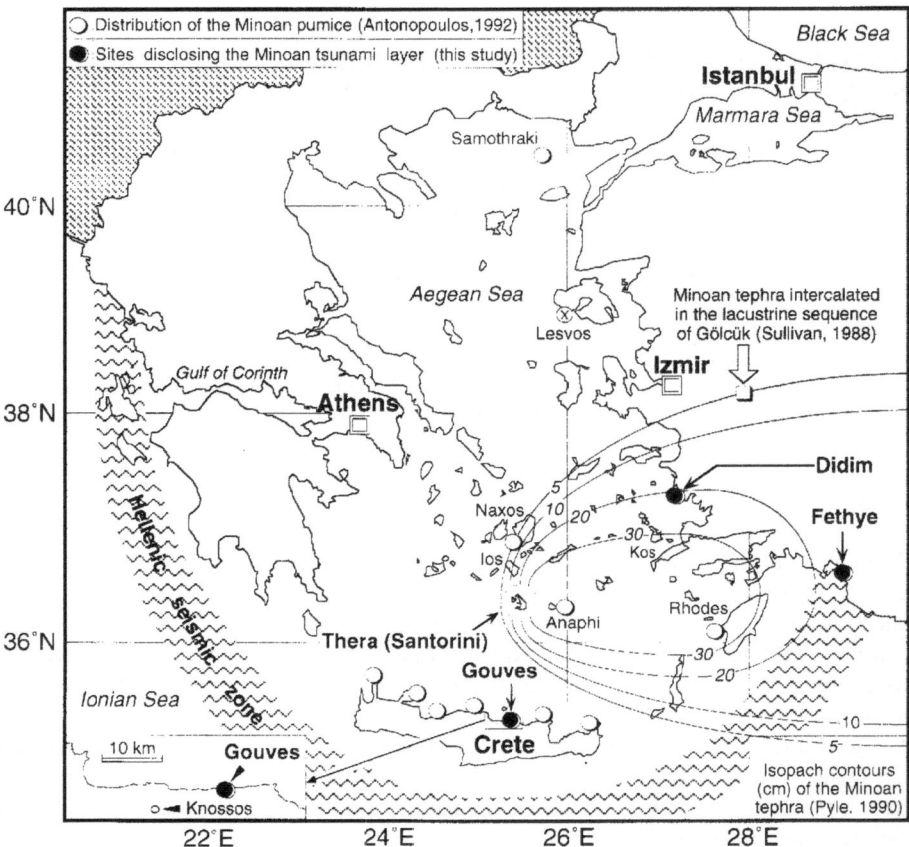

Figure 1. Map of the Aegean Sea and the adjacent region showing areas and sites mentioned in the text. Felsic volcanic products of the Minoan eruption are found on Aegean Sea coasts and in Eastern Mediterranean deep-sea cores. Tsunamigenic sediment layers were discovered in Didim and Fethiye (western Turkey) and Gouves (Crete).

At the archaeological site of Amnissos on Crete, Marinatos [7] found the floors of Minoan ruins to be covered by a thick layer of seaborne pumice from Thera, and he hypothesized that a destructive tsunami caused by the Thera eruption invaded the Aegean Sea coast of Crete. The Marinatos' theory has not been involved in archaeological debates, however, because of the lack of sufficient scientific evidence. The discovery of the Thera tephra in deep-sea sediment cores from the eastern Mediterranean [8] and in lacustrine sediments of western Turkey [2] has presented a new opportunity for estimating the effect of the eruption on human activities.

2. Numerical Simulation of Tsunami Occurrence

Heiken and McCoy [9] reconstructed explosion-collapse processes of the Stronghyle dome complex on Thera. They postulated that volcanic collapse and subsequent emergence of a submarine caldera resulted in the occurrence of tsunamis. Sudden collapse of the Stronghyle volcano probably generated large waves, which formed the Minoan tsunami. Inrush of seawater into the caldera and collision of water masses with the caldera wall could have oscillated sea level on a large scale, resulting in the generation of a tsunami. [10] has carried out the investigation on the generation and propagation of tsunamis in Aegean Sea by mathematical modeling. We carried out a numerical simulation of tsunami generation and propagation, employing TUNAMI N2 that incorporates the shallow-water theory consisting of nonlinear long-wave equations. Figure 2 is the computational result, which indicate that a train of waves with an elevation of 5 to 8 m reached the Aegean Sea shore within 1 to 2.5 h and the great surf set up by the arrival of waves attacked the coast of Crete and western Turkey. The highest water level (~21 m) in the model is found to the south of Ios, and it appears that the Minoan harbor of Amnissos was attacked by waves with an elevation exceeding 10 m.

3. Traces of Tsunami Invasion

To verify the values of tsunami runup, we used trenching to try to detect traces of tsunami invasion on the coast of Didim and Fethiye, western Turkey where human activities have not been carried out since the Hellenic period. On the trench wall of Didim, it was clear that the fine sand layer composed of carbonate grains underlies a 10-15-cm-thick yellowish white layer of felsic tephra (see Figure 4). No erosional contact of the sand layer with the overlying tephra implies that the fallout of airborne ash followed subsequently the deposition of marine materials. Silty mud rich in fossils of shallow-marine benthic foraminifera underlies the carbonate sand layer, and grades landward into nonmarine organic mud that is ideal for the preservation of fossil plant roots and their impressions. Within the sediment sequence of the Fethiye section, it was found that the sand layer consisting of siliciclastic grains is sharply covered by 5-10-cm-thick white felsic tephra. The sand layer, overlying the nonmarine sandy silt rich in plant debris, yields abundant shell fragments of shallow marine gastropods of indeterminate genus.

At the time of the excavation in the archaeological site of Gouves, located about 15 km to the east of Knossos, one of us (Papadopoulos) found that the floor of the late Minoan potter's workshop is covered by a thin veneer of carbonate sand and an overlying 10-20-cm-thick pumice layer. The carbonate sand, composed of unsorted grains of skeletal fragments, is marine in origin. The excavated site is situated 30 to 90 m inland from the Minoan harbor installation and 2 to 3 m high above present sea level, and it is interpreted that the sand layer was deposited during seawater flooding in late Minoan time.

232

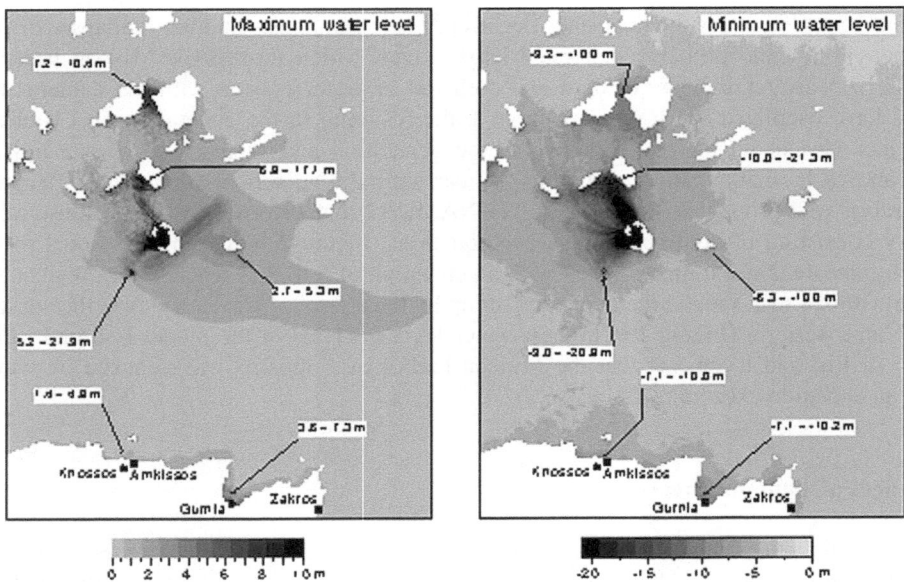

Figure 2. Results of numerical simulation, showing maximum (left) and minimum (right) water levels. First wave reached Aegean Sea coasts of western Turkey about 150 minutes after volcanic collapse. It is suggested that on the coast of Gurnia (Crete) maximum runup height was 6 to 11 m and inundation distance was 500 m at most.

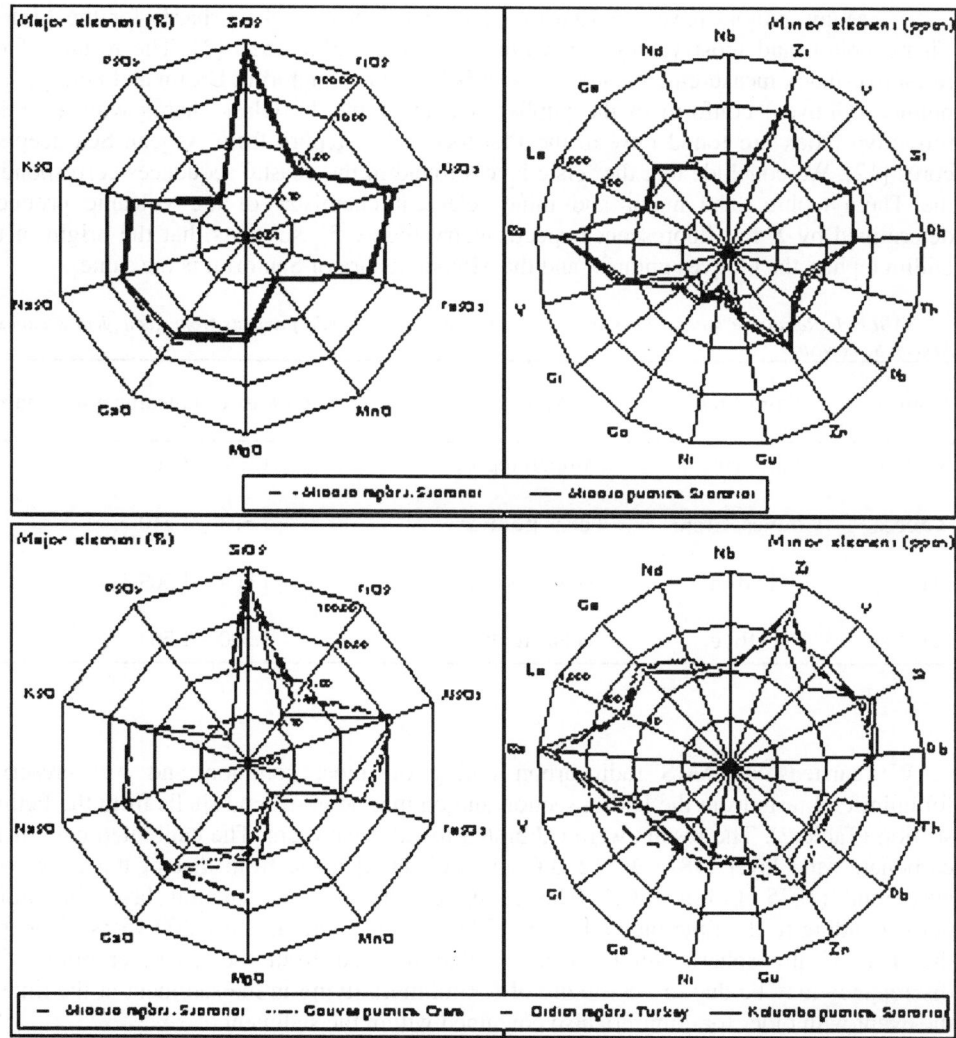

Figure 3. Comparison of analytical data of volcanic samples on circular diagrams. The upper two diagrams represent the chemical trend of major and minor elements of the Minoan effusives in Santorini, showing that the Minoan tephra has the very same origin as the Minoan pumice. The distribution patterns of major and minor element chemistry of volcanic samples from other geological sites are compared with that of the Minoan tephra in the lower two circular diagrams. The Didim tephra and the Gouves pumice have nearly the same pattern as the Minoan tephra. The geochemistry of the Kolumbo pumice originating from a submarine eruption of Kolumbo Volcano in 1650 AD is largely different from that of other volcanic samples.

The geological results show that waves penetrated into the coastal zone of Didim and formed fast-flowing currents associated with the rapid lateral translation of seawater and suspended sediments of offshore origin. Observation of storm surges on the coast of the Aegean Sea suggests that they are generally agents of erosion and do not produce regionally extensive deposits on land areas. Thus the unusual process of sediment transport together with the landward-tapering sedimentation of offshore materials is best explained by interpreting the carbonate sand layer to have been deposited by a tsunami [11].

Many authors have reported the refractive index of glass shards for the Minoan Thera tephra, and most values are within the range 1.506-1.510 [2]. The results of our refractive-index measurements on glass shards in tephra (Santorini, Didim and Fethiye) and pumice (Gouves) conform to the published data (Table 1). Glass shards with a similar refractive index are found only in the Pleistocene V-1 tephra from Aegean Sea deep-sea cores [12]. We conclude that the felsic layers found in the coastal sequences correspond to the Thera tephra. The major and minor element chemistry of the volcanic products, determined by X-ray fluorescence spectrometry Figure 3), suggests that the origin of the Didim tephra, the Gouves pumice, and the Minoan tephra of Santorini is the same.

TABLE 1. *Refractive index values of glass shards in felsic volcanic products from geological sections of Aegean Sea coasts.*

Country	Sampling	Material	Refractive index of volcanic glass shards
Greece	Santorini	Minoan tephra	1.504 - 1.509
Greece	Gouves, Crete	Felsic pumice	1.503 - 1.510
Turkey	Didim	Felsic tephra	1.504 - 1.509
Turkey	Fethye	Felsic tephra	1.500 - 1.504

We carried out AMS radiocarbon dating on most common and well-preserved foraminiferal tests from the Didim section and on marine gastropod shells from the Fethiye section (Table 2). The results were calibrated to calendar years. The calibrated date of the carbonate sand layer (1818 ± 112 B.C.) is indistinguishable from that of the underlying silty mud (1875 ± 116 B.C.). The calibrated date of the tsunami layer in Didim, corresponding to the time interval 1930-1706 B.C., is approximately 200 years older than the date of atmospheric acidity increase that resulted from the Thera eruption. This discrepancy may be due to the mixing of foraminifera living in the sediment at the time of the event with older foraminifera also contained within the sediment.

TABLE 2. *AMS radiocarbon dates of calcareous shells preserved in tsunamigenic sediments of Didim, western Turkey.*

Country	Sampling	Material	$\delta^{13}C$ PDB (‰)	^{14}C age (yr B.P.)	Error (1σ)	Calibrated age (B.C.)
Turkey	Didim	Benthic foraminifera in calcareous sand	-2.2	3837	88	1930 - 1706
Turkey	Didim	Benthic foraminifera in silty mud	+0.1	3886	86	1991 - 1759

4. Discussion and Concluding Remarks

It has been made clear that the Thera eruption and the following tsunami occurrence are recorded in the coastal sequences of western Turkey and Crete. The fallout of tephra carried by the prevailing winds in the troposphere was preceded by the invasion of tsunamis, and the duration between the arrival of tsunamis and the ash fall was probably very short. A sequence of volcanic phases starting with the Plinian activity has been discriminated within the stratigraphic record of the Minoan eruption of Santorini, and it is estimated that the time span of the sequence was on the order of tens of hours [1]. The ejecta probably reached over the Eastern Mediterranean within a few of days. A train of tsunami waves is numerically inferred to have arrived at the Aegean Sea shore of western Turkey about 2.5 h after the caldera collapse. Figure 4 is the schematic model for the explanation of the Minoan tectono-volcanic event. The airborne tephra conformably overlies the tsunami layer, and no erosional structure was observed just below the tephra layer. It follows, therefore, that the time scale of the Minoan event in the Aegean Sea region is on the order of 24 h.

Our numerical results and the sediment distribution show that the seawater flooding due to the tsunami invasion was restricted to the coastal zone of Crete and all along the Aegean Sea coast. Despite its large wave height at the harbor of Amnissos and the Gulf of Mirambelo (6-11 m) of Crete, it is estimated that the run-up distance of waves was only several hundred meters from the coast. The destruction of boats and harbor installation, however, might have damaged the maritime trade and the fishing economy. In this context, the Minoan civilization was influenced by the tsunami hazards associated with explosion collapse processes of a dome complex on Minoan Thera.

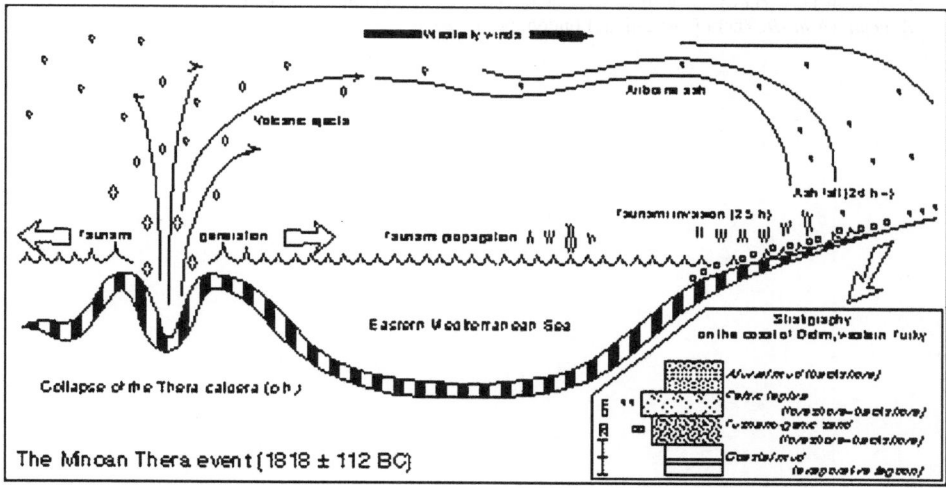

Figure 4.. Explanation model of the Minoan event. The fallout of tephra carried by the prevailing winds in the troposphere was preceded by the invasion of tsunamis, and the duration between the arrival of tsunamis and the ash fall was probably very short. A sequence of volcanic phases starting with the Plinian activity has been discriminated within the stratigraphic record of the Minoan eruption of Santorini, and the time span of the sequence is estimated on the order of tens of hours.

236

Acknowledgements

This research was supported financially by the Ministry of Education, Science and Culture, Japan, and partly supported by TUBITAK with the Project Grants INTAG-827 and YDABCAG-60 in Turkey.

References

1   Bond, A., and Sparks, R.S.J. (1976) The Minoan eruption of Santorini, Greece, *geological Society (London) Journal* **132**, 1-16.
2   Sullivan, D.G. (1988) The discovery of Santorini Minoan tephra in western Turkey, *Nature* **333**, 552-554.
3   Doumas, C., and Papazoglou, L. (1980) Santorini tephra from Rhodes, *Nature* **287**, 322-324.
4   Pyle, D.M. (1990) New estimates for the volume of the Minoan eruption, in D.A. Hardy, J. Keller, V.P. Galanopoulos, N.C. Flemming, and T.H. Druitt, (eds.), *Thera and the Aegean world III*, Thera Foundation, London, pp. 113-121.
5   Antonopoulos, J. (1992) The great Minoan eruption of Thera volcano and the ensuing tsunami in the Greek Archipelago, *Natural Hazards* **5**, 153-168.
6   Page, D. (1970) The Santorini volcano and the destruction of Minoan Crete, *The Society for the Promotion of Hellenic Studies, Supplement*, International University Booksellers, London.
7   Marinatos, S. (1939) The volcanic destruction of Minoan Crete, *Antiquity* **13**, 425-439.
8   Ninkovich, D., and Heezen, B.C. (1965) Santorini tephra, London, *Proceedings of 17th Symposium of Colston Research Society*, pp. 413-453.
9   Heiken, G., and McCoy, F., Jr. (1984) Caldera development during the Minoan eruption, Thira, Cyclades, Greece, *Journal of Geophysical Research* **89**, 8441-8462.
10 Yalciner, A.C., Kuran, U., Akyarlı, A., Imamura, F., 1995. An investigation on the generation and propagation of tsunamis in the Aegean Sea by mathematical modelling. In: Tsuchiya, Y., Shuto, N. (Eds.), Tsunami: Progress in Prediction, Disaster Prevention and Warning, Book Series of Advances in Natural and Technological Hazards Research by Kluwer Academic Publishers, pp. 55-70
11 Minoura, K., and Nakaya, S. (1991) Traces of tsunami preserved in inter-tidal lacustrine and marsh deposits: Some examples from northeast Japan, *Journal of Geology* **99**, 265-287.
12 Keller, J., Rehren, T.H., and Stadlbauer, E. (1990) Explosive volcanism in the Hellenic arc: a summary and review, in D.A. Hardy., J. Keller, V.P. Galanopoulos, N.C. Flemming, and T.H. Druitt (eds.) *Thera and the Aegean world III*, Thera Foundation, London, pp. 13-26.

POSSIBLE TSUNAMI DEPOSITS DISCOVERED ON THE BULGARIAN BLACK SEA COAST AND SOME APPLICATIONS

B.K. RANGUELOV
Geophysical Institute, BAS.
Acad. G.Boncev str. bl.3, Sofia 1113, Bulgaria.

Abstract

The recently discovered tsunami deposits on the Northern Bulgarian Black Sea coast supports the idea that these shore lines have been flooded by local tsunamis in the past. The dating of the discovered deposits and inclusions in them as well as the levelling of their recent positions show that these deposits have been formed may be during the significant earthquake (M>7.0, approximate coordinates: 28.3E, 43.3N) generated the giant landslide occurred in I-st century BC. The effects of this landslide can be still observed on the seashore. The obtained data combined with others (such as tsunami catalogue for the Black sea, the recurrence graphs, the refraction models, the zoning of the tsunami generating sources and vulnerable areas, etc.) can be useful for the tsunami hazard assessment of the Black sea. This possibility is slightly mentioned using the previous investigations of the author and the results obtained about the newly discovered deposits.

1. Introduction

Many new investigations have been performed during the mid 90s of the last century for the tsunami research of the Black Sea region. A lot of new data, maps and interpretations have been made for the tsunami zoning purposes. For the first time tsunami deposits have been discovered in the outcrops located on the northern part of the Bulgarian coast. Different observations have been made. Geodesy positioning according to the recent sea level helps to estimate the supposed tsunami height [1]. A large amount of data has been collected and processed during the performance of the GITEC Project with EU [2]. Calculations and modelling of different parameters of the tsunamis as well the tsunamigenic sources and vulnerable areas locations have been performed [3, 4]. The travel times of the tsunamis from different sources, zones of shadows due to the refraction, tsunami energy distribution according to the geometry of the coast and bottom and other important values have been calculated [5,6]. Many calculations of the probabilistic tsunami risk assessment and the expected negative consequences from the tsunami flooding have been done and the average repeatability estimated for the different time periods and the different run-up's [7, 8]. New data and interpretations about the fractal properties of the coastal line and the respective observed run-ups have been obtained using the method of fractal analysis as well [9]. All these calculations and models have been verified by the data of observed tsunamis described in old chronics or registered by the recent equipment. Several observed cases

A. C. Yalçıner, E. Pelinovsky, C. E. Synolakis, E. Okal (eds.),
Submarine Landslides and Tsunamis 237-242.
©2003 Kluwer Academic Publishers.

support the recognition of the estimated tsunami run-ups [10, 7]. All these data can be combined and used in case of the tsunami hazard assessment of the seashore.

2. Observations

The discovery and the observations of the well-preserved possible tsunami deposits have been made during the field expedition - summer time of 1996 [1]. The area under investigations is located on the northern part of the Bulgarian coast near Varna City. The discovery has been made by chance following the digging works of the local house builders. Several points with the visible tsunami deposits have been recognized as the probable sites in a small bay located north of Varna City and south from the most powerful seismic source at the region (fig.1.).

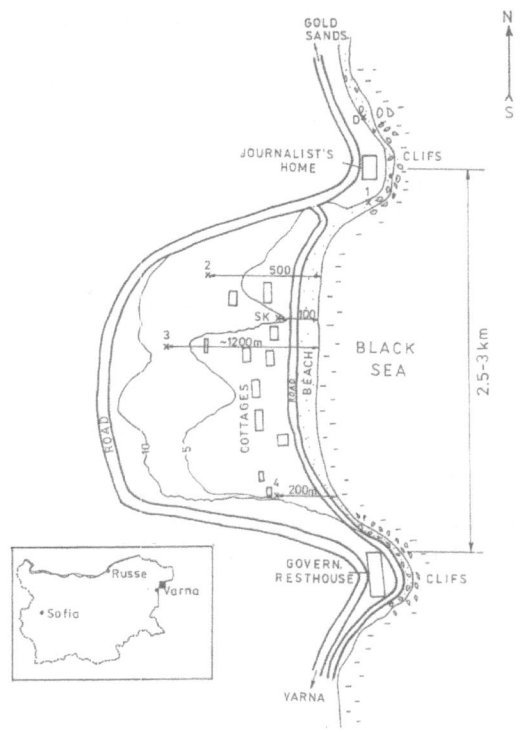

Figure 1. Sketch of the location of the tsunami deposits' sites pointed by vertical crosses and numbered consequently. Site 1 – distance 4 meters from the sea shore line (height – 4.25 meters above the water level); site 2 – 503 meters from the seashore (height – 5.15 meters); site 3 – 1215 meters from the seashore (height – 8.50 meters); site 4 – 204 meters from the seashore (height – 2.75 meters)

The layers have been carefully investigated. Usually the deposited sands over the hard-carbonized sandstone compose them. The sands include bigger sandstone pieces (size between 2-3 and 10-15 cm.) brought by the visible lower levels of the beach gravel. Often in this suspected tsunami deposits the shells of different mollusks can be observed. The sites

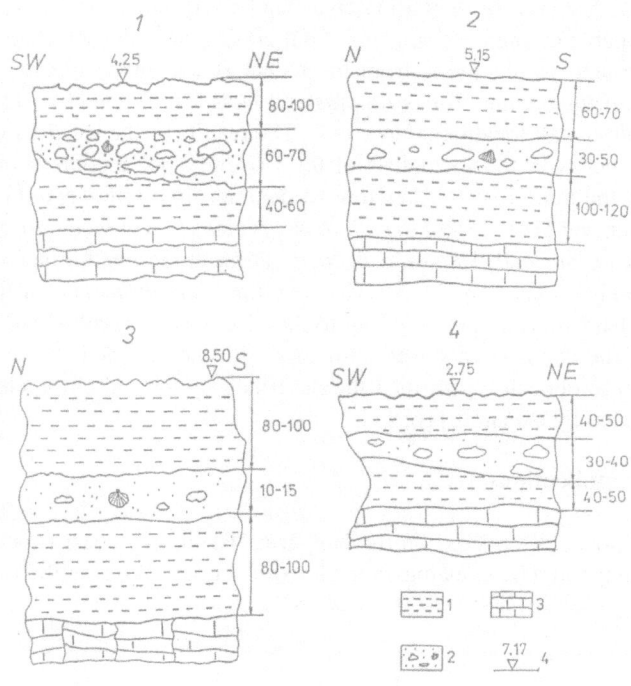

Figure 2. Sketches of the cross-sections of some selected sites (1-4) of fig.1. Stratigraphy and the position of the discovered tsunami layers are presented:1-soil layers; 2 - marine sediments of gravel, sands, some shells of Mollusks and microfossils (tsunami deposits!); 3 - hard rock basement (carbonized sandstone); 4 - measuring point above sea level. The thickness is in cm.

are well visible, with clear stratification of the layers. At the lower part – gravel with rough sands can be seen followed by the rough grains and the finest fraction over all. At some places the preserved back drive channels of the water can be observed (for example on site 2). The correlation of the positioned level of the deposits can be follow to all sites. The deposits have different thickness according their space position – distance from the recent seashore and the respective heights. The interpolation of the level's positions performed for the different sites fits very well most of them. This shows that the deposits have not been moved from their original places. Some samples from the located sites have been collected. The sampling points have been positioned by the leveling measurements to the recent water level (fig.2).

Photographs have been taken of almost all discovered sites. After the samples have been collected and investigated in the laboratory of Coventry Polytechnic, the microfossil and radiocarbon analysis have been performed. The results show that these deposits can be recognized as a result of the local tsunamis generated by a big earthquake occurred in I-st century BC. This event is famous, with the subsided underwater city of Bizone (old Greek colony) and the heavy very big monolith blocks slide down in the water. They are visible even at the recent seashore. The estimated magnitude is M>7.0 and approximate coordinates – 28.3E, 43.3N [11, 10, 12, 13]. The radiocarbon dating method C14 was performed in the Laboratory of Radioactivity and environment in Sofia, on three samples and gives reasonable dating in average of about 2000 years. The selected samples consist of many fragments and whole shells from the recent Black Sea *molluscs*. The main species belong to the *Cardium edule, Tapes rugatus* and *Donax julianae* [14]. The specimens have been found at almost all localities (No's 1-3). The results of the special microfossil analysis performed together with the specialists of the Coventry polytechnic show the presence of the skeletons of the microfossils. Skeletons of the microfossils *Foraminifera* and *Ostracoda* have been recognised [15]. "The *Foraminifera* species present were *Haynesina depressula, Elphidium williamsoni* and *Anunonia becarii*. These three benthic species in general are tolerant of brackish conditions" [15]. The levelling measurements on the described sites established the tsunami run-ups according to the recent water level reached up to 7-8 meters [7]. The age of the deposits very well coincides with the reported strong earthquake in I-st century BC. These deposits confirmed the described by the old chroniclers large flooding, probably generated by tsunamis [16, 11].

3. Results and applications

The tsunami deposits discovered on the Bulgarian Black Sea coast support the idea of the local flooding generated by tsunamis in the historical times [11]. On the fig.3 the

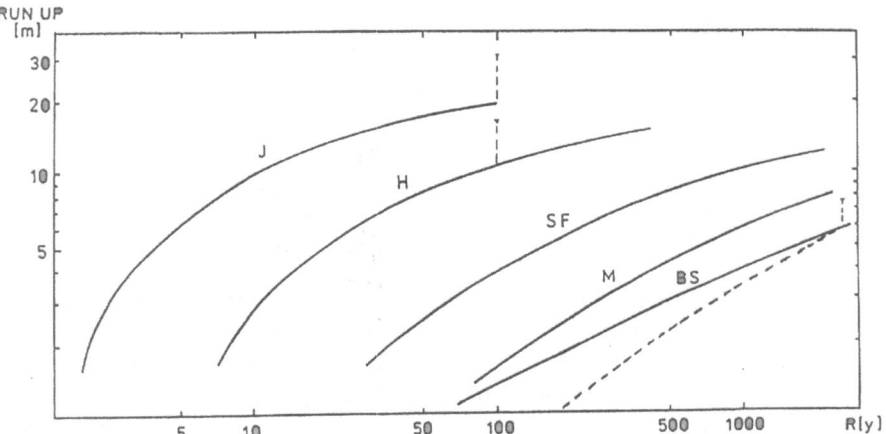

Figure 3. Approximate run-up tsunami heights according the return period for various places (according [17]): J - Japan, H - Hilo bay (Hawaii), SF - San Francisco, M - Mediterranean, BS - Black Sea. The dashed line show the same for the tsunami sources in the aquatory of BS, the solid one - including offshore tsunami sources. The dashed vertical bars show the extreme values observed for a certain period. For the Black Sea, the extreme value is obtained due to the new discovered paleotsunami deposits. The position of the vertical bars is according the appearance of the extreme heights of tsunamis.

extreme value has been added on the repeatability graph established earlier [7, 17] for the Black Sea.

The data from this graph have been used about the model calculation for the empirical Black Sea tsunami zoning [8]. The results of the investigations of the tsunami deposits and the tsunami hazard for the Bulgarian Black Sea coast are reflected on the created map of the Geological Hazards in Bulgaria (scale 1:500 000) [18]. All tsunami effects on the Bulgarian part of the Black sea coast have been outlined and possible tsunami influence such as generating sources as well as flooding areas from near and far-field tsunami sources are reflected [5]. A selection of some most tsunami vulnerable areas of the coast have been made for target calculations. The tsunami influence due to the refraction is estimated [3]. The results show that the most dangerous for the coastal facilities are the local tsunami sources. The far field sources can produce huge tsunamis, but their energy dissipates in the sea and only the local sources can produce significant influence on the coast [19]. This is due to the specific location of the tsunami generating sources (mostly earthquakes and very rare other generators such as landslides (both - surface and submarine), stonefalls, etc., which can generate tsunamis. On the other hand the geometry of the bottom and the coastal line and the reflections due to those bathimetry peculiarities can create strong auto focusing effects to the local specific sites attacked by the waves [3]. As it was shown in the previous investigations, the tsunami energy is concentrated by the local geometry peculiarities of the coast and the sea bottom [6, 10, 9]. The newest tsunami zoning made by an empirical approach [8] shows the possibility of combined influence - tsunami plus sashes (or meteosurges) which can amplify the tsunami flooding effects. May be the discovered tsunami deposits also reflects such combined influence. The established height from the tsunami deposits position reaches 7-8 meters up and thus provides the data to suggest the big sea surges, which can occur in the Black sea. The newest achievements can help the reliability of the adopted or suggested solutions. Up to now there is no systematic search for similar deposits. They are intended in a near future. It seems very probable that some other promising places (for example near Crimea coasts and near Varna City south coast) also contain tsunami deposits [13]. Several deposits of sapropel breccia have been discovered according the oil prospecting in the regions. Usually they are indicators of the fast and massive flooding. The necessity to try to find them is obvious, for future science investigations. They may help the dating of other local tsunamis as well. Unfortunately the activated landslides on the Bulgarian coast during April 1997 disturbed the observational points and changed dramatically all observed levels of the discovered deposits' sites. This means that no more real position of the deposits can be recognised [12]. This fact shows that similar investigations must be performed with highest accuracy on the collected data, materials and samples.

4. Conclusions

Using new and sophisticated methods of microfossil analysis, precise levelling and radiocarbon dating the new discovered probable tsunami deposits are reported. The age of the deposits estimated by the carbon 14 method correlated with the time consequences of the strongest earthquake and tsunami reported in the I-st century BC. The depth origin of the deposits is confirmed by the microfossil analysis. This new fact applied to the existing repeatability relationships can help the tsunamizoning of the Black Sea coasts and the justification of the existing tsunami catalogue of the same sea.

References

1. Ranguelov, B. (1998) Tsunami investigations in the Black Sea - new data from discovered tsunami deposits., *Abst. Intl. Symp. HAZARDS'98., Hania, Crete - May 17-22*, 125.
2. Ranguelov, B. (1997) Tsunami catalogues for the Black Sea - a Comprehensive Study., *Abst, XXIX. Gen. Ass. IASPEI, Tessaloniki, 18-28 Aug.*, 431.
3. Ranguelov, B. (1996) Tsunami Vulnerability in the Black Sea., *Proc. I-st Int. Conf. SESURB'96., 12-16 Feb., , Petropavlovsk - Kamchatsk.*, 39-45.
4. Ranguelov, B. and Gospodinov, D.(1995) Tsunami Vulnerability Modelling for the Bulgarian Black Sea Coast., *Wat. Sci. Tech*, v. 32, No 7, , 47-53.
5. Ranguelov, B. (1994) New Investigations of the Tsunami Danger in the Black Sea., *Proc. XXIV Gen. Ass. ESC., 19-24 Sept., Athens*, v. III, 1806-1807.
6. Ranguelov, B. and Gospodinov, D. (1994) Tsunami Energy Distribution According to the Black Sea Geometry., *Proc. XXIV Gen. Ass. ESC.,19-24 Sept., Athens*, v. III, 1808-1813.
7. Ranguelov, B. (1997) Tsunami repeatability for the Black Sea., *Proc. Intl. Conf. "Port - Coast - Environment" (PCE'97)., 30.06 - 4.07., Varna*, **2**, 293-297.
8. Ranguelov, B. (2000) Tsunami investigations in the Black Sea - empirical relationships and practical applications., *Proc. Moscow tsunami risk workshop,* (in press)
9. Ranguelov, B. (1997) Fractal dimensions and tsunami run-ups in the Black Sea *Compt. Rend. de'lAcad. Sci.*,vol.50, No 6, 47-50.
10. Ranguelov, B. (1996) Seismicity and Tsunamis in the Black Sea., in Stefanson R. (ed.), *Seismology in Europe,* Reykjavik, pp.667-673.
11. Guidoboni, E. (1989) *I terremoti prima Mille in Italia e nell'area mediterranea.*, ING, Bologna. (in Italian).
12. Ranguelov, B. (1998) Earthquakes, tsunamis and landslides on the Bulgarian Northern coast of the Black sea., in Mariynsky L. (ed.) *Coastal safety series*, Academic press, Sofia., pp. 46-51. (in Bulgarian).
13. Nikonov, A. (1977) Tsunamis observed on the coasts of the Black and the Azov Seas, *Physics of the Earth*, 1 , 1-11.(in Russian)
14. Kuneva-Abadjieva, V. (1960) *Black Sea mollusca.*, Gov. edition, Varna. (in Bulgarian).
15. Dawson, A. (1996) Personal communication. Ref. AGD140/GV. Coventry Polytechnic University.
16. Chrsitoskov, L. and Tupkova-Zaimova, V. (1979) Possible tsunami genesis of earthquake sources for the Bulgarian Black Sea coast, *Bulg. Geophys J.*,**V,4**, 98-99. (in Bulgarian)
17. Tiedemann, H. (1992) Earthquakes and Volcanic eruptions - A handbook on risk assessment, Swiss Re, Zurich.
18. Brouchev, I. et al. (1995) Geological Hazards Map of Bulgaria, Scale 1:500 000, in I. Brouchev (ed.), *Geological Hazards in Bulgaria*, Academic Press, Sofia. (in Bulgarian with English summary).
19. Pelinovsky, E. (1999) Preliminary estimates of tsunami danger for the northern part of the Black Sea., *Phys.Chem.Earth (A)*, **24**, No 2, 175-178.

INFLUENCE OF THE ATMOSPHERIC WAVE VELOCITY IN THE COASTAL AMPLIFICATION OF METEOTSUNAMIS

M. MARCOS[1,2], S. MONSERRAT[1,2], R. MEDINA[3], C. VIDAL[3]
[1] Grupo de Oceanografía Interdisciplinar/ Oceanografía Física, IMEDEA (CSIC-UIB)
[2] Departamento de Física, Universitat de les Illes Balears
[3] Grupo de Ingeniería Oceanográfica y de Costas, Universidad de Cantabria, E.T.S. Ingenieros de Caminos

Abstract

Extremely large seiche oscillations are regularly observed in some specific areas around the world even in the absence of any seismic forcing. These seiches have been successfully associated with strong atmospheric pressure perturbations inducing sea level oscillations at the open ocean, before entering the inlet, which are in turn resonantly amplified by the geometric characteristics of the inlet. The coastal behaviour of such waves, although of different origin, is similar to tsunami waves behaviour and is sometimes referred to as meteotsunamis. In some specific places, such as Ciutadella inlet, Balearic Islands, western Mediterranean, these seiche oscillations are stronger than expected, even taking into account the large amplification factor by resonance of the inlet. For these cases, some external amplification, before entering the inlet, is necessary to explain the phenomenon. In this paper, it is numerically shown how the phase speed of the atmospheric pressure disturbance generating the surface waves is a critical factor in the energy transfer between the atmosphere and the ocean.

1. Introduction

Large amplitude seiche oscillations with periods in the tsunami frequency range (several minutes) are periodically observed in some specific areas around the world even in absence of seismic forcing. These waves have been successfully related to atmospheric forcings (mainly atmospheric pressure oscillations) and therefore referred to as meteotsunamis. Strong seiche oscillations associated with atmospheric forcing have been reported in Japan [1, 2], China [3], the Adriatic Sea [4], the Aegean Sea [5], etc.

Waves of this kind are periodically observed in Ciutadella harbour, located at the end of an elongated inlet in Menorca Island, western Mediterranean (Fig. 1), where are locally known as rissaga. It has been demonstrated that rissagas are due to the resonant amplification of the inlet normal mode (10 min period) when this is externally forced by open ocean long waves generated by rapid atmospheric pressure fluctuations [6, 7]. The

A. C. Yalçıner, E. Pelinovsky, E. Okal, C. E. Synolakis (eds.)
Submarine Landslides and Tsunamis 243-249.

characteristics of these atmospheric waves are well established [8]. They are non-dispersive waves, travelling from the SW to the NE and with a phase speed always ranging between

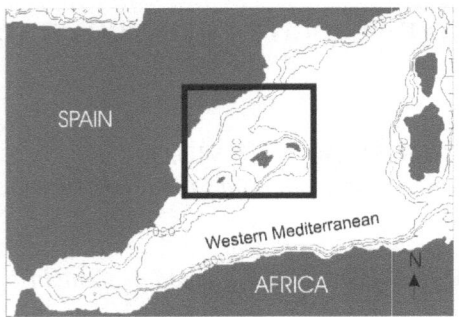 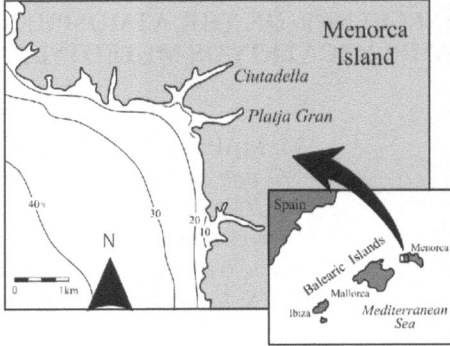

Figure 1. Situation of Ciutadella inlet in the western Mediterranean

20 and 30 m/s. The elongated dimensions of Ciutadella inlet (1 km long, 50 m wide) is a key point conferring to this inlet a particularly high amplification factor by resonance. However, the large sea level oscillations recorded at this site (up to 3 m wave height) can only be explained if some amplification of the ocean long wave occurs prior to its arrival to the inlet mouth. A finite difference 2D numerical model, able to simulate ocean long waves generated by atmospheric pressure fluctuations, is used to show how the platform characteristics allow this amplification and how the phase speed of the atmospheric pressure waves is critical in this amplification.

2. Numerical model

The numerical model is a finite difference, 2D long wave model developed by the Ocean and Coastal Research Group at the University of Cantabria. It has been modified to include an atmospheric pressure term in the momentum equation. The model integrates the depth-averaged equations of continuity, momentum and diffusion over a finite difference grid. The equations, in Cartesian coordinates have the form:

- Mass conservation:

$$\frac{\partial(HU)}{\partial x} + \frac{\partial(HV)}{\partial y} + \frac{\partial \eta}{\partial t} = 0$$

- Momentum conservation:

$$\frac{\partial(UH)}{\partial t} + \frac{\partial(U^2 H)}{\partial x} + \frac{\partial(UVH)}{\partial y} = fVH - \frac{1}{\rho_0} \cdot \frac{\partial P_a}{\partial x} - gH\frac{\partial \eta}{\partial x} - \frac{g}{2\rho_0}H^2\frac{\partial \rho_0}{\partial x}$$

$$-\frac{g}{\rho_0}\int_{-h}^{\eta}\left[\frac{\partial}{\partial x}\int_{z}^{\eta}\rho'\cdot dz\right]dz + \tau_{xz(\eta)} - \tau_{xz(-h)} + H\varepsilon_h\left[\frac{\partial^2 U}{\partial x^2} + \frac{\partial^2 U}{\partial y^2}\right] + 2H\frac{\partial \varepsilon_h}{\partial x}\frac{\partial U}{\partial x} + H\frac{\partial \varepsilon_h}{\partial y}\left[\frac{\partial U}{\partial y} + \frac{\partial V}{\partial x}\right]$$

$$\frac{\partial(VH)}{\partial t}+\frac{\partial(V^2H)}{\partial y}+\frac{\partial(UVH)}{\partial x}=-fUH-\frac{1}{\rho_0}\cdot\frac{\partial P_a}{\partial y}-gH\frac{\partial\eta}{\partial y}-\frac{g}{2\rho_0}H^2\frac{\partial\rho_0}{\partial y}$$

$$-\frac{g}{\rho_0}\int_{-h}^{\eta}\left[\frac{\partial}{\partial y}\int_{z}^{\eta}\rho'\cdot dz\right]dz+\tau_{yz(\eta)}-\tau_{yz(-h)}+H\varepsilon_h\left[\frac{\partial^2 V}{\partial x^2}+\frac{\partial^2 V}{\partial y^2}\right]+2H\frac{\partial\varepsilon_h}{\partial y}\frac{\partial V}{\partial y}+H\frac{\partial\varepsilon_h}{\partial x}\left[\frac{\partial U}{\partial y}+\frac{\partial V}{\partial x}\right]$$

- Diffusion equations for temperature, T and salinity, S (here both are denoted as C):

$$\frac{\partial C}{\partial t}+U\cdot\frac{\partial C}{\partial x}+V\cdot\frac{\partial C}{\partial y}=\frac{1}{H}\frac{\partial}{\partial x}\left(HD_x\cdot\frac{\partial C}{\partial x}\right)+\frac{1}{H}\frac{\partial}{\partial y}\left(HD_y\cdot\frac{\partial C}{\partial y}\right)$$

where, x, y, z form the right-handed Cartesian coordinate system, U and V are the depth-averaged velocity and $H=h+\eta$, where η is the free surface and h is the depth. The term f is the Coriolis parameter, P_a is the atmospheric pressure, ε_h, and ε_z are, respectively, the horizontal [9] and vertical [10] eddy viscosity coefficients, D_x and D_y are the horizontal diffusivity coefficients and $\rho=\rho_0+\rho'$ is the water density, with ρ_0 being the reference density. Density is obtained from the values of T and S using the UNESCO equation of state, as adapted by Mellor [11].

Model equations are written on a staggered grid (Arakawa C) and are solved by means of an implicit finite difference method, except for non-linear terms, which are treated explicitly. The finite difference algorithm is a centered, two time levels scheme, resulting in a second order approximation in space and time.

3. Model inputs

The actual bathymetry between Mallorca and Menorca islands (Fig 2a) may be schematically modelled as shown in Fig. 2b, taking a platform depth of 60m bounded by a 1000m ocean. Both actual and simplified bathymetries are used in separate simulations. A set of atmospheric perturbation propagating upwards with phase speeds ranging between 15 and 50 m/s has been used as the boundary condition along the lower boundary of the computational domain. The atmospheric perturbation employed in simulations is the actual record measured during 22-23 July, 1997, corresponding to a rissaga event In numerical computations, the grid size is 0.5 km and the time step for all the simulations has been 5 s. Boundary conditions in both bathymetry cases are radiation conditions in the upper and lower boundaries and reflection condition on the left and right ones.

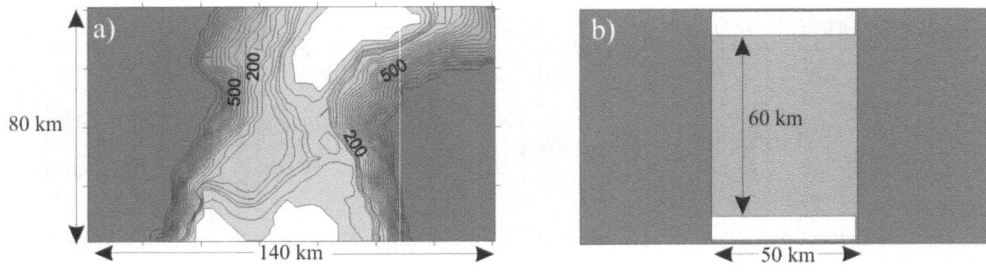

Figure 2. Actual (a) and idealized (b) bathymetries of Menorca Channel

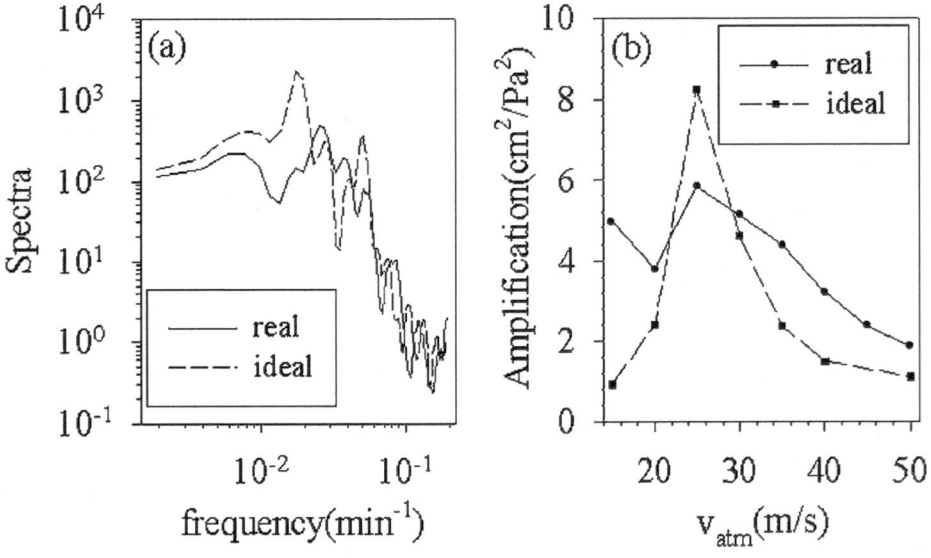

Figure 3. Sea level spectra for 25 m/s atmospheric phase speed forcing for real (full line) and idealized (dashed line) bathymetries (16 degrees of freedom) a). Spectral ratio between sea level (in cm) and atmospheric pressure (in Pa) versus phase speed for real (full line) and idealised (dashed line) bathymetries b.)

4. Results

Time series of modelled sea level height at a point close to the upper coast have been analysed by means of spectral techniques and compared with in-situ measurements. A Kaiser-Bessel window of 512 points has been used for segments of 2400 data, corresponding to 40 hours of simulation, resulting in 16 degrees of freedom. In both spectra, several peaks, probably associated with standing waves, are clearly identified (Fig. 3a). The spectral ratio between sea level height (cm) and atmospheric pressure (Pa) for

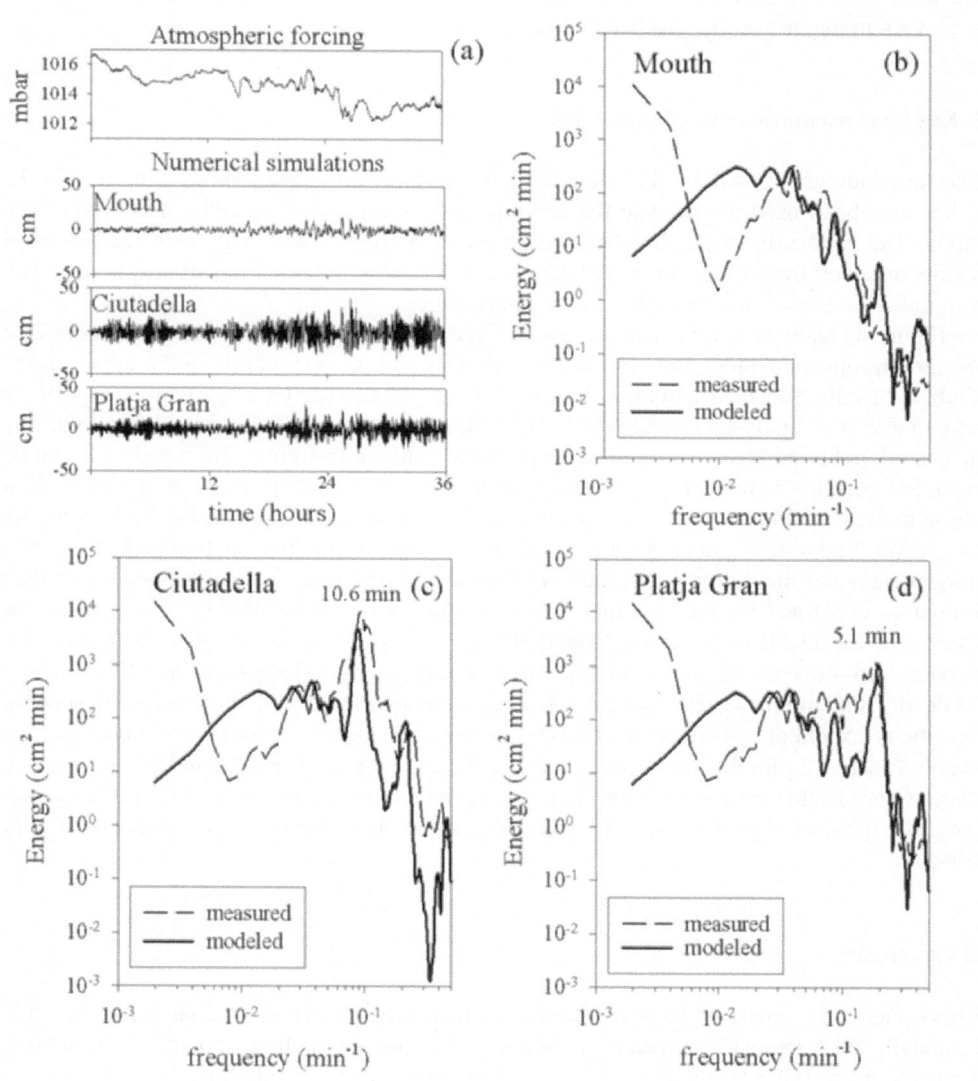

Figure 4. Atmospheric forcing and computed sea level at three different points a). Comparison between the measured and computed spectra of the free surface at these three points (14 degrees of freedom): outside Ciutadella inlet b),atthe end of line Ciutadella inlet c) and inside the neighbouring inlet of Platja Gran d).

different values of the atmospheric perturbation phase speed and for both bathymetries are shown in Fig. 3b.

These results suggest that the presence of the two islands in the wave track is a favourable point to the generation of standing waves, whose periods are basically controlled by the platform length and depth. However, their energy is deeply depending on the atmospheric waves phase speed. The optimal energy transfer between the atmosphere and the ocean occurs for phase speeds of about 25 m/s, value which is close to $(gH)^{1/2}$ in the ideal case (24.24 m/s) and coincides with the measured velocity of the atmospheric waves when major events have been observed.

5. Sea level oscillations inside the inlet

The same numerical model, with a smaller grid of 10 m and corresponding time step of 0.5 s, has also been used to simulate the wave propagation inside Ciutadella and Platja Gran inlets. The finer grid model has been fed through the lower boundary with the sea level results obtained from the outer model which was forced with the 25 m/s atmospheric wave. The radiation condition is used in the open boundaries.

Figure 4a shows the atmospheric pressure record and the modelled sea level records at the mouth (outside Ciutadella inlet) in approximately 30 m water depth, in the upper end of Ciutadella inlet and in a neighbouring inlet, Platja Gran (see Fig. 1). Time series of 36 hours have been analysed as described above, now resulting in 14 degrees of freedom. The measured and computed spectra at these points are shown in figures 4b, c and d. It can be seen that outside the inlet (Fig. 4b), the oscillation is very weak, peaking at 24min, both in the numerical and in the measured spectra. Other smaller peaks, appearing both in higher and lower frequencies, have poorer match due to the limitations of the grid (for lower frequencies) and the definition of the boundaries. The disagreement for periods larger than 1 hour is explained by the fact that the used domain is too small. Figure 4c shows the spectra of the oscillations inside Ciutadella, in its upper end, where the oscillations are maximal. In this case, the peak corresponds to the resonant period of the inlet fundamental mode of 10.5 minutes. The matching between the measured and the computed spectral density is excellent. Also the first mode, of about 5 minutes, shows an almost perfect match. Finally, figure 4d shows the spectra of the oscillations inside the neighbouring cove Platja Gran. In this case, the major peak corresponds to the resonant period of this smaller cove, 5.1 minutes. Again, the matching between the measured and computed spectra is very good.

6. Conclusions

Atmospherically generated large amplitude seiche oscillations (meteotsunamis) observed in Ciutadella inlet (Balearic Islands) are numerically simulated. It is shown that travelling atmospheric pressure fluctuations generate standing waves in the platform between the two islands. Their periods are basically controlled by the platform length and depth, but their energy depends on the atmospheric waves phase speed. The optimal energy transfer between the atmosphere and the ocean occurs for phase speeds of about 25 m/s, value which coincides with the measured velocity of the atmospheric waves when major events have been observed. Finally, the complete process inside the inlet is simulated by using a fine grid model with boundary forcing obtained from the coarse grid model on the shelf. The simulated sea level oscillations inside the inlets compare very well with observed data.

References

1. Honda, K., Terada, T., Yoshida, Y. and Isitani, D. (1908) An investigation on the secondary undulations of oceanic tides, J. College Sci., Imp. Univ. Tokyo, 108 p.
2. Hibiya, T. and Kajiura, K. (1982) Origin of `Abiki' phenomenon (a kind of seiches) in Nagasaki Bay, *J. Oceanogr. Soc. Japan*, **38**, 172-182.
3. Wang, X., Li, K., Yu, Z. and Wu, J. (1987) Statistical characteristics of seiches in Longkou Harbour, *J. Phys. Oceanogr.*, **17**,1063-1065.
4. Hodvzic, M. (1979) Exceptional oscillations in the Bay of Vela Luka and meteorological situation on the Adriatic, *International School of Meteorology of the Mediterranean, 1 Course, Erice, Italy.*
5. Papadopoulos, G. A. (1993) Some exceptional seismic (?) sea-waves in the Greek Archipelago, *Science of Tsunami Hazards*, **11**, 25-34.
6. Gomis, D., S. Monserrat, J. Tintoré (1993) Pressure-forced seiches of large amplitude in inlets of the Balearic Islands, *J. Geophys. Res.*, **98**, 14437-14445
7. Monserrat, S., A.B. Rabinovich, B. Casas (1998) On the Reconstruction of the Transfer Function for Atmospherically Generated Seiches, *Geophys. Res. Let.*,**25**, 2197-2205
8. Monserrat, S., A.J. Thorpe (1992) Gravity wave observations using an array of microbarographs in the Balearic Islands, *Q. J. R. Meteor. Soc.*, **118**, 259-282.
9. Smagorinsky, J. (1963) General circulation experiments with the primitive equations, *Mon. Weather Rev.*, **91**, 99-165
10. Jin, X., Kronenburg, C. (1993) Quasi-3D numerical modelling of shallow water circulation, *Journal of Hydraulic Engineering*, Vol 119, **4**, 458-472
11. Mellor, G.L. (1991) An equation of state for numerical models of oceans and estuaries, *J. Atmos. Oceanic. Technol.*, **8**, 601-611

References

IMPACT OF SURFACE WAVES ON THE COASTAL ECOSYSTEMS

S.R. MASSEL[1], E.N. PELINOVSKY[2]
1 Institute of Oceanology of the Polish Academy of Sciences
Powstańców Warszawy 55, 81-712 Sopot, Poland
2 Institute of Applied Physics of the Russian Academy of Sciences
Ulyanov 46, 603600 Nizhniy Novgorod, Russia

Abstract

Wave motion in vegetated coastal zones is relatively poorly understood when comparing with non-vegetated coasts. This is mainly due to lack of good quality data as well as to lack of the theoretical (or numerical) solutions. In this paper an attempt to model the wave propagation in two vegetated coastal environments is demonstrated. The attenuation of short surface waves as well as tsunami waves in the mangrove forest is determined by treating the mangrove forest as a random media with certain characteristics depending on the geometry of mangrove trunks and their locations. In the second example, the transformation of surface waves on sandy beaches, their breaking, set-up and run-up are determined and resulting fluctuations of the water table and groundwater flow are discussed.

1. Introduction

We are living in the age of "environmental awareness", with increasing demands on minimizing negative impacts of human activity on the environment. The environmental factor becomes move and move important in traditional coastal management and engineering. The necessity of inclusion of the environmental requirements is very clearly seen at the vegetated coasts, especially in the tropical climate. In contrast to the non-vegetated coasts, the understanding of the physical processes at vegetated coasts is very poor and it is not adequate to develop effective management plans or engineering designs which are today subjected to very stringent requirements to minimize their impact on the environment.

The role of waves in the coastal marine environment can not be overestimated. Waves approaching the shoreline break and dissipate their energy in the very shallow water. Tsunami, cyclone or storm waves impose large forces on natural coastal and man made structures. Knowledge of wave motion and the sediment budget provides the key to the proper selection of protecting structures and methods of coastal management [1]. However, hydrodynamics of the vegetated coastal zones, especially the wave motion, are still very poorly understood. Various types of ecological models are used to simulate the impact of the specific project on the vegetated environment to provide the physical and biological

A. C. Yalçıner, E. Pelinovsky, E. Okal, C. E. Synolakis (eds.)
Submarine Landslides and Tsunamis 251-258.

consequences of the alternative solution for the coastal regions [2] [3]. Most of the existing ecological models are concentrated on the flow pattern and transport and dispersion of the pollutants.

For the purpose of this paper, two examples of the surface wave impact on the ecology of the coastal system are given. The first example deals with the surface waves attenuation in the mangrove forest. Using the linearised governing equations, the rate of wave energy attenuation is obtained in the closed form for a given geometry of mangrove trees and roots. In the second example the main factors contributing to fluctuations in the water table and groundwater flow induced by surface waves in the sandy beach are discussed.

2. Wave Propagation in Mangrove Forests

Mangroves are densely vegetated mudflats that exist at the boundary of marine and terrestrial environments. Inherent in this habitat is their ability to survive in a highly saline environment [4]. In recent years it has been realized that mangroves may have a special role in supporting fisheries, stabilizing the coastal zone and protecting the lives and properties of the people living near the sea and offshore islands [5][6][7]. For example, after the 1998' tsunami in Papua New Guinea, the International Tsunami Survey Team recommended to plant the *Casuarina* mangrove species in front of coastal communities. They argued that the local *Casuarina* species withstood the wave attack significantly better than coconut trees [8].

Hydrodynamic factors play a major role in the structure and function of mangrove ecosystems. Biogeochemical and trophodynamic processes, and forest structure and growth are intimately linked to water movement. However, studies of physical processes in tropical mangrove swamps and mangrove-fringed estuaries are few, and far behind compared to those of temperate estuaries.

During tsunami and tropical cyclones, however, energy of waves substantially exceeds tidal energy. Due to the complexity of mangrove systems, the transmission of waves through mangrove areas is still poorly understood and number of papers dedicated to the mangrove hydrodynamics is very limited [9][10][11][12][13][14].

In the paper [15], the theoretical prediction model for attenuation of random surface waves propagated through mangrove forests was developed. A full boundary value problem was solved and attenuation of wave spectrum was predicted. Assuming that the diameter of particular mangrove trunks is very small in comparison with wavelength, wave energy is dissipated mostly due to drag forces induced on trunks by waves. Therefore within the mangrove forest (Region II - see Fig. 1), the momentum equation for motion with dissipation can be written as follows:

$$\frac{\partial \vec{u}}{\partial t} = \frac{1}{\rho} \nabla (p + \rho g z) - \frac{1}{\rho} \vec{F},$$

(1)

in which $\vec{u} = (u, w)$ is a wave-induced velocity vector, p is a corresponding dynamic pressure and $\vec{F} = 0.5 \rho\, C_d D \vec{u} |\vec{u}|$ is the drag force vector per unit volume [15].

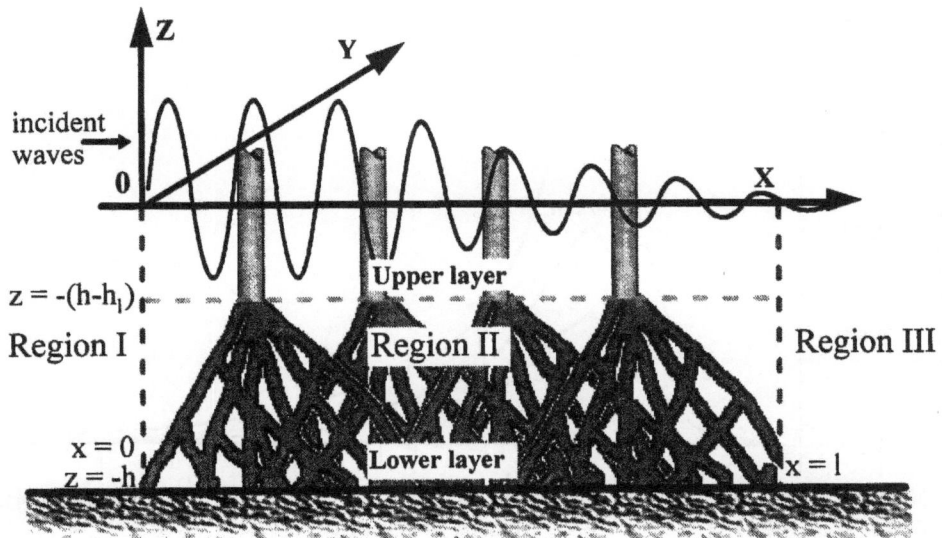

Figure 1. Reference scheme

We assume that in a unit control area of mangrove there are N_u trunks piercing the sea surface (usually N_u is of order of 1-10 per m²), each of the mean diameter $\vec{D}_u$. In the bottom layer of thickness h_l (usually thickness h_l is of order 0.3 m - 1.0 m) mangroves are very dense and smaller trunks and roots are randomly oriented. It is assumed that the number of trunks, N_l, each of the mean diameter $\vec{D}_l$, is of the order of 10-30 per m² (Fig. 1). The control area has to be selected sufficiently large to accommodate N_u and N_l trunks, where $N_u > 1$ and $N_l \gg 1$. On the other hand, this area has to be sufficiently small in order to neglect the variation of wave velocity within the control area and to neglect the exact location of each trunk within the control area.

Equation (1) is the nonlinear equation due to a quadratic drag term involving absolute value of local random orbital velocity. This equation cannot be solved exactly for a mangrove forest of arbitrary density and for arbitrary forcing by a random wave field. To get a practical solution, the linearization procedure, widely used in ocean engineering for determination of the forces on offshore structures, was applied [16]. The nonlinear term was replaced by the linear one under the condition that the mean error of this substitution becomes minimal. In physical terms, it means that instead of the mangrove forest with a complicated spatial net of trunks and roots, we are dealing with a medium for which the energy dissipation is characterised by term $f_e \omega_p \vec{u}(x,z)$, where f_e is the linearization coefficient and ω_p is the peak frequency. Hence we have:

$$\frac{\partial \vec{u}}{\partial t} = \frac{1}{\rho} \nabla (p + \rho g z) - f_e \omega_p \vec{u}$$

(2)

The details of solution of the boundary value problem and linearization procedure are given by in [15].

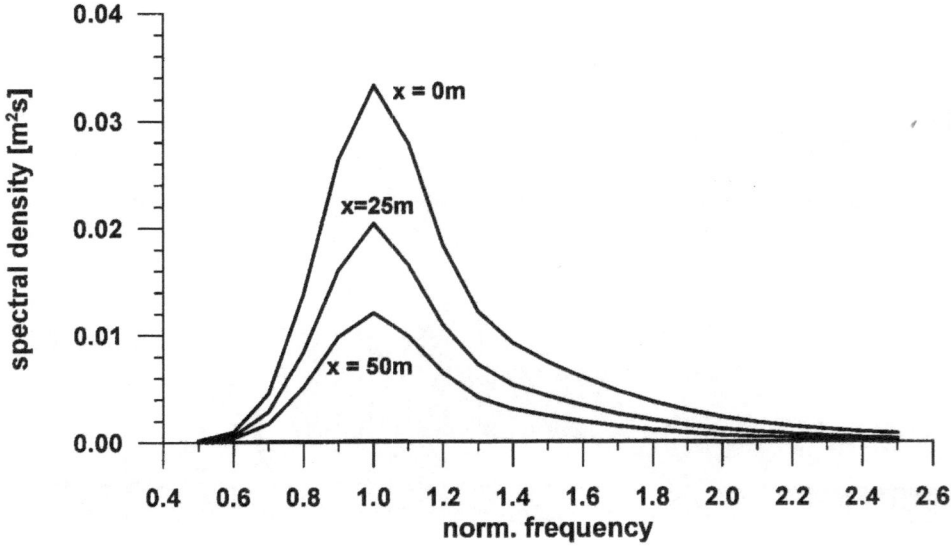

Figure 2. Wave spectrum at three cross-sections (x = 0, 25, 50 m) in densely populated forest

To illustrate developed model let us assume the following parameters of the mangrove forest: forest width l = 50 m, water depth h = 1 m, number of trunks in upper layer N = 16/m^2, number of trunks in lower layer N_l = 49/m^2, mean diameter of upper layer trunks $\vec{D}_u = 0.08$ m m and mean diameter of lower layer trunks $\vec{D}_l = 0.02$ m. The mangrove forest is subjected to surface waves characterized by a typical spectrum for shallow water [2]. In the calculation, the significant wave height H_s = 0.6 m and the peak period is T_p = 5 s. In Fig. 2, the frequency spectra at three cross-sections (x = 0, 25, 50 m) from the mangrove front are shown. Numerical calculations indicate that for a given incident wave spectrum and given mangrove density 99% of energy is dissipated within mangrove forest [17][15].

3. Groundwater Circulation due to Wave Set-up and Run-up

Transformation of waves on the sandy beach, their breaking, set-up and run-up are the main factors contributing to fluctuations in the water table and groundwater flow. Sandy beaches are highly exploited but very dynamic and fragile environments. The beach system is driven by the physical energy induced by waves and tides. The water flow through the beach body is of great importance in introducing water, organic materials and oxygen to the ground environment. Moreover, it controls the vertical and horizontal, chemical and biological gradients, and nutrient exchange in the beach [18]. Also water filtration through a sandy beach is considered to be significant for water improvement as the beach retains and processes organic matter and pollutants. Some pollutants, like hydrocarbons and heavy metals are sorbed on the surface of microbes and diatoms the more numerous and diverse the organisms, the more effective the area of sorption [19].

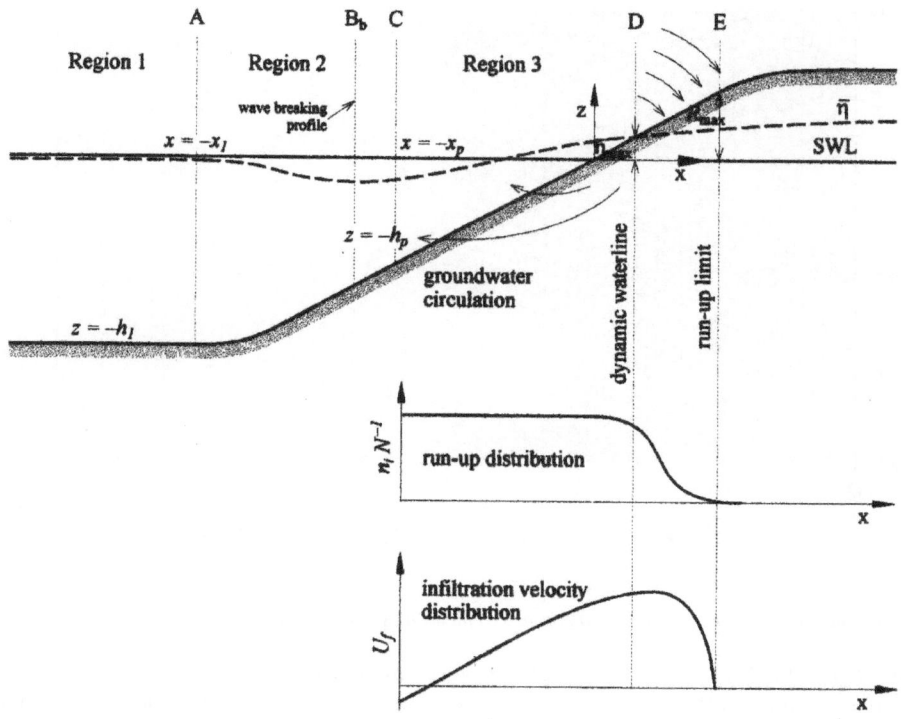

Figure 3. Relationships between wave run-up, infiltration and coastal watertable

Wave motion on the beach is very complex and the groundwater flow is different in different beach regions. In Region 3, between points D and E (see Fig. 3), the wave run-up infiltration contributes mainly to the raising of the coastal water table. The hypothetical distribution of the infiltration velocity U_f induced by the run-up is shown in the same figure, and the vertical axis $n_i\, N^{-1}$ denotes the ratio of the events when the beach surface is covered by water to the total number of events.

The beach groundwater flow in the set-up region, between points B_b and D (see Fig. 3), induces groundwater circulation which contributes to the submarine groundwater discharge [20][21]. Little is known about the groundwater flow in this region. One of the models of the groundwater flow is the Longuet-Higgins analytical solution for the circulation induced by wave set-up in a semi-infinite domain [22].

In the run-up region, between points D and E, wave uprushes on the beach and the water level reaches the height R_{max}, defined as the maximum vertical height above still water level. The run-up height is always greater than the wave set-up. The wave run-up limit and induced water infiltration into beach body is a response to the instantaneous flow of the surface water. Massel and Pelinovsky [23] developed the analytical model for the run-up of dispersive breaking and non-breaking waves. Waves approaching the shallow water area were modelled by the mild-slope equation [2] [3]. The wave run-up during

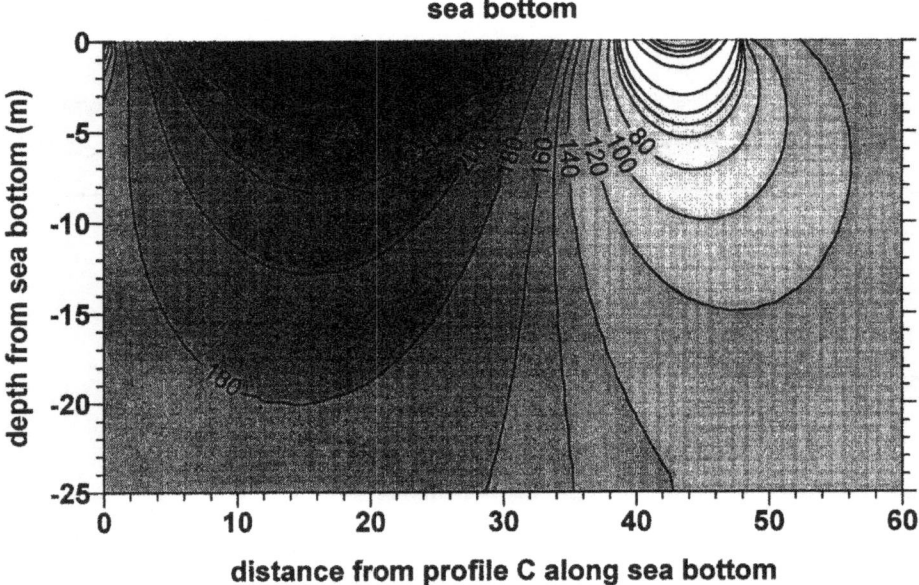

Figure 4. Streamlines contours (in cm²/s) in a porous bed induced by wave set-up

tsunami waves attack is particularly high. At very small water depths, the non-linear and linear equations for shallow water waves were considered and the dissipation due to wave breaking was included, providing a more realistic estimation of run-up characteristics [24].

The experimental data on the wave run-up are numerous. However, they are mostly related to the long or solitary waves propagating on steep slopes, especially on marine revetments (see for example [25]). On the other hand, the data on run-up of short waves on the gentle slopes of the natural beaches are rather rare. In paper [23], the predicted wave run-up heights have been compared with data reported in [26] [27] [28] showing a good agreement.

However, the beaches consisting sand or unconsolidated sediment are porous and any changes of pressure associated with the wave set-up produce flow of sea water within the beach itself (see Fig. 3). Moreover, the wave motion percolated in a permeable bottom influences the wave forces on the hydraulic structures founded on or extended into the bottom. Using set-up and run-up values, calculated in [23], the streamlines of flow in a porous medium beneath the sea bottom in Region 3 were calculated by the approximate method suggested by Longuet-Higgins [22]. Example of streamlines distribution for the incident deepwater wave height $H_0 = 1.95$ m and period $T = 6$ s propagating on beach with the slope 0.1 is shown in Fig. 4.

The pattern of streamlines indicates that the induced circulation begins with infiltration close to profile D and ends with exfiltration at the lower part, close profile C, as was shown schematically in Fig. 3. Move detail description of the water infiltration due to wave set-up, based on the conformal mapping approach will be published in a separate paper.

4. Conclusions

The following major conclusions can be drown from this short paper:

1. Tsunami waves or tropical cyclone induced waves are extraordinary events and their energy substantially exceeds the usual energy level at a given location. These waves are the dominant cause of water movement, sedimentation and coast erosion.

2. Wave motion in vegetated coasts subjected to the tsunami waves or tropical cyclones is relatively poor understood, when comparing with non-vegetated coasts. This paper presents some results on the mangrove forest which is a good example of the vegetated coast. The energy dissipation is determined by treating the mangrove forest as a random media with certain characteristics determined using the geometry of mangrove trunks and their locations. Example given in this paper as well as in other papers by one of the authors, showed that mangrove species withstood the wave attach significantly better than for example coconut trees.

3. The wave run-up on a beach face is one of the basic factors that determines beach stability and sediment transport, induces beach groundwater flow and raises the groundwater table. Driven by waves, the water flow through the beach body is able to transport oxygen, and hence helps to maintain biological activity in the porous media. In particular, the pressure changes associated with the wave set-up, through small, produces effects that, because they are cumulative in time, may be more far-reaching. Two systems of circulations have been discovered, related to different gradients of the set-up height. The pattern of groundwater flow is of vital importance to the status of the organisms inhabiting the beach sand.

References

1 Matsumoto, T., Kimura, M., Nakamura, M. and Ono T. (2001) Large-scale slope failure and active erosion occurring in the southwest Rynkyn fore-are area, *Natural Hazards and Earth System Science* 1, 203-211.

2 Massel, S.R. (1996) *Ocean Surface Waves: their Physics and Prediction*, World Scientifc Publ., Singapore, 491pp.

3 Massel, S.R. (1999) *Fluid Mechanics for Marine Ecologists*, Springer Verlag, Heidelberg, Berlin, 565pp.

4 Robertson, A.I. and Alongi, D.M. (eds). (1992) *Tropical mangrove ecosystems*, American Geophysical Union, Washington, 329pp.

5 Jenkins, S.A. and Skelly, D.W. (1987) Hydrodynamics of artificial seaweed for shoreline protection, *Scripps Inst. Oceano.*, 87-16.

6 Qureshi, T.M. (1990) Experimental plantation of rehabilitation of mangrove forests in Pakistan. *Mangrove Ecosystems Occasional Papers*, UNESCO, 4.

7 Siddigi, N.A. and Khan, M.A.S. (1990) Two papers on mangrove plantations in Bangladesh. *Mangrove Ecosystems Occasional Papers*, UNESCO, 8.

8 Kawata, Y., Benson, B.C., Borrero, J.C., Borrero, J.L., Davies, H.L., de Lange, W.P., Imamura, F., Letz, H., Nott, J. and Synolakis, C.E. (1999) Tsunami in Papua New Guinea was as intense as first thought, *Eos, Transactions* 80, 101-105.

9 Asano, T., Deguchi, H. and Kobayashi, N. (1992) Interaction between water waves and egetations, *Proc. 23th Coastal Eng. Conf.* 3, 2710-2723.

10 Dubi, A. and Torum, A. (1994) Wave damping by kelp vegetation, *Proc. 24th Coastal Eng. Conf.* 1, 142-156.

11 Lovas, S.M. (2000) *Hydro-physical conditions in kelp forest and the effect on wave damping and dune erosion*, The Norwegian University of Science and Technology, Trondheim, 187pp.

12 Mazda, Y., Magi, M., Kogo, M. and Hong, P. N. (1997a) Mangroves as a coastal protection from waves in the Tong King Delta, Vietnam, *Mangrove and Salt Marches* 1, 127-135.

13 Mazda, Y., Wolanski, E., King, B., Sase, A. and Ohtsuka, D. (1997b) Drag force due to vegetation in mangrove swamps, *Mangroves and Salt Marches* 1, 193-199.

14 Wolanski, E., Mazda, Y. and Ridd, P. (1992) Mangrove hydrodynamics. In: Robertson, A.I. and Alongi, D.M. (eds), *Tropical mangrove ecosystems*, American Geophysical Union, Washington, 43-62.

258

15 Massel, S.R., Furukawa, K. and Brinkman, R.M. (1999) Surface wave propagation in mangrove forests, *Fluid Dynamics Research* 24, 219-249.

16 Gudmestad, O.T. and Connor, J.J. (1983) Linearization methods and the influence of the nonlinear hydrodynamic drag force, *Applied Ocean Research* 5, 184-194.

17 Brinkman, R.M., Massel, S.R., Ridd, P.V. and Furukawa, K. (1997) Surface wave attenuation in mangrove forests, *Proc. 13th Australasian Coastal and Ocean Engineering Conf.* 2, 941-946.

18 McLachlan, A. (1989) Water filtration by dissipative beaches, *Limnology and Oceanography* 34, 774-780.

19 Węsławski, J.M., Urban-Malinga, B., Kotwicki, L., Opalińnski, K., Szymelfenig, M. and Dutkowski, M. (2000) Sandy coastlines - are there conflicts between recreation and natural values ?, *Oceanological Studies* XXIX, 5-18.

20 Kang, H.Y. and Nielsen, P. (1996) Watertable dynamics in coastal areas, *Proc. Coastal Eng. Conf.* 3, 4601-4612.

21 Li, L. and Barry, D.A. (2000) Wave-induced beach groundwater flow, *Advances in Water Resources*f 23, 325-337.

22 Longuet-Higgins, M.S. (1983) Wave set-up, percolation and undertow in the surf zone, *Proc. Roy. Soc.* A390, 283-291.

23 Massel, S.R. and Pelinovsky, E.N. (2001) Run-up of dispersive and breaking waves on beaches, *Oceanologia* 43, 61-97.

24 Pelinovsky, E. and Mazova, R. (1992) Exact analytical solutions of nonlinear problems of tsunami wave run-up on slopes with different profiles, *Natural Hazards* 6, 227-249.

25 Titov, V.V. and Synolakis, C.E. (1995) Modeling of breaking and nonbreaking long-wave evolution and run-up using VTSC-2, *Proc. ASCE, Jour. Waterway, Port, Coastal and Ocean Eng.* 121, 308-316.

26 Führböter, A. (1984) Model and prototype tests for wave impact and run-up on a uniform slope 1:4, *Proc. Conf. on Water Wave Research. Theory, Laboratory and Field*, Hannover, 1-51.

27 Gourlay, M.R. (1992) Wave set-up, wave run-up and beach water table: Interaction between surf zone hydraulics and groundwater hydraulics, *Coastal Eng.* 17, 93-144.

28 Saville, T. (1958) Wave runup on composite slopes. *Proc. 6th Coastal Eng. Conf.*, ASCE, 691-699.

Part 5

Mitigation

ENGINEERING STANDARDS FOR MARINE OIL TERMINALS AND OTHER NATURAL HAZARD THREATS

M. L. ESKIJIAN[1], R.E. HEFFRON[2], T. DAHLGREN[3]
[1]*California State Lands Commission 330 Golden Shore, Suite 210
Long Beach, CA 90802 USA*
[2]*Han-Padron Associates 100 Oceangate, Suite 650
Long Beach, CA 90802*
[3]*Ben C. Gerwick, Inc 20 California Street
San Francisco, CA 94111*

Abstract

Marine Oil Terminals, when inadequately designed or maintained can pose a significant risk to public safety and the environment. Natural hazard threats such as earthquakes, high winds, and excessive currents from runoff and tsunamis should be considered for both new and existing Terminals Damage to or collapse of a wharf or pier carrying pipelines may result in an oil spill. In order to protect the public health, safety and the environment, the California legislature passed the Lempert-Keene-Seastrand Oil Spill Prevention and Response Act in 1990 (the Act). The Marine Facilities Division was formed to implement the Act. It soon became obvious that there were no rehabilitation or upgrade standards for these aging facilities. Significantly, seismic standards did not exist when the vast majority of the facilities were built. The new comprehensive engineering standards developed specifically for Marine Oil Terminals in the U.S. The standards fulfill the mandate of providing regulations for the performance standards of marine oil terminals, to minimize the possibilities of a discharge of oil. In this study the criteria defined by standards, seismic analysis of design criteria, mooring loads for the marine oil terminals are presented. However, much of the criterion is equally applicable to other types of marine structures. It is expected that these standards will form the basis for similar standards developed for other state, federal, and international regulatory agencies.

1. Introduction

Marine Oil Terminals (MOTs), when inadequately designed or maintained can pose a significant risk to public safety and the environment. Natural hazard threats such as earthquakes, high winds, excessive currents from runoff and tsunamis should be considered for both new and existing MOTs. Damage to or collapse of a wharf or pier carrying pipelines may result in an oil spill. In addition to the obvious environmental damage and public health concerns, an oil spill may be expensive to clean up. Recent experience in the U.S. has shown that the cost of cleaning up oil spills in waterways can range from $10,000/bbl up to $250,000/bbl.

A. C. Yalçıner, E. Pelinovsky, E. Okal, C. E. Synolakis (eds.)
Submarine Landslides and Tsunamis 259-266.
©2003 Kluwer Academic Publishers.

In order to protect the public health, safety and the environment, the California legislature passed the Lempert-Keene-Seastrand Oil Spill Prevention and Response Act in 1990 (the Act). The Act defines a new mission for California State Lands Commission (SLC), in part stating:

" the commission shall adopt rules, regulations, guidelines and commission leasing policies for reviewing the location, type, character, performance standards, size and operation of all existing and proposed marine terminals within the state, whether or not on lands leased from the commission, and all other marine facilities on land under lease from the commission to minimize the possibilities of a discharge of oil..." (Public Resources Code (PRC) Section 8755 (a))

"The commission shall periodically review and accordingly modify its rules, regulations, guidelines and commission leasing policies to ensure that all operators of marine terminals within the state and marine facilities under the commission's jurisdiction always provide the best achievable protection of the public health and safety, and the environment..." (PRC Section 8756).

The Marine Facilities Division (MFD) was formed to implement the Act. Initial oversight of MOTs was operational in nature, but by 1991 engineering inspections had begun. It soon became obvious that there were no rehabilitation or upgrade standards for these aging facilities. In other words, a geriatric structure could be "rehabilitated" to the same design criteria as was used originally. The average age of the MOTs in California is about 50 years, which is typically considered past the design life of a marine facility. Many were constructed in the 1901 to 1925 period. The newest MOT in California, an exception to the rule, was built in 1987. Significantly, seismic standards did not exist when the vast majority of the facilities were built. The need for definitive engineering standards for MOTs became obvious and has resulted in the "Marine Oil Terminal Engineering and Maintenance Standards," or MOTEMS [1].

2. Project Participants and Overview

An overview of the current state of practice will help define the need for and urgency of the engineering standards. One of the biggest problems with the current state of practice is the situation of "grandfathering" existing operations. When the MFD was formed in 1990, each operator was required to submit an Operations Manual, which defined the maximum size of vessel currently calling at the terminal. The MFD then allowed the operator to continue the present practice. However, if the operator wished to increase the size of vessels calling at the terminal, an engineering justification was required to verify the capacity of the terminal. The fundamental problem with grandfathering of existing operations is that fairly large vessels were allowed to call at geriatric facilities simply because historically this had been the practice. There was no engineering justification for the increase in vessel size from the original construction time, when vessels were much smaller. In some cases, large tank vessels, over 125,000 DWT were calling at geriatric timber wharves over 60 years old. Terminals such as this may have inadequate mooring, berthing, fendering and seismic capabilities by today's standards. To compound this problem, many of these facilities have had inadequate maintenance such as routine inspections above and below the water line. Some have not been inspected after significant earthquakes.

The MFD started to inspect these facilities, and document structural deficiencies, knowing that the operator could repair to any standard that he desired. With this as a background, the Marine Oil Terminal Engineering and Maintenance Standards (MOTEMS) project was initiated by the MFD, to develop comprehensive engineering standards for the analysis, design and inspection/maintenance of MOTs. Funding for this project is through a hazard mitigation grant from the U.S. Federal Emergency Management Agency (FEMA), and was a result of the 1994 Northridge earthquake. Additional funding came from the California State Lands Commission.

The prime consultant for the development of these standards is a joint venture of Han-Padron Associates of Long Beach, California (CA) and Ben C. Gerwick, Inc. of San Francisco, CA. The U.S. Naval Facilities Engineering Service Center (NFESC), through a "Cooperative Research and Development Agreement" has also been an active participant. The Civil Engineering Department of the University of Southern California has also participated in the work by providing research and criteria in the area of tsunami risk assessment and also has contracted for the offshore seismic hazard risk assessment. The Civil Engineering Department of the University of California, San Diego has contributed to the structural performance criteria for seismic loading. This process has also included strong participation by the regulated community (marine oil terminal operators), engineering consultants, port engineers and academics. Peer review has been extensive, involving participants in workshops and technical advisory group meetings. The review process has even involved an industry group, the Western States Petroleum Association. The SLC has been careful to avoid imposing onerous or unnecessary restrictions.

3. The Engineering Standards

The standards will require each MOT to conduct an audit (detailed structural, mechanical and electrical evaluation) to determine the continuing fitness-for-purpose of the MOT and the level of compliance. The operators must then determine what actions are necessary and provide a schedule for implementation. The standards define criteria in the following areas:

- Audit and Inspection Procedures
- Loading Combinations
- Seismic Analysis and Design
- Mooring and Berthing Analysis and Design
- Geotechnical Hazards and Foundations
- Structural Analysis and Design of Components
- Fire Prevention, Detection and Suppression Systems
- Piping and Pipelines
- Electrical and Mechanical Equipment Systems

4. Audits and Inspections

The standards define three distinct types of inspections and audits that are required. The first is an Annual Inspection which consists of a walk-down of the facility conducted by the MFD along with the operator.

A second inspection type, called the Audit, is the primary focus of the standards. It consists of a comprehensive evaluation of all structural, mooring, electrical and mechanical systems relative to specific criteria defined in the standards. The Audit is conducted every three years for above water structure components, as well as electrical and mechanical systems. The frequency for underwater inspection varies with the construction type and condition of the facility, and may range from six months to six years. The audit requires a seismic assessment, a comprehensive mooring/berthing analysis, an evaluation of the fire plan, and a review of the geotechnical features, including the possibility of liquefaction or slope failure. The MOT is rated from good to critical, with numerical grades of 6 to 1, respectively, for both the structural and non-structural operational features. Scores of 4 (fair) and below may require some corrective actions prior to the restoration of full operational status. Deficiencies require the submittal of a schedule for rehabilitation that must be approved by the MFD. The frequency and scope of underwater inspections are consistent with the new ASCE Underwater Investigation Standard Practice Manual [2].

The Post-Event Inspection is conducted following a significant event such as an earthquake, flood, fire, vessel impact or similar event. The requirement to perform the Post-Event Inspection may be initiated by the MFD or the operator, depending on the level of ground motion or the specific incident. MFD regulations currently mandate that the operator must inform the MFD if the damage is greater than $50,000. The Post-Event Inspection serves to determine if the facility is safe to continue operations and to determine if remedial action is necessary.

5. Loading Criteria

Loading criteria are defined for both new and existing MOTs. The requirements for existing MOTs are lower than for new facilities, but rehabilitation may still be required. In addition to defining loading criteria for dead loads, live loads, earthquake loads, mooring loads, and berthing loads, the standards also define required loading combinations and associated safety factors for structural components and mooring lines.

6. Seismic Analysis and Design Criteria

The seismic analysis requirements of the standards differentiate facilities into high, moderate and low risk classifications. These classifications are used to define earthquake ground motion to be applied to the facility, as well as to determine the level of sophistication required for the structural analysis. The existing MOT classification for seismic risk is based on the number of barrels of oil at risk (flowing and stored), the number of transfers per year and the vessel size in dead weight (long) tons. The classification levels are as follows:

TABLE 1. EXISTING FACILITY SEISMIC CLASSIFICATION

Classification Level	Exposed Oil (bbl)	Transfers per Year/Facility	Vessel Size (DWTx1000)
High	≥ 1200	N.A.	N.A.
Moderate	< 1200	≥ 90	≥ 30
Low	< 1200	< 90	< 30

Two different levels of design earthquake motion are defined for various risk classifications. The design spectral acceleration, design peak ground acceleration, and design earthquake magnitude may be determined, once the return period of the event is established. The Level 1 and Level 2 performance criteria are defined as follows [3]:

Level 1 Earthquake: No or minor structural damage without interruption in service or with minor temporary interruption in service.

Level 2 Earthquake: Controlled inelastic behavior with repairable damage resulting in temporary closure in service restorable within months (prevention of structural collapse) and the prevention of a major spill, defined as 1200 barrels of petroleum product.

Currently, the initial earthquake parameters are taken from the U.S. Geological Survey (USGS) web site, http://geohazards.cr.usgs.gov/eq/html/canvmap.html. They are available as peak ground acceleration and spectral acceleration values at 5 percent damping for 10, 5, and 2 percent probability of exceedance in 50 years, which correspond to Average Return Periods of 475, 975, and 2,475 years, respectively. The spectral acceleration values are available for 0.2, 0.3, and 1.0 second spectral periods. In obtaining peak ground acceleration and spectral acceleration values from the USGS web site, the site location can be specified in terms of site longitude and latitude or zip code when appropriate. It should be noted that for existing MOTs located on weak soils or near a known, active fault, a site-specific hazard assessment is required. The seismic load criteria and the performance-based requirements are similar to those described in FEMA 356 [4].

The seismic criteria for the three existing facility classifications (high, moderate and low), for the two performance levels, are as follows:

TABLE 2. DESIGN EARTHQUAKE MOTIONS FOR EXISTING MOTs

Facility Classification		Probability of Exceedance	Return Period
High	Level 1	50% in 50 years	72 years
	Level 2	10% in 50 years	475 years
Moderate	Level 1	64% in 50 years	50 years
	Level 2	14% in 50 years	333 years
Low	Level 1	75% in 50 years	36 years
	Level 2	19% in 50 years	238 years

For new marine oil terminals, a site-specific seismic hazard assessment is required, using the "high" classification for all cases (50% probability in 50 years and the 10% probability in 50 years, for Level 1 and 2 respectively).

The analytical procedures used for the seismic analysis range from simple to very complex. Minimum levels of sophistication are required by the standards, with the procedures becoming more complex for higher risk classification levels. The minimum procedure requirements are summarized in Table 3. Complex analytical procedures, such as full nonlinear time-history analysis, are not required for any facility, but it should be noted that the more complex analytical procedures may result in less rehabilitation, since the procedures account for the residual capacity of structures in the inelastic range of material properties.

TABLE 3. MINIMUM REQUIRED ANALYTICAL PROCEDURES

Classification	Substructure Material	Demand Procedure	Capacity Procedure
High/Moderate Irregular	Concrete/Steel	Linear Modal Procedure	Nonlinear Static Procedure
High/Moderate Regular	Concrete/Steel	Nonlinear Static Procedure	Nonlinear Static Procedure
Low	Concrete/Steel	Nonlinear Static Procedure	Nonlinear Static Procedure
High/Moderate/Low	Timber	Nonlinear Static Procedure	Nonlinear Static Procedure

After conducting the structural analysis, the component capacity is established on a performance basis, using limiting levels of displacement or strain. Some of the strain levels prescribed in MOTEMS are as follows:

Description	Strain

Unconfined concrete piles
(no confining steel, or spacing greater than 12 inches)   0.005

Ultimate reinforcing steel tensile strains (Existing structures)

Grade 40 steel   0.15

Grade 60 steel   0.12

Ultimate prestress strand strain   0.06

Confined Concrete Piles:

Pile/deck hinge (Level 1 earthquake)   <0.005
Pile/deck hinge (Level 2 earthquake)   <0.025
In-ground hinge (Level 1 earthquake)   < 0.005
In-ground hinge (Level 2 earthquake)   <0.008

Other current activities, in addition to the MOTEMS project are directly related to the seismic vulnerability of marine oil terminals. One MOT in Northern California recently completed a major seismic rehabilitation, using the procedures of MOTEMS, and is placing accelerometers along the wharf. The hardware will be tied into California's Strong Motion Instrumentation Program (CSMIP) and will quickly provide in-structure acceleration records. Within minutes, the time history records would be available on the Internet. In addition to this marine oil terminal, the Port of Oakland has agreed to instrument both new and rehabilitated wharves.

These records can be used to calibrate the new analysis methodologies presented in MOTEMS, correlating finite element models to actual behavior. With the maximum in-structure acceleration known, the operator and regulator can quickly evaluate whether or not a facility should remain operational. An additional benefit to having accelerometers is that they may provide a date/time stamp for major impact loads from vessels or barges.

7. Mooring Loads

Since its inception in late 1990, the MFD has been concerned about the mooring of vessels at marine oil terminals. Most of these facilities were designed and built long ago, to handle much smaller vessels. Larger vessels have greater sail areas resulting in greater wind loads, as well as larger hull areas below the water line resulting in higher current loads. Aging structures with older design fender types may not be adequate for higher impact loads. The problem is further complicated by the new generation of double-hulled vessels, which have even larger sail and hull areas, with the same cargo capacities.

Two load cases are specified for both new and existing facilities:
(1) Survival Condition:
25 year return period wind, with 30 second duration

For an existing facility, the vessel may alternatively choose to leave berth when the operational wind speed is exceeded, within 30 minutes. In such cases, a suitable operational wind envelope can be developed.

(2) Operating Condition:

The operating condition is determined by the mooring analysis, using appropriate safety factors for lines and mooring points. An operational wind rose is required, with limiting speed/direction. When the operational wind speed is exceeded, the vessel must terminate the transfer operation.

The analytical procedures required for the mooring assessment are consistent with those of the Oil Companies International Marine Forum publications [5].

8. Tsunami and Seiche

In addition to determining mooring and berthing forces due to wind, waves, and current, the standards also address requirements for seiche and tsunamis. Run-up heights will be added to the MOTEMS, as the research at the University of Southern California is completed. Information about near-field subsea landslide induced tsunamis is also being investigated, along with triggering mechanisms and postulated run-up heights. In the interim, while the research is being completed, the tsunami hazards are addressed by operational measures. A tsunami generated by a distant source (far field event) may allow for adequate warning time for operators to request tank vessels to depart the MOT and move into deep water. The MOT is required to have a plan with specific actions for responding to far-field events.

9. Conclusions

Upon completion, the new standards will be the first set of comprehensive engineering standards developed specifically for MOTs in the U.S. The background for the standards is based on the MFD's experience of inspecting and reviewing mooring and structural upgrades during the past nine years. The standards fulfill the mandate of providing regulations for the performance standards of marine oil terminals, to minimize the possibilities of a discharge of oil. However, much of the criterion is equally applicable to other types of marine structures. It is expected that these standards will form the basis for similar standards developed for other state, federal, and international regulatory agencies. The potential application of this effort is worldwide, for any new or existing marine oil terminal, regardless of the seismic hazard classification of the region [6]

References

1. California State Lands Commission. *Marine Oil Terminal Engineering and Maintenance Standards,* Draft #6, December 2000.
2. American Society of Civil Engineers, Ports and Harbors Committee, *Underwater Investigations Standard Practice Manual*, May 2001.
3. Ferritto et al., *Seismic Criteria for California Marine Oil Terminals.* Port Hueneme, CA, Naval Facilities Engineering Service Center, 1999.
4. "*Prestandard and Commentary for the Seismic Rehabilitation of Buildings*", Federal Emergency Management Agency, FEMA 356, November 2000.
5. *Mooring Equipment Guidelines.* Oil Companies International Marine Forum, 2nd Edition, London, England: Witherby & Co, LTD, 1997.
6. "*Seismic Design Guidelines for Port Structures*", Working Group No. 34 of the Maritime Navigation Commission International Navigation Association", A.A. Balkema Publishers, Lisse, 2001.

A TSUNAMI MITIGATION PROGRAM WITHIN THE CALIFORNIA EARTHQUAKE LOSS REDUCTION PLAN

RICHARD J. MCCARTHY and ROBERT L. ANDERSON
California Seismic Safety Commission
1755 Creekside Oaks Drive, Suite 100
Sacramento, California 95833, USA

Abstract

Tsunami assessment, education, warning, and mitigation efforts are intended to reduce losses related to tsunamis. California is participating in an effort with the states of Alaska, Hawaii, Oregon and Washington as well as the Federal Emergency Management Agency, the National Oceanic and Atmospheric Administration, and the United States Geological Survey in developing tsunami hazard risk reduction techniques under the National Tsunami Hazards Mitigation Program. This program supports the California Earthquake Loss Reduction Plan through the geosciences, landuse, and (indirectly) education elements of the Plan. This paper briefly describes the relationship between the National Tsunami Hazard Mitigation Program and the California Earthquake Loss Reduction Plan and highlights tsunami hazards assessment, education, warning, and mitigation in California.

1. Introduction

1.1 WHY WORRY ABOUT TSUNAMIS?

California has been subjected to fourteen known locally generated tsunamis dating back to 1800 which have not been assessed and mitigated to the same level, as have earthquakes. This is because over the last 200 years earthquakes have caused more damage and loss of life than tsunamis in California. Tsunamis, however, have caused significant damage in California in both Santa Barbara and Crescent City.

1.2 FOUNDING OF THE CALIFORNIA SEISMIC SAFETY COMMISSION AND THE ESTABLISHMENT OF THE CALIFORNIA EARTHQUAKE LOSS REDUCTION PLAN

California's Seismic Safety Commission (Commission) was established, as a direct result of the 1971 San Fernando earthquake, by legislation that took effect on January 1, 1975. The legislation directed the commission to engage in the following activities:
- Set mitigation goals and priorities in the public and private sectors
- Request state agencies to devise criteria to promote earthquake and disaster safety

A. C. Yalçıner, E. Pelinovsky, E. Okal, C. E. Synolakis (eds.)
Submarine Landslides and Tsunamis 267-276
©2003 Kluwer Academic Publishers.

- Recommend program changes to state agencies, local agencies, and the private sector to further seismic safety
- Provide incentives for research
- Coordinate earthquake safety activities of government at all levels

Because of its desire to maintain the momentum of a goal and policy-setting process, the State established the *California Earthquake Hazard Reduction Act of 1986*. Enactment of the Act followed the devastating Mexico City earthquake of 1985, which brought home the specter of massive urban losses. The 1986 law also required the Commission to develop a series of five-year strategic plans that focused on reducing earthquake losses and speeding recovery.

The *California Earthquake Loss Reduction Plan* (Plan) continues a planning process that began over 25 years ago. Although the Commission has taken an appropriate new look and a somewhat different emphasis, it has done so with a continued commitment to the original goals and the intent that the Plan serves multiple purposes:

- First, it continues to be the Commission's policy statement about what needs to be done to reduce earthquake risk over the long term
- Second, it is the state's strategic plan guiding the Executive Branch agencies in their overall implementation strategies and priorities for seismic safety
- Third, it complies with the Federal Emergency Management Agency's *National Hazards Mitigation Strategy* and serves as the state's federally required hazard mitigation plan for earthquakes.

The Plan is a roadmap to achieve a safer California. The Plan contains 11 elements, each addressing a distinct but interrelated area of concern. The Plan sets forth statewide objectives and strategies to support the goals of learning from earthquakes, advancement in building for earthquakes, and advancement in living with earthquakes. The goals, objectives, and strategies presented in the Plan address the state's most pressing seismic issues through the following 11 elements:

- Geosciences
- Research and Technology
- Education and Information
- Economics
- Land Use
- Existing Buildings
- New Construction
- Utilities and Transportation
- Preparedness
- Emergency Response
- Recovery

Until the early 1990s, the tsunami risk to the California coast focused on the hazards presented by distant sources (Japan, Alaska, and South America). However, in 1992, the Commission began to identify potential tsunami hazards from coastal California sources and, after reviewing historical data, determined that a policy must be developed and implemented to address hazards presented from near shore tsunami sources. In response to the determination, the Commission included tsunami hazard in the *California Earthquake Loss Reduction Plan* and proposed specific questions to be addressed. The questions listed below were selected to be addressed so that a frame work for assessing tsunami hazard and preparing strategy for tsunami risk mitigation could be developed. The answers to the questions are still being developed.

Risk Issues
- What is the current risk to California from distant and locally generated tsunamis?
- Are the current mitigation actions by federal and state government appropriate to the level of risk?
- Are there better or more effective ways to assess tsunami risk given the scarcity of data?
- How can acceptable levels of tsunami risk be determined?
- What should be the highest priority for run-up studies? (Ports and harbors or developed beach areas with high populations)?

Preparedness Issues
- Does the current tsunami warning system operated by the National Oceanic and Atmospheric Administration meet California's needs?
- Could the present warning system provide sufficient notice to alert coastal residents that a tsunami has been generated from a local offshore source?
- What is the least amount of time that a warning can be issued?
- How is a tsunami warning issued?
- How is the information released to the public?
- Could the present seismic networks in California be modified to function as or form the foundation of a local tsunami warning center?

Education Issues
- What should people know about tsunamis?
- What methods have been most effective in educating the public to the tsunami hazard in the United States and other countries?
- In countries with public information programs, do residents resist evacuating? If so, why?

Recovery Issues
- Are there unique circumstances presented by tsunamis that impact recovery?
- What federal programs and insurance coverage apply to tsunami damage?
- If tsunami run-up areas were identified for California, how would the insurance industry respond?

Land Use Issues
- If tsunami run-up areas were identified for California, should there be specific design requirements for structures located in the inundation zone? How have other states and countries approached this problem?
- Design or avoidance, which is the best tsunami mitigation for a development proposed in an inundation zone?
- Are there classes of use that should be prohibited from locating within an inundation zone (hospitals, schools, essential services)?
- Are there classes of use that cannot avoid being located in inundation zones (ports, marinas, water dependent industries)?

Addressing the questions presented above represents a major step in moving the State's tsunami mitigation program in a direction similar to land use laws that require mitigation for development located within liquefaction and earthquake induced landslide zones. This paper is not intended to address all of the questions.

Regrettably, tsunami inundation areas were not included along with liquefaction, and earthquake induced landslides in the development of seismic hazard zone maps mandated by the State in the early 1990s. Incorporation of all three hazards into one overall land use policy is the most practical, efficient, and cost effective way to identify and mitigate these hazards. Today, the tsunami hazard and associated mitigation activities of the National Tsunami Hazard Mitigation Program are a major part of the *California Earthquake Loss Reduction Plan*. This fact, along with recognition and funding assistance by the Federal Emergency Management Agency and the National Oceanic and Atmospheric Administration has fostered a major tsunami risk reduction program for California and other states in or along the Pacific Ocean.

2. Discussion

2.1 STATE AND FEDERAL ACTIVITIES REGARDING TSUNAMI AND LOSS REDUCTION

The West Coast and Alaska Tsunami Warning Center and the Pacific Tsunami Warning Center, have been providing tsunami warnings for California from distant and local tsunamis. The current tsunami warning system operated by the National Oceanic and Atmospheric Administration does not fully meet California's needs. This is due to California being vulnerable to both local and distant source tsunamis. There are areas along the California Coast that are considered to be tsunami generation regions (see table 1 and figure 1). A locally generated tsunami may not be detected in sufficient time for one of the tsunami warning centers to provide adequate warning time for local authorities to evacuate areas impacted by the tsunami. The National Tsunami Hazards Mitigation Program (NTHMP) has identified the need for additional data collection capability (filling of data gaps), data analysis, tsunami modeling, and continued development, deployment, and implementation of tsunami warning guidance tools.

Near-Shore Potential Tsunami Sources Along the California Coast

Figure 1. Near-Shore Potential Tsunami Sources Along the California Coast [7]

The NTHMP tsunami detection and data collection efforts include: deploying additional tidal gauges, or replacing tidal gauges, deep water tsunami buoys, seismometers, and tsunami damage observations. The collection and analysis/interpretation of tsunami data and tsunamigenic data is needed in order to develop new or revise existing tsunami models. With better data, models, and mapping of tsunami inundation zones, area and site specific mitigation

TABLE 1. Local Tsunami Source Regions of California [7]

SOURCE ZONE	MAJOR OFFSHORE FAULTS	MAJOR SUBMARINE CANYONS	EARTHQUAKE MAGNITUDE AND YEAR	HISTORICAL TSUNAMI RUNUP AND YEAR
Crescent City to Cape Mendocino	Little Salmon Fault(T) Mad River Fault Zone(T) Mendocino Fault(S) Cascadia Subduction Zone(T)	Trinity, Eel, Mendocino, Mattole	Ms=7.4 (1923) M=7.2 (1923)	1.1-m (1992)
Cape Mendocino to San Francisco	San Andreas Fault(S) Point Reyes Fault(T)	Spanish, Delgada, Vizcaino, Noyo, Navarro, Arena, Bodega	M=7.7 (1906)	0.1-m (1906)
San Francisco to Monterey	San Gregorio Fault(S)	Pioneer, Ascension, Monterey	M=7.1 (1989)	0.3-m (1989)
Monterey to Point Arguello	Hosgri Fault Zone(RS) Santa Lucia Bank Fault(RS?)	Sur, Lucia	Ms=7.3 (1927)	0.6-m (1927)
Point Arguello to Los Angeles (Santa Barbara Channel and Santa Monica Bay)	Santa Barbara Channel Faults(T) Anacapa-Dume Fault Zone(RS) Santa Monica Fault(T)	Arguello, Hueneme, Mugu, Dume, Santa Monica, Redondo	MI=7.7 ½ (1812)	3-4m (1812)
Los Angeles to San Diego (Inner Borderland)	San Clemente(R) Catalina – San Diego Trough(S-RS?) Palos Verdes(RS) – Coronado Bank(NS) Newport-Inglewood – Rose Canyon(S)	San Gabriel, Newport, Carlsbad, La Jolla, Coronado	M=6.25 1933	Uncertain (1862, 1933)
Northern Channel Islands to San Nicolas Island (Northern Outer Borderland)	East Santa Cruz Basin Fault Zone(S) Ferrelo Fault Zone(S) San Nicolas Island Escarpment(T?)	Santa Cruz	ML=5.1 (1969)	?
San Nicolas Island to Mexican Border Southern Outer Borderland	East Santa Cruz Basin Fault Zone(S) Ferrelo Fault Zone(S)	Unnamed	ML=5.3 (1948)	?

T = Thrust Fault; RS = Reverse Oblique Fault; NS = Normal Oblique Fault; S = Strike Slip Fault; Ms = Surface Wave Magnitude; M = Moment Magnitude; MI = Seismic Intensity Magnitude; ML = Local Magnitude; = Magnitude Unknown

needs can be better assessed. The NTHMP is an excellent vehicle to continue to use to develop and deploy tsunami hazards monitoring and warning systems and to develop tsunami hazard mitigation schemes.

The California Public Resources Code also includes tsunamis as a subset of seismic hazards (Public Resources Code Section 2692.1) within the State of California Seismic Hazards Mapping Act. Development of tsunami inundation maps for California is underway by the University of Southern California Tsunami Research Center and the Tsunami Inundation Mapping Effort (TIME) Center in Newport, Oregon. The State of

California, the National Oceanic Atmospheric Administration, and the Federal Emergency Management Agency are supporting the development of these tsunami inundation maps. Tsunami inundation maps for California outside of Crescent City and Humboldt Bay have not been developed for use for other than evacuation planning in California.

California is a subscriber to the West Coast and Alaska Tsunami Warning Center, located in Palmer, Alaska and in the Pacific Tsunami Warning Center in Ewa Beach, Hawaii. Other states subscribing to the Alaska Tsunami Warning Center and the Pacific Tsunami Warning Center include Oregon, Washington, Alaska, and Hawaii. California is also a partner in the NTHMP. California does not have a separate tsunami hazard mitigation plan. Instead of developing a stand-alone tsunami hazard assessment, warning and mitigation program the State elected to partner with the NTHMP. The NTHMP consists of three interdependent components [1]:

- Hazard assessment
- Warning guidance
- Mitigation.

The three components fit into the Plan under the geosciences and land use elements. The geosciences element calls for the development of data to provide accurate and useful planning scenarios to reduce the risk from the hazards of seiches and tsunamis. The land use element of the Plan calls for the identification of all areas subject to potential inundation from dam or levee failure or tsunami run-up, and the incorporation of appropriate loss reduction strategies to be incorporated into general plans ([2], page 12). The tsunami run-up identification effort is well underway by the University of Southern California Tsunami Research Center and the TIME Center.

2.1.1 California Tsunami Assessment, Warning Guidance, and Mitigation Activities

The following tsunami assessment, warning guidance, and mitigation activities are underway or in planning and research stages in California:

2.1.1.1 Assessment: The geosciences initiative is directly related to tsunami assessment activities. The geosciences initiative regarding tsunamis is carried out by the collection and processing of data, the development of tsunami generation and propagation models, and the development of tsunami inundation maps. The collection, processing, and analysis of data is done not only for tsunamis that have affected California but is also reviewed for tsunamis in other parts of the world. Recent tsunamis generated by the Manzanillo, Mexico earthquake of 1995, the Papua New Guinea earthquake of 1998, and the Kocaeli, Turkey earthquake of 1999 have been evaluated. The installation of tidal gauges, seismometers, and collection of first hand accounts, has also helped to fill in data gaps. Modeling of tsunamis has led to a beginning in the understanding of tsunamigenic events. A better understanding of sea floor topography has led to a refinement in the estimate of California coastal areas that are prone tsunami inundation. So far, the majority of information regarding marine geology has been provided by the USGS. Future activities may also include data provided by the United States Navy, oil companies, and project Earthscope (if funded).

2.1.1.2: Several tsunami inundation maps cover portions of the Southern California Coast have been completed and several more are under development in both Southern and Northern California.

2.1.2.1 The warning guidance effort of California follows that of the NTHMP. However, not all of the 15 coastal counties of California have adopted the NTHMP tsunami warning guidance in their general plans. Areas of California such as Crescent City, Del Norte County, and Humboldt County have adopted a more visible warning guidance stance. Tsunami hazard inundation posting has been established in Crescent City and in the Eureka/Arcata Bay areas of Northern California. The California Department of Transportation is assessing the use of the national tsunami hazard and evacuation signs. The Crescent City draft publication entitled "Tsunami" is under consideration for revision by OES for applicability on a statewide basis. OES, the California Coastal Commission, and Humboldt State University are planning to develop a tsunami curriculum for kindergarten through twelfth grade students. The curriculum will enable children to understand what tsunamis are, what hazards are associated with tsunamis, in general, when they may be expected, and finally how to protect oneself from tsunami hazards. The curriculum is to be partially based upon a similar program developed by the State of Washington. Another useful education tool is United States Geological Survey Circular No. 1187 [3] entitled "Surviving a Tsunami-Lessons from Chile, Hawaii, and Japan". This document contains strategies that an individual may follow when a tsunami is expected.

2.1.3 Mitigation

2.1.3.1 The completed maps are to be used in the planning of evacuation routes and posting of areas that are in a tsunami hazard inundation area. Tsunami mitigation has included the development of tsunami evacuation plans for several communities. So far, no significant County or City ordinances have been adopted restricting coastal development in California with respect to tsunami hazard. There are four documents that are useful in understanding the disposition of the tsunami risk reduction activities implemented in California. The documents are briefly described below.

Designing for Tsunamis Seven Principles for Planning and Designing for Tsunami Hazards, National Tsunami Hazard Mitigation Program, dated March 2001. The document is a guideline to help coastal communities understand tsunami their hazard potential and risks in addition to providing guidance for the mitigation of tsunami risk through land use planning and building design.

Local Planning Guidance on Tsunami Response, Second Edition, A Supplement to the Emergency Planning Guidance of Local Governments, dated May 2000; State of California Office of Emergency Services [4] The guidance document contains a template for developing planning activities to mitigate tsunamis and a sample tsunami warning checklist based upon efforts of the County of San Mateo, California.

Findings and Recommendations for Mitigating the Risks of Tsunamis in California, dated September 1997, State of California Office of Emergency Services [5]. This document called for the development of tsunami inundation maps for California which are to be used for planning evacuation routes.

Planning Scenario in Humboldt and Del Norte Counties, California for a Great Earthquake on the Cascadia Subduction Zone, California Division of Mines and Geology, Special Publication No. 115, dated 1995 [6]. The scenario contains information regarding the 1964 Crescent City tsunamis as well as graphics regarding potential tsunami runup zones.

3. Summary

3.1 The 1992 Cape Mendocino earthquake demonstrated that in addition to being susceptible to distant source tsunamis, California is also susceptible to locally generated tsunamis and to tsunamis generated from earthquakes along the Cascadia Subduction zone. Although California had been subjected to other more destructive tsunamis in the past, the Cape Mendocino event led to the inclusion of tsunami hazard initiatives in California's Earthquake Loss Reduction Plan. Instead of implementing a stand alone tsunami hazard assessment and mitigation program, California's Office of Emergency Services has been able to share in the collection, processing, and analysis of data, the development of tsunami modeling, the cost of mapping tsunami inundation zones, the development of tsunami hazard warning guidance materials, as well as the development of tsunami hazard mitigation techniques by working with universities, federal, state (including Washington, Oregon, Alaska, and Hawaii) and local agencies. The NTHMP supports the Plan's initiative with respect to geosciences, land use and education. State activities regarding tsunami education are being planned and is expected to be partially based on the State of Washington's tsunami hazard education program.

The successful interaction of the NTHP and *California's Earthquake Loss Reduction Plan* points to the value of incorporation of tsunami risk reduction activities into existing or developing earthquake loss reduction programs or national hazard reduction programs. It is the intent of the authors to encourage the incorporation of tsunami risk reduction activities into national and or local earthquake risk reduction or natural hazard reduction programs for countries that have potential tsunami exposure.

4. Acknowledgements

4.1 The authors acknowledge the invaluable assistance of Mr. James Godfrey from the State of California, Office of Emergency Services provided invaluable background information regarding current tsunami hazard assessment, warning guidance, and mitigation activities and Mr. Vincent Vibat of the California Seismic Safety Commission provided graphics and formatting support for the paper.

5. References

1. Bernard, E. N. (1998) Program Aims to Reduce Impact of Tsunamis on Pacific States, Eos Transactions, American Geophysical Union, Vol. 79, No. 22, June 2, 1998, pp 258, 262-263.
2. California Seismic Safety Commission (1997) California Earthquake Loss Reduction Plan, 1997-2001, 49 pages.
3. Atwater, B.F., Cisternas, M.V., Bourgeois, J., Dudley, W.C., Hendley, J.W., and Stauffer, P.H. (1999) United States Geological Survey Circular No. 1187, 19 pages.
4. California Office of Emergency Services (2000) Local Planning Guidance on Tsunami Response, Second Edition, A Supplement to the Emergency Planning Guidance for Local Governments, 195 pages.
5. California Office of Emergency Services (1997) Findings and Recommendations for Mitigating the Risks of Tsunamis in California, 30 pages.
6. California Division of Mines and Geology (1995) Planning Scenario in Humboldt and Del Norte Counties, California for a Great Earthquake on the Cascadia Subduction Zone, Special Publication No. 115.

7. McCarthy, R. J., Bernard, E.N. and Legg, M.R. (1993) The Cape Mendocino Earthquake: A Local Tsunami Wakeup Call? In Coastal Zone '93 Volume 3, Proceedings of the Eight Symposium on Coastal and Ocean Management, American Society of Civil Engineers, Pages 2812-2828.

SHORT-TERM INUNDATION FORECASTING FOR TSUNAMIS

V.V. TITOV[1,2], F.I. GONZÁLEZ[2], H.O. MOFJELD[2] and J.C.NEWMAN[1,2]

[1]*Joint Institute for the Study of the Atmosphere and Ocean, University of Washington*
[2]*NOAA/Pacific Marine Environmental Laboratory, Seattle, Washington*

Abstract

Since 1997, PMEL has been involved in the R&D effort to provide tsunami-forecasting capabilities for the Pacific Disaster Center (PDC) in Hawaii. As a part of this effort, modeling tools for the short-term forecasting and assessing the risk of tsunami inundation have been developed. The Short-term Inundation Forecasting for Tsunamis (SIFT) will involve gathering information from several observation system – seismic network, Deep-ocean Assessment and Reporting of Tsunamis (DART) and coastal tide-gages –, "sifting" through the information that will be used for tsunami modeling and, finally, provide inundation forecast for selected communities based on simulation results.

The modeling part of the SIFT project will employ a two-step procedure. The first step will estimate the offshore wave heights using a database of the pre-computed tsunami propagation runs. This phase will utilize linearity of the tsunami propagation to construct a solution that matches observations for a particular event. This offshore forecasting methodology has been implemented for the PDC [1] to predict tsunami amplitudes in deep ocean for tsunamis originated in Alaska. The second step of the tsunami forecasting procedure will include model estimates of tsunami inundation for specified coastal sites. The inundation modeling will use the offshore estimates from the first step as input to obtain amplitude and current velocity estimates of tsunami inundation for selected sites.

1. The Need for SIFT

Emergency managers and other officials are in urgent need of operational tools that will provide Short-term Inundation Forecasting for Tsunamis (SIFT) as guidance for rapid, critical decisions in which lives and property are at stake. These decision-makers must issue warnings, direct vessels to put to sea, order the evacuation of coastal communities, send search-and-rescue teams into the disaster area, and sound an "all clear" that officially declares to citizens and vessel operators that it is safe to return to homes, businesses, coastal ports and harbors. Because such decisions must be made throughout the life of the emergency, forecasts must be constantly updated to assess the current hazard and provide continual guidance during the entire duration of the event.

A. C. Yalçıner, E. Pelinovsky, E. Okal, C. E. Synolakis (eds.)
Submarine Landslides and Tsunamis 277-284.
©2003 Kluwer Academic Publishers.

2. Exploitation of Measurement and Modeling Technologies

Recent advances in tsunami measurement and numerical modeling technology can be exploited, combined and integrated to create an optimal tsunami forecasting system. Neither technology can do the job alone. Observational networks will never be sufficiently dense; the ocean is vast, and establishing and maintaining monitoring stations is costly and difficult, especially in deep water. Numerical model accuracy is inherently limited by errors in bathymetry and topography and uncertainties in the generating mechanism, especially in the near-field; furthermore, case studies that compare model results with observations generally demonstrate that model accuracy degrades with successive tsunami waves.

Weather forecasting is an obvious example in which models and real-time data are combined to improve predictions by using data assimilation techniques to optimize agreement between model simulations and observations. Essentially, the model functions as a sophisticated space-time interpolator between measurement stations.

3. Time as a Fundamental Constraint

Even a perfect SIFT capability requires a finite period of time to acquire seismic and tsunami data and execute a forecast algorithm. And it can be expected that the more time available to prepare a forecast, the more effective the forecast -- more observational data can be incorporated into the forecast and more time is available for evacuation or other appropriate community responses. Thus, in emergency management terms, SIFT will be more effective in the case of a "distant tsunami" than the case of a "local tsunami."

This can also be stated in simple mathematical terms. If t_M is the tsunami travel time to a measurement station, t_C is the arrival time at a coastal community, and T is the (effective) period of the first tsunami wave, then the time available to acquire adequate tsunami data and provide a forecast before the tsunami strikes is approximately

$t_F = t_C - (t_M + T/4)$.

Here, it is assumed that the forecast algorithm execution time is negligible, that the magnitude of the first tsunami extremum is essential input to the algorithm, and $T/4$ is the time delay for arrival of the first extremum at the measurement station. This can be written

$t_F = (t_C - t_M) - T/4 = t_{CM} - T/4$,

where t_{CM} is the time difference between arrival of the first wave at the coastal community and the measurement station. Thus, the greater the inequality $t_{CM} > T/4$, the more effective we can expect the forecast to be.

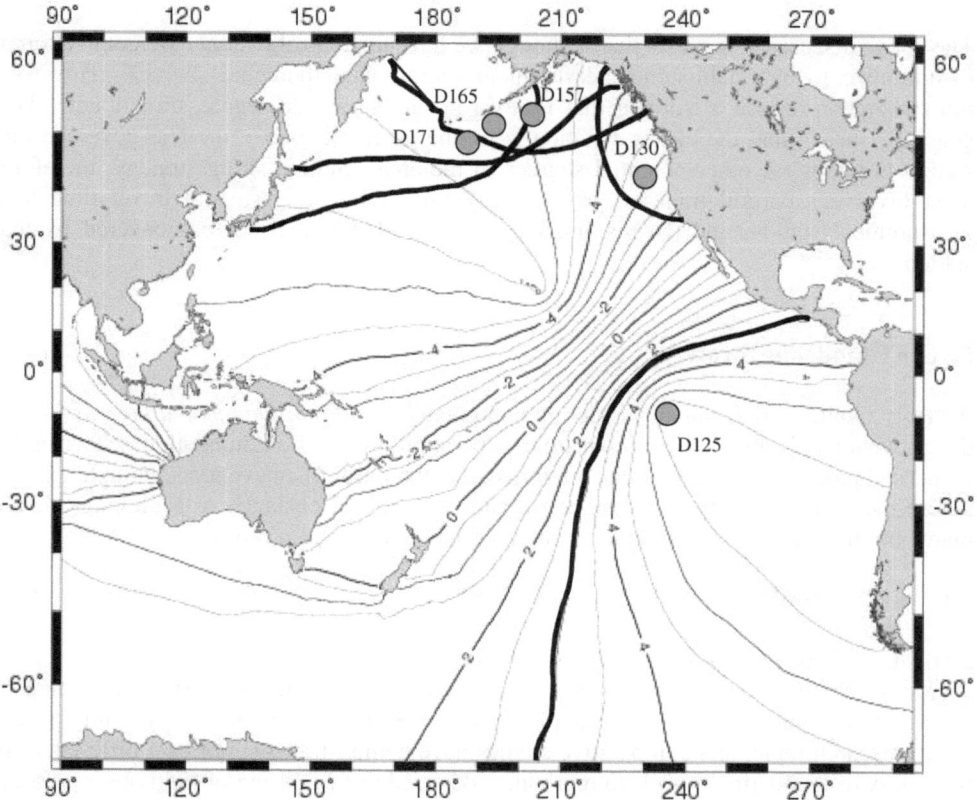

Figure1. Time difference (t_{CM}) between tsunami arrival at Hilo and at D125 DART station (thin lines). Thick lines outline sources of tsunamis that reach at least one DART location 3 hours before Hilo.

Physically, t_{CM} is maximized if the measurement station is near the source and the threatened community is far from the source i.e., if t_C is large and t_M is small. It is minimized when either (a) the community is very near the source, i.e., t_C is small, or (b) the station is very near the community, i.e., $t_C \sim t_M$, regardless of the source location. We also note that the wave period, T, is typically in the range 5-45 minutes, but is generally longer for large tsunamis and shorter for small tsunamis. In addition, for a given tsunami, it can be expected that dispersive effects will result in a shorter T in the near-field.

Figure 1 illustrates this concept using Hilo as an example of coastal community. The plot shows the difference between tsunami travel time (t_{CM}) to Hilo and to D125 DART location for every point in the Pacific. The thick line is the 3-hour contour, which outlines sources of tsunamis that would arrive at D125 three hours earlier than at Hilo, leaving enough time for evacuation decision. The 3-hour contours are also shown for other DART locations. The envelope of thick contours outlines sources of tsunamis that would be detected by at least one DART system in time to decide on evacuation at Hilo.

4. SIFT Measurement Strategy

The above discussion suggests that, although we have no control over t_C, we could attempt to minimize t_M by establishing measurement stations near potential sources. However, most sources produce a tsunami that is, to a greater or lesser degree, directional. This generally means that the closer a station is to one potential source, the less coverage that station provides for other potential sources in the area. Station siting strategy, therefore, must involve a careful trade-off between (a) minimizing t_M, tsunami travel time to a measurement station, and (b) maximizing the area of potential sources covered by that station.

5. SIFT Modeling Strategy

A "Holy Grail" of the tsunami modeling community is the *real-time model forecast* -- i.e., the provision of site- and event-specific generation/propagation/inundation scenarios by numerical model computations that are initiated at the onset of an event and completed well before the first wave arrives at threatened sites. Technical obstacles to achieving this are many, but three primary requirements are *accuracy, speed,* and *robustness*.

5.1. ACCURACY

Errors in two broad categories are especially important:

Model Physics Error. The physics of propagation are better understood than that of generation and inundation. For example, landslide generation physics is currently a very active research topic, and comparative studies have demonstrated significant differences in the ability of inundation models to reproduce idealized test cases and/or field observations.

Model Input Error. This issue is also known as the "garbage in, garbage out" problem, i.e., model accuracy can be degraded by errors in (a) the Initial Conditions set for the sea surface and water velocity, due to inadequate physics and/or observational information, and (b) the bathymetry/topography computational grid, due to inadequate coverage, resolution and accuracy, including the difficult issues encountered in merging data from different sources.

Generally speaking, the practical manifestations of these errors are *temporal* and *near-field degradation* of model accuracy. *Temporal degradation* refers to the fact that only the first few wave computations can be considered reliable. Thus, forecasting later tsunami waves with existing deterministic numerical models is generally not possible. However, near-site tsunami observations can be subjected to efficient, newly-developed statistical algorithms that provide estimates of the largest future tsunami wave expected during a particular event. *Near-field degradation* refers to the fact that near-field computations are generally very sensitive to spatio-temporal details of the source and computational grid. In contrast, far-field solutions are much less sensitive to spatio-temporal details of the source; for example, numerical experiments demonstrate that the far-field solution depends primarily on the magnitude and location of the epicenter and is relatively insensitive to other earthquake parameters. Furthermore, hindcast studies indicate that if existing numerical models are exercised and interpreted with care by an experienced tsunami modeler, they can produce results that adequately match measurements of the first few waves in the far-field.

5.2. SPEED.

We refer here to *forecast speed*, relating to the time taken to make the first forecast product available to an emergency manager for interpretation and guidance. This process involves at least two important, potentially time-consuming, steps:

Source Specification. Seismic wave data are generally available first, but finite time is required to interpret these signals in terms of descriptive parameters for earthquakes, landslides, and other potential source mechanisms. Tsunami waves travel much slower and, as noted above, time on the order of *T/4* will be needed to incorporate these data into a forecast. Seismic networks are much denser than tsunami monitoring networks, but inversion algorithms for both must be developed that provide greater source detail, more rapidly.

Computation. Currently available computational power can provide useful, far-field, real-time forecasts before the first tsunami strikes a threatened community, if the time available for forecasting, t_F , is sufficiently large, the source can be quickly specified, and an accurate computational grid is available. In fact, if powerful parallel computers and/or pre-computed model results are exploited, model execution time can be reduced almost to zero so that the minimum t_F required for an effective forecast might be very small, at least in principle.

In practice, of course, there will always be situations for which $t_F < 0$, i.e. in which the source-site geometry makes it impossible to provide a warning forecast. But even a late forecast will still provide valuable assessment guidance to emergency managers responsible for critical decisions regarding response, recovery, and search-and-rescue.

5.3. ROBUSTNESS.

With lives and property at stake, reliability standards for a real-time forecasting system are understandably high, and the development of such a system is a difficult challenge. It is one thing for an experienced modeler to perform a hindcast study and obtain reasonable, reliable results; such exercises typically take months to complete, during which multiple runs can be made with variations in the model input and/or the computational grid that are suggested by improved observations and/or speculative experimentation, and the results examined for errors and reasonableness. It is quite another matter to design and develop a system that will provide reliable results in real-time, without the oversight of an experienced modeler.

The previous discussion suggests that critical components of SIFT technology exist now that could provide rapid, usefully accurate forecasts of a limited but important category -- the first few waves of a far-field, earthquake-generated tsunami. In particular, it seems feasible to develop a forecast system which combines real-time seismic and tsunami data with a forecast database of pre-computed event- and site-specific scenarios that have been thoroughly tested and scrutinized for reasonableness and sensitivity to errors. Later waves could also be usefully forecast by processing real-time tsunami data with a statistical/empirical model. Implementation of this technology requires integration of these components into a unified, robust system. This is the strategy adopted by the NOAA/PMEL SIFT Project.

6. The NOAA/PMEL SIFT Project.

In the event of a Pacific Rim earthquake, the National Oceanic and Atmospheric Administration (NOAA) bears primary responsibility in the United States for real-time assessment of the tsunami hazard and, if warranted, the issuance of a warning. Similarly, the Department of Defense's Pacific Disaster Center (PDC) is responsible for providing timely and accurate information to emergency managers during a disaster. The PMEL Tsunami Program is developing a SIFT methodology for both early and later tsunami waves. Different modeling solutions are used for different stages of tsunami evolution: generation, propagation and inundation.

6.1. GENERATION

Generation process of tsunami wave may never be understood in all the details needed for short-term inundation forecasting. Many source models have been shown to produce very good agreement with observation. However, most of the data used for these source models are collected over long period of time after the event and often include tsunami inundation observations – the very data we are trying to forecast. This problem has been long recognized in the tsunami community and real-time tsunami observations have been identified as a key for better source determination [2].

Direct tsunami observation, however, may never be dense enough to obtain all the parameters of the tsunami source. Sensitivity studies have been performed to identify and reduce the number source parameters that are important for tsunami generation and for the short-term forecasting. Numerical experiments demonstrate that far-field solutions depend primarily on the magnitude and location of the epicenter and are relatively insensitive to other earthquake parameters. For local forecast, on the other hand, site-specific source characteristics – such as local asperities or earthquake induced landslides – are most important for tsunami forecast. In either case, only partial source characterization is needed for the tsunami forecast and limited tsunami observation can provide the data needed.

6.2. PROPAGATION

Sensitivity studies suggest that rapid, useful forecasts of the first few waves of a far-field tsunami may be provided by a limited-size database of pre-computed scenarios that have been thoroughly tested and scrutinized for reasonableness and sensitivity to errors. Web-based user interface has been developed to exploit the database approach for the offshore forecasting. The interface allows creating a variety of offshore tsunami propagation scenarios by combining pre-computed propagation solutions. Combined with data assimilation scheme, this approach can provide quick estimates of tsunami offshore propagation that conform to real-time buoy observations.

Data assimilation scheme is under development to combine real-time DART data of tsunami offshore amplitude with the forecast database to improve accuracy of an offshore tsunami scenario. This is a very important component of SIFT methodology, since it provides corrections for the initial imperfect "guesses" of the generation model.

6.3.   INUNDATION

Once the offshore scenario is obtained, the results of the propagation run are used for the site-specific inundation forecast. That is achieved in two ways. Fast preliminary estimates of inundation amplitudes can be obtained by a one-dimensional runup model (1 spatial dimension). The method uses results of the offshore simulation at a point near the target coastline as initial conditions for one-dimensional runup computation along selected bathy-topo transect. The approach assumes that the wave has little long-shore dissipation. This technique was pioneered by Titov & Synolakis ([3], [4]) and has been tested by modeling inundation amplitudes for a number of historical events ([5]; [1]; [4]). Although the method uses simplified assumptions about tsunami inundation it can provide timely preliminary estimates of averaged tsunami runup along uniform portion of a coast. The advantage of having this capability is the quickness of the prediction, which can be obtained virtually at the same time as the offshore forecast. Using the 1-D runup approach, the site-specific inundation forecast can be completed in a matter of seconds after receiving observation data. The speed and robustness of the 1-D inundation prediction can be exploited for a local tsunami forecasting, using direct input from a real-time gage measurement.

Inundation modeling in two spatial dimensions demands more computer power, time. A 2-D model needs much more input data to perform inundation forecast. At the same time, it is a more realistic simulation of the event and, therefore, can produce more reliable forecast. Again, input for the 2-D inundation computations are the results of the offshore forecast – tsunami parameters along the perimeter of 2-D inundation study area. Our tests show that inundation forecast for first several waves at Hilo, Hawaii for a tsunami originated near Alaska can be obtained in about 20 minutes from the time when the offshore forecast is complete.

In summery: to forecast inundation by early tsunami waves, seismic parameter estimates and tsunami measurements are used to *sift* through a pre-computed generation/propagation forecast database and select an appropriate (linear) combination of scenarios that most closely matches the observational data. This produces estimates of tsunami characteristics in deep water which can then be used as initial conditions for a site-specific (non-linear) inundation algorithm. A statistical methodology has been developed to forecast the maximum height of later tsunami waves that can threaten rescue and recovery operations. The results are made available through a user-friendly interface to aid hazard assessment and decision-making by emergency managers.

The focus of this initial effort is on forecasting inundation at selected sites in Hawaii by tsunamis generated by earthquakes in North Pacific. Computations of generation-propagation scenarios for the forecast database are performed by the Method of Splitting Tsunami (MOST) model. Fast, preliminary inundation forecasts are computed by a 1-dimensional model; slower, but more detailed and accurate forecasts are provided by a 2-dimensional model. Tsunami measurements are provided by local Hawaii tide gauge stations and by real-time reporting tsunami measurement stations established by the PMEL Deep-ocean Assessment and Reporting of Tsunamis (DART) Project.

284

Acknowledgements

This work is jointly funded by NOAA, the National Tsunami Hazard Mitigation Program to develop tsunami inundation maps for U.S. coastal communities and by the DoD/NASA Pacific Disaster Center Program to improve short-term tsunami inundation forecasting, NOAA's contribution number 2487. The study was supported in part by the Joint Institute for the Study of the Atmosphere and Ocean under NOAA Cooperative Agreement #NA17RJ11232, contribution number 940.

References

1. Titov, V.V. and F.I. Gonzalez, (1999): Numerical study of the source of the July 17, 1998 PNG tsunami, *IUGG'99 Abstracts*, Birmingham, 19-30 July 1999.
2. Synolakis, C.E., P. Liu, G. Carrier and H. Yeh (1997): Tsunamigenic Sea-floor Deformations. 1997, *Science*, 278, 598-600.
3. Titov, V.V. and Synolakis, C.E. (1995): Modeling of Breaking and Nonbreaking Long Wave Evolution and Runup using VTCS-2, *Journal of Waterways, Ports, Coastal and Ocean Engineering*, Nov./Dec., 121, 6, 308-316.
4. Titov V.V. and Synolakis C.E. (1993): A numerical study of the 9/1/92 Nicaraguan Tsunami, *Proc. of the IUGG/IOC Int. Tsunami Symposium*, Wakayama, Japan. Published by the Japan Society of Civil Engineers, pp. 585-598.
5. Bourgeois, J. C. Petroff, H. Yeh, V. Titov, C. Synolakis, B. Benson, J. Kuroiwa, J. Lander, E. Norabuena (1999): Geologic setting, field survey and modeling of the Chimbote, northern Peru, tsunami of 21 February 1996. *Pure and Appl. Geophys*, 154 (3/4), 513-540.
6. Titov, V. V., H.O. Mofjeld, F.I. González and J.C. Newman, 1999, Offshore forecasting of Alaska-Aleutian Subduction Zone tsunamis in Hawaii, *NOAA Tech. Memo*. ERL PMEL-114, 22pp.

QUANTIFICATION OF TSUNAMIS: A REVIEW

GERASSIMOS A. PAPADOPOULOS
Institute of Geodynamics
National Observatory of Athens, 11810Athens, Greece

Abstract

The efforts made since 1923 to quantify tsunami size in terms of either intensity or magnitude are critically reviewed. The existing 6-point intensity scales need a drastic revision and replacement by modern, detailed, 12-point scales in analogy to earthquake intensity scales. A new tsunami intensity scale proposed by Papadopoulos and Imamura [1] seems to meet these requirements. Among the existing tsunami magnitude scales even the most sophisticated ones need either better calibration of formulas based on more wave height data or significant improvement in the tsunami source energy calculation.

1. Introduction

Efforts towards a quantification of tsunamis started about seventy-five years ago by the pioneering work of Sieberg [2, 3] who defined the first tsunami intensity scale. However, the tsunami quantification is still a puzzling aspect in the tsunami research since the several scales proposed to measure tsunami size often either are confusing as for the quantity they represent, that is intensity or magnitude, or lie under serious difficulty in their applicability. After several attempts made by many researchers to quantify tsunamis in terms of either intensity or magnitude it is extremely useful to reexamine critically not only the various definitions given but also their practical implementation. Particularly it is shown the general need for (1) to develop detailed, pure tsunami intensity scales, established on standard principles and on modern, well - elaborated criteria, and (2) to improve drastically calibration of magnitude scales.

2. Intensity and Magnitude Scales of Tsunami

The earthquake magnitude is an objective physical parameter that measures either energy radiated by, or moment released in, the earthquake source and does not reflect macroseismic effects. On the contrary, the earthquake intensity is a rather subjective estimate of the macroseismic effects. In every earthquake event only one magnitude or moment on a particular scale corresponds. However, every earthquake is characterized by different intensities in different locations of the affected area.

Okal [4] showed that source depth and focal geometry plays only a limited role in controlling the amplitude of the tsunami, and that more important are the effects of directivity due to rupture propagation along the fault and the possibility of enhanced

A. C. Yalçıner, E. Pelinovsky, C. E. Synolakis, E. Okal (eds.),
Submarine Landslides and Tsunamis 285-291.
©2003 Kluwer Academic Publishers.

tsunami excitation in material with weaker elastic properties, such as sedimentary layers. Therefore, a tsunami can be considered as a particular case of seismic wave and problems related to tsunami quantification could be approached in analogy to seismology.

Sieberg [2, 3] is very likely the first to present a 6-point tsunami intensity scale which, in analogy to earthquake intensity scales, was based not on the measurement or estimation of a physical parameter, e.g. the wave height, but it was established on the description of tsunami macroscopic effects, like damage etc. Ambraseys [5] published a modified version of Sieberg's scale known as Sieberg-Ambraseys tsunami intensity scale. In the Japanese tsunami literature one may find a long tradition in the effort for tsunami quantification. Imamura [6, 7] introduced and Iida [8, 9] and Iida [10] developed further the concept of tsunami magnitude, m, defined as

$$m = \log_2 H_{max} \tag{1}$$

Where H is the maximum tsunami wave height (in m) observed in the coast or measured in the tide gages. Practically, the so-called Imamura – Iida scale is a 6-point scale ranging from −1 to 4 giving the impression of a rather intensity than a magnitude scale. However, m does not estimate effects but it measures by definition $H_{max\ that}$ is a physical quantity. In this sense it may represent magnitude in a primitive way since it does not calibrate the wave height with the distance. In his attempt to improve the Imamura – Iida's definition, Soloviev [11] proposed to define tsunami intensity, i_S, by

$$i_S = \log_2 \sqrt{2}\ (H) \tag{2}$$

where H (in m) is the mean tsunami height in the coast. However, this is still a primitive magnitude scale since it is also based on the physical quantity H. Tsunami magnitude M_t [12, 13, 14, 15] or m [16] was defined by the general form

$$M_t = a \log_{10} H + b \log \Delta + D \tag{3}$$

where H = maximum single (crest or trough) amplitude of tsunami waves (in m) measured by tide gages, Δ is the distance (in km) from the earthquake epicenter to the tide station along the shortest oceanic path (in km), and a, b, D are constants. Expression (3) is similar to the Prague formula [17] used since 1960's for the measurement of the surface-wave earthquake magnitude. A different approach for the calculation of the tsunami magnitude was introduced by Murty and Loomis [18]. Their tsunami magnitude, ML, is defined by

$$ML = 2\ (\log_{10} E - 19) \tag{4}$$

where E is the tsunami potential energy (in ergs). Definition of ML is in close analogy to the Kanamori's [19] definition of moment magnitude

$$M_w = 2/3\ (\log_{10} M_0 - 16.1) \tag{5}$$

as well as to the mantle magnitude [20]

$$M_m = \log M_0 - 20$$

Where M_0 is the seismic moment.

A particular scale measuring tsunami size is that proposed by Shuto [21] who considered it as an intensity scale:

$$i = \log_2 H \tag{6}$$

Where H is the local tsunami height (in m). Obviously by definition it is still a magnitude scale. However, in order to use it as an intensity scale for the tsunami damage description, Shuto [21] proposed to define H according to its possible impact. A 6-point classification of tsunami effects ranging from 0 to 5 is tabulated for the description of the expected damage or destruction as a function of H.

3. Possibilities and Limitations of the Tsunami Size Scales

All the tsunami magnitude scales that are based on measurements of tsunami wave heights at coastlines, from the primitive ones, like those of Imamura - Iida and Soloviev, to the more recent and more sophisticated scales of Abe and Hatori are very sensitive to local effects like coastal topography, near-shore bathymetry, refraction, diffraction and resonance. However, better calibration of formulas, based on more tide-gage and measured in the field wave heights, may drastically improve the applicability of such scales for the tsunami magnitude determination in the future.

TABLE 1. Time evolution of tsunami size scales proposed.

Tsunami Scale	Type of Tsunami	Analogy to Earthquake Scales
Intensity Scales		
Sieberg [2, 3]	primitive 6-point intensity scale	early intensity scales
Ambraseys [5]	improved 6-point intensity scale	improved intensity scales
Shuto [21]	developed 6-point intensity scale	developed intensity scales
Papadopoulos and Imamura [1]	new 12-point intensity scale	new EMS '92 and '98 12-point intensity scale
Magnitude Scales		
Imamura –Iida (40's, 50's and 60's)	primitive magnitude scale	local Richter magnitude scale
Soloviev [11]	primitive magnitude scale	local Richter magnitude
Abe [12, 13, 14, 15]	magnitude scale	surface-wave magnitude scale
Murty –Loomis [18]	magnitude scale	moment – magnitude scale

On the other hand, the Murty-Loomis tsunami magnitude, which is directly based on the total tsunami energy, E, at the source, provides a wider magnitude range but is not easily applicable at the moment because of serious difficulties involved in the calculation of energy E . Better esimates of tsunami energy in the future certainly will result in the magnitude determination of a more and more increasing number of tsunamis. Table 1 summarizes a classification of the several tsunami size scales proposed and their analogy to earthquake size scales.

The tsunami intensity scale proposed by Sieberg [2, 3] and modified by Ambraseys [5] is a 6-point scale constructed in such a way that its divisions are not detailed enough and certainly do not incorporate the experience gained from the impact of large destructive tsunamis occurring in the last decades. Shuto's [21] tsunami scale is by definition a magnitude scale because H is simply a physical parameter. On the other hand, its description of tsunami impact is a 6-point tsunami intensity scale, ranging from 0 to 5, the division of which, however, is a function of H. Therefore, the scale under discussion is a mixture of magnitude and intensity. Apparently, Shuto [21] tried rather to produce a predictive tool that describes expected tsunami impact as a function of H, than to create a new tsunami intensity scale describing tsunami effects independently from physical parameters that control the type and extent of the effects. The overall approach is a useful tool for the tsunami size quantification.

The lack of a pure tsunami intensity scale with a detailed description of its divisions that incorporate recent experience from large, catastrophic tsunamis of the Pacific Ocean, creates serious problems in the standardization of the estimation of the tsunami effects, as well as in the comparisons of the effects from site to site for a given tsunami and from case to case for different tsunami events. Following the long seismological experience, Papadopoulos and Imamura [1] proposed the establishment of a new tsunami intensity scale based on the next principles: (a) independency from any physical parameter ; (b) sensitivity, that is incorporation of an adequate number of divisions (or points) in order to describe even small differences in tsunami effects; (c) detailed description of each intensity division by taking into account all possible tsunami impact on the human and natural environment, the vulnerability of buildings and other engineered structures on the basis of recent experiences gained from large, catastrophic tsunamis of the Pacific Ocean. The new tsunami intensity scale incorporates twelve divisions and is consistent with the several 12-point seismic intensity scales established and extensively used in Europe and North America in about the last 100 years. The new scale is arranged according to (a) the effects on humans, (b) the effects on objects, including vessels of variable size, and on nature, and (c) damage to buildings and other engineered structures.

4. Conclusions

The present review implies that the time evolution of the tsunami quantification follows the steps made for the earthquake quantification with a time shift of about 30 years. For a drastic improvement of the tsunami quantification some further, drastic developments are needed. In the field of tsunami intensity scaling, new, detailed and sensitive scales are needed with the intensity to be estimated independently from the wave heights or any other physical parameter observed. The intensity scale proposed recently by Papadopoulos and Imamura [1] seems to meet these requirements. As for the tsunami magnitude scales that are based on measurements of wave heights in tide-gages, there is a general need for better calibration of the formulas in use which strongly depends on the improvement of both the quality and quantity of instrumental data collected. Tsunami magnitude scales based on the energy at the source need improvement of the methods in use for the energy calculation that is improvement of our understanding of the tsunami generation mechanisms.

References

1. Papadopoulos, G.A. and F. Imamura (2001). A proposal for a new tsunami intensity scale *Internat. Tsunami sympocium 2001 Proc., Seattle, Washington, Aug. 7 –10, 2001,* 569- 577.
2. Sieberg, A. (1923). Geologische, physicalische und angewandte Erdbebenkunde . Jena, Verlag von G. Fischer.
3. Sieberg, A. (1927). Geologische einführung in die Geophysik . Jena,Verlag von G. Fischer, 374pp.
4. Okal, E.A. (1988). Seismic parameters controlling far-filed tsunami amplitudes: a review *Natural Hazards* **1,** 67 – 96.
5. Ambraseys, N.N. (1962). Data for the investigation of seismic sea waves in the Eastern Mediterranean *Bull. Seism. Soc. Am.* **52,** 895 – 913.
6. Imamura, A. (1942). History of Japanese tsunamis *Kayo-No-Kagaku (Oceanography)* **2,** 74 –80 (in Japanese).
7. Imamura, A. (1949). List of tsunamis in Japan *J. Seismol. Soc. Japan* **2,** 23 – 28 .
8. Iida, K. (1956). Earthquakes accompanied by tsunamis occurring under the sea off the islands of Japan *J. Earth Sciences Nagoya Univ.* **4,** 1 – 43.
9. Iida, K. (1970). The generation of tsunamis and the focal mechanism of earthquakes. In : Adams, W.M., ed. *Tsunamis in the Pacific Ocean, Honolulu: East-West Center Press,* 3-18.
10. Iida, K., D.C. Cox and G.Pararas-Carayannis (1967). Preliminary Catalog of Tsunamis Occurring in the Pacific Ocean, *Data Rep. 5, HIG-67-10, Hawaii Inst. of Geophys., Univ. of Hawaii* .
11. Soloviev, S.L. (1970). Recurrence of tsunamis in the Pacific. . In : Adams, W.M., ed. *Tsunamis in the Pacific Ocean, Honolulu: East-West Center Press,* 149-163.
12. Abe, K.(1979). Size of great earthquakes of 1837-1974 inferred from tsunami data. *J. Geophys. Res.* **84,** 1561- 1568.
13. Abe, K. (1981). Physical size of tsunamigenic earthquakes of the northwestern Pacific *Phys. Earth Planet. Inter.* **27,** 194 – 205.
14. Abe, K. (1985). Quantification of major earthquake tsunamis of the Japan Sea *Phys. Earth Planet. Inter.* **38,** 214 – 223.
15. Abe, K. (1989). Quantification of tsunamigenic earthquakes by the M_t scale *Tectonophysics* **166,** 27 – 34.
16. Hatori, T. (1986). Classification of tsunami magnitude scale *Bull. Earthq. Res. Inst.* **61,** 503-515.
17. Vaněk, J., Kárník, V., Zátopek, A., Kondorskaya, N.V et al. (1962) . Standardization of magnitude scales *Izvest. Acad. Sci. U.S.S.R., Geophys. Ser.* . **2,** 153 – 158.
18. Murty, T.S. and H.G. Loomis (1980). A new objective tsunami magnitude scale *Marine Geodesy* **4,** 267 – 282.
19. Kanamori, H. (1977). The energy release in great earthquakes *J. Geophys. Res.* **82,** 2981- 2987.
20. Okal, E.A. and J. Talandier (1988). M_m : A variable –period mantle magnitude *J. Geophys. Res.* **94,** 4169 – 4193.
21. Shuto, N. (1993). Tsunami intensity and disasters In : Tinti S., ed. *Tsunamis in the World, Kluwer,* 197 – 216.

Appendix: A New Tsunami Intensity Scale

The new tsunami intensity scale proposed by Papadopoulos and Imamura [1] incorporates twelve divisions and is consistent with the several 12-grade seismic intensity scales established and extensively used in Europe and North America in about the last 100 years. The new scale is arranged according to the effects on humans, the effects on objects, including vessels of variable size, and on nature,and damage to buildings:

I. *Not felt*
Not felt even under the most favourable circumstances.
No effect. No damage.

II. *Scarcely felt*
Felt by few people on board in small vessels. Not observed in the coast. No effect. No damage.

III. *Weak*
Felt by most people on board in small vessels. Observed by few people in the coast. No effect. No damage.

IV. *Largely observed*
Felt by all on board in small vessels and by few people on board in large vessels. Observed by most people in the coast. Few small vessels move slightly onshore. No damage.

V. *Strong*
Felt by all on board in large vessels and observed by all in the coast. Few people are frightened and run to higher ground.
Many small vessels move stronlgy onshore, few of them crash each other or overturn. Traces of sand layer are left behind in grounds of favourable conditions. Limited flooding of cultivated land.
Limited flooding of outdoors facilities (e.g. gardens) of near-shore structures.

VI. *Slightly damaging*
Many people are frightened and run to higher ground.
Most small vessels move violently onshore, or crash stronly each other, or overturn.
Damage and flooding in a few wooden structures. Most masonry buildings withstand.

VII. *Damaging*
Most people are frightened and try to run in higher ground.
Many small vessels damaged. Few large vessels oscillate violently. Objects of variable size and stability overturn and drift. Sand layer and accumulations of pebbles are left behind. Few aquaculture rafts washed away.
Many wooden structures damaged, few are demolished or washed away. Damage of grade 1 and flooding in a few masonry buildings.

VIII. *Heavily damaging*

All people escape to higher ground, a few are washed away.

Most of the small vessels are damaged, many are washed away. Few large vessels are moved ashore or crashed each other. Big objects are drifted away. Errosion and littering in the beach. Extensive flooding . Slight damage in tsunami control forest, stop drifts. Many aquaculture rafts washed away, few partially damaged.

Most wooden structures are washed away or demolished. Damage of grade 2 in a few masonry buildings. Most RC buildings sustain damage, in a few damage of grade 1 and flooding is observed.

IX. *Destructive*

Many people are washed away.

Most small vessels are destructed or washed away. Many large vessels are moved violently ashore, few are destructed. Extensive errosion and littering of the beach. Local ground subsidence. Partial destruction in tsunami control forest, stop drifts. Most aquaculture rafts washed away, many partially damaged.

Damage of grade 3 in many masonry buildings, few RC buildings suffer from damage grade 2.

X. *Very destructive*

General panic. Most people are washed away.

Most large vessels are moved violently ashore, many are destructed or collided with buildings. Small bolders from the sea bottom are moved inland. Cars overturned and drifted. Oil spill, fires start. Extensive ground subsidence.

Damage of grade 4 in many masonry buildings, few RC buildings suffer from damage grade 3. Artificial embankments collapse, port water breaks damaged.

XI . *Devastating*

Lifelines interrupted. Extensive fires. Water backwash drifts cars and other objects in the sea. Big bolders from the sea bottom are moved inland.

Damage of grade 5 in many masonry buildings. Few RC buildings suffer from damage grade 4, many suffer from damage grade 3.

XII. *Completely devastating*

Practically all masonry buildings demolished. Most RC buildings
suffer from at least damage grade 3.

*Table 2. Possible correlation between the intensity domains, **I**, proposed here and the quantities **H** and **i** introduced in formula (5) by Shuto [21].*

I	H (m)	i
I-V	<1.0	0
VI	2.0	1
VII-VIII	4.0	2
IX-X	8.0	3
XI	16.0	4
XII	32.0	5

RUBBLE MOUND BREAKWATERS UNDER TSUNAMI ATTACK

C.E. BALAS[1] and A. ERGİN[2]
1 Gazi University,Faculty of Engineering and Architecture, Civil Engineering Department, Ankara, Turkey.
2 Middle East Technical University, Engineering Faculty, Civil Engineering Department, Ankara, Turkey.

Abstract

In the past applications of risk assessment for coastal structures; only wave characteristics, tidal range, storm surge, wave set-up, surf beat and structural system parameters were considered, but the tsunami risk could not be incorporated to the reliability based design in the literature. The reliability model REBAD introduced in this study primarily enabled the risk assessment of breakwaters subject to tsunami risk. The Second-Order Reliability Method (SORM) was applied to determine the safety of Haydarpaşa Port, Sea of Marmara, Turkey. The failure probability was forecasted by approximating the Van der Meer failure surface with a second-degree polynomial having an equal curvature at the design point. Inclusion of tsunami risk that has an extended return period when compared to storm waves, increased the failure risk of the structure in its lifetime. For Haydarpaşa port main breakwater, the failure risk of the structure was not sensitive to the tsunami occurrence. However, in places with great seismic activity, tsunami risk may be very significant depending on the occurrence probability and the magnitude of the tsunami.

1. Introduction

The safety of coastal structures is evaluated by modelling random resistance and load variables with probability distributions at the limit state. The implementation of REBAD model [1] that can be employed both for design and safety evaluation intentions, to the design of rubble-mound breakwaters has a foremost importance especially in countries such as Turkey, where the tsunami risk is significant in Marmara and Aegean regions. REBAD enabled the design to be accomplished for several damage levels and failure consequences, leading to an optimal design where initial and maintenance costs of the structure were optimised [2]. In the model, the hydraulic failure mode of the armor layer is described in terms of its limit state to compare the effects of failure mode response functions on the preliminary design of rubble-mound breakwaters including tsunami risk [3].

A. C. Yalçıner, E. Pelinovsky, E. Okal, C. E. Synolakis (eds.)
Submarine Landslides and Tsunamis 293-302.
©2003 Kluwer Academic Publishers.

2. Structural Risk Assessment

REBAD deals with the problems of uncertainty that affect most of the variables in structural reliability, since the design of coastal structures incorporates a considerable extent of uncertainty in the resistance and potential load intensities [4]. The safety of the structure relies on the joint influence of loads acting on the structure and the available strength. The limit state function defined for a specific failure mode consists of load and strength variables that are random in nature. The primary variable vector $\mathbf{z}$ in the normalized space indicates these random variables. The serviceability limit-state was implemented for the safety evaluations, as the exceedance of the failure damage level that may not result in complete breakdown of the structure, but may cause an interruption in the achievement of its functions. The functional form of the basic variables used in the limit state is defined as failure function by: $g(\mathbf{z})=(z_1,z_2,...,z_n)$. The safety of the structure can be assured by designating an admissible value of the probability of achieving the limit state defined by: $g(\mathbf{z})=0$. The failure probability of the coastal structure P_f, i.e. the probability of reaching the limit state that is influenced by the uncorrelated extreme value distribution of wave height can be expressed in universal form:

$$P_f = \int\int\int_{g(Z)=0} ... \int f_{Z_1,...,Z_n} (z_1,...,z_n)\,d z_1,...,d z_n \qquad (1)$$

where, $f_{Z_i}(z_i)$ is the joint probability density function of standardized basic variables Z_i. For a definite damage level, this failure probability can be comprehended as the exceedance probability of that damage level. In system reliability, the second order estimate to the failure probability is obtained by approaching the joint failure set by the set bounded by hyperplanes at the design point on the joint failure surface closest to the origin [5]. In the applications for a selected design wave, tidal level, storm surge and tsunami set-up were randomly generated (on average 30,000 times) from the probability distributions to obtain the design load of the breakwater. Its reliability was investigated (again on average 30,000 times) by the SORM method at the limit state. Consequently for each randomly generated load combination of the computer, the joint damage probability reflected both the occurrence probability of loading conditions and the exceedance probability of the limit state which is the damage level of the Haydarpaşa main breakwater under tsunami attack.

3. Case Study: Haydarpaşa Port Breakwater

The Haydarpaşa port situated on the Anatolian side of the Bosphorus (latitude 41 00 N and longitude 29 01) has a wide hinterland connected to inland by means of highways and railroads which are the shortest route connecting Europe to Middle East countries. The port has two breakwaters of 1700 m. and 600 m. long protecting the sea area of 62 hectares. The main breakwater was designed at a depth of 15 m. with armor weights of 4 tonnes on a slope of 1:2.5 (Table 1). The significant design wave height and period were H_s= 4m. and T_s=6 sec, respectively, obtained from extreme value statistics [6].

The reliability risk assessment model was applied in order to evaluate the structural safety of the breakwater by modelling random design and structural variables, i.e. wave height, tidal range, storm surge, wave set-up and the structural system parameters by probability distributions (Table 2). The wave set-up and surf beat were taken as 6% and 9% of the significant wave height, respectively [7]. Since the breakwater was constructed at a depth of 15 m tsunami height is taken as the design parameter as defined in Figure (1) [8]. The available information concerning tsunamis associated with the İstanbul and the eastern Marmara earthquakes have been used as the tsunami data [9]. Data documented for İstanbul between the years of 358-1999 (1641 years) gives the number of major tsunamis in İstanbul and on the coasts of Marmara Sea as 32, where the tsunami height exceeds 0.5 m, as descriptively listed in Table (3).

There are several valuable magnitude and intensity definitions, classifications and statistical approaches for the occurrence probabilities of tsunami. Efforts towards the quantification of tsunami in terms of intensity scale started with Sieberg [10]. Since then several investigators put a great effort to grade the tsunami in terms of intensity [11] and magnitude scales [12] [13] [14] [15] [16] [17]. Based on these studies Table (4) was prepared for structural risk assessment, with the possible tsunami height ranges judged by the intensity scale and the descriptions related to the major earthquakes in the Marmara region tabulated in Table (3) which is adopted from Altınok and Ersoy [9]. With the limited data, the tsunami height ranges were identified based on the tsunami information in Table (3).

TABLE 1. Design parameters for Haydarpaşa main breakwater.

Design Variable X_I	Value
Nominal diameter D_n (m.)	1.15
Weight W_{50} (tones)	4.0
Design wave height H_s (m.)	4.0
Tidal range R_T (m.)	±0.25
Surf Beat (m.)	0.42
Storm surge S_S (m.)	0.22
Relative density Δ_ρ	1.64
Height of structure (m.)	10
Structure slope $\mathrm{Cot}\theta$	2.5

The lack of tsunami intensity and magnitude scale with detail description in the data utilized, forced the investigators to make a decision on the tsunami height range by using Modified Sieberg Seismic Sea-Wave Intensity Scale (i) [11] in Table (3). In this table, H_{Tm} is the maximum tsunami wave height (m), D is the distance that the water penetrated inland (m) and NTI designates that there is no tsunami information. As a pioneering work, a simple statistical analyses of the classified tsunami heights in İstanbul and the adjacent coasts of Marmara is analysed in terms of probability of occurrences of the mean values of the ranges as given in Figure (2). In this Figure, the regression line, presenting the statistical characteristic of the mean values, was provided with a certain confidence limit indicating the lowest and the highest tsunami height ranges. The purpose of this attempt was to introduce a simple conceptual statistical model of tsunami occurrence and the probability distribution of tsunami height for İstanbul. Based on the number of occurrences of tsunamis in 1641 years, the return period of tsunami heights (for the mean values of ranges) were estimated as given in Table (5).

TABLE 2. Probability parameters of variables used in reliability-based risk assessment.

X_1	Distribution	Parameters
Y	Beta	a=2.9 ; b=2.1 c=1.4
D_n	Beta	a=3.1 b=1.6 c=1.5
H_s	Gumbel	a=1.88 b=0.78
Tide	Triangular	min=-0.25 max=0.25
Tsunami height	Log-normal	a=2.3 b=8.3
Δ_ρ	Normal	μ_{xi}=1.64 σ_{xi}=0.13
Cotθ	Beta	a=9 ; b=1.8 c=2.8

Climatic and geomorphologic changes in the future may alter the statistical characteristics of tsunamis in the Marmara region. Hence it is well known that, modeling of tsunami in general depends on the number and accuracy of data, time period studied and the statistical analysis techniques utilized. Although, this study is based on the limited descriptive data available, results were regarded as representative for the population.

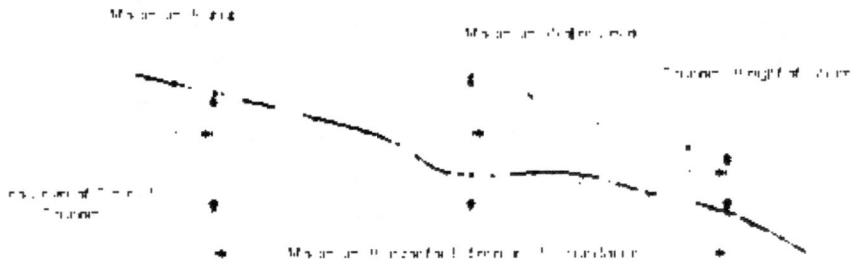

Figure 1. Maximum run-up of tsunami inundation, adopted from Farreras [8].

The structural performance function $g(z_i)$ for the rubble mound breakwater was simulated under design conditions using the probability distributions of the load and strength parameters for the following two cases: Case I: Tsunami risk not included and Case II: Tsunami risk included. The scatter range of the randomly generated values was between $g_{min}= -44.2$ m and $g_{max}=3.68$ m for Case I; $g_{min}=-44.3$ m and $g_{max}=3.64$ m, for Case II. This signifies the effect of uncertainties inherent in the design parameters of the limit-state functions having the simulated mean values of $\mu_g=1.1$ m (Case I) and $\mu_g=1.08$ m (Case II) (Table 6).

Simulation results for Case I and Case II are illustrated in Figure (3a-b), respectively, where the occurrence probability of structural performance function in 100 years is given. From Figure (3-a), where the tsunami risk was not included (Case I), structural performance function $g(z_i)$ was safe with an annual probability of 99.95%, signifying that the failure risks will be 1.65% and 3.25% in 50 years and 100 years, respectively.

As for the Case II (Figure 3b) where the tsunami risk was included, it is seen that, the structural performance function is safe with an annual probability of 99.98% signifying the failure risks as 2.2% and 4.2% in 50 years and 100 years, respectively.

It is clearly seen that, including the tsunami risk increased the failure risk of the breakwater and it is observed that:

I.   The failure probability is relatively higher in Case II (tsunami included) than Case I (not included).

II.   Longer the duration in years, the failure probability and the impact of tsunami risk on the failure mechanism are increased.

III. For Haydarpaşa port main breakwater, tsunami was not the key design parameter when compared to storm waves; therefore, the difference in the failure risk was not very significant.

IV. However, in places with great seismic activity, tsunami risk may be very significant depending on the occurrence probability and the magnitude of the tsunami.

298

TABLE 3. Major tsunamis in İstanbul and Marmara region, adopted from Altınok and Ersoy [9].

Date	Place	Tsunami Information
24.08.358	İzmit Gulf, Iznik, İstanbul	NTI
11.10.368	Iznik and its surroundings	NTI
01.04.407	İstanbul	NTI
08.11.447	Marmara Sea, İstanbul, İzmit Gulf, Marmara Islands	i=3
26.01.450	Marmara Sea, İstanbul	i=3
26.09.488	İzmit Gulf	NTI
Winter 529	Thracian coasts of Marmara	NTI
Winter 542	West coast of Thracia	i=4
06.09.543	Kapidag Peninsula, Erdek Bandirma	NTI
15.08.553	İstanbul, İzmit Gulf	D=3000 m.
15/16.08.555	İstanbul, İzmit Gulf	NTI
14.12.557	İstanbul, İzmit Gulf	D=5000 m.
715	İstanbul, İzmit Gulf	NTI
26.10.740	Marmara Sea, İstanbul, Iznik Lake	i=3/i=4
26.19.975	İstanbul, Thracian coast of Marmara	i=3
989	İstanbul, Marmara coast	NTI
990	İstanbul, Marmara coast	NTI
02.02.1039	İstanbul, Marmara coast	NTI
23.09.1064	Iznik, Bandirma, Murefte, İstanbul	NTI
12.02.1332	Marmara Sea, İstanbul	i=3
14.10.1344	İstanbul, Marmara coast, Thracian coast, Gelibolu	i=4
10.09.1509	İstanbul, Marmara coast	i=3 H_{Tm} >6m.
17.07.1577	İstanbul	NTI
05.04.1646	İstanbul	i=3/i=4
15.08.1551	İstanbul	NTI
02.09.1554	İzmit Gulf, İstanbul	NTI
22.05.1766	İstanbul, Marmara coast	i=2
23.05.1829	İstanbul, Gelibolu	i=2
19.04.1878	İzmit, İstanbul, Marmara coast	i=3
10.05.1878	İzmit, İstanbul	40 people killed by tsunami
09.02.1894	İstanbul	i=3 H_{Tm} <6 m.
18.09.1963	Eastern Marmara, Yalova, Gemlik Gulf	H_{Tm} =1m.
17.09.1999	İzmit Gulf	i=3

TABLE 4. Tsunami height range defined for the structural risk assessment.

Intensity	Tsunami height range	Description
I = very light ii =light	H_T=0.1-1 m.	Minor
iii = rather strong iv = strong	H_T=1-3 m.	Moderate
V= very strong vi =disastrous	H_T=3-6 m.	Major

TABLE 5. Return periods based on the number of occurrences of tsunamis in 1641 years.

Tsunami Height (m.)	Return Period (years)
H_T=0.5	R_p=50
H_T=2.0	R_p=100
H_T=4.0	R_p=200

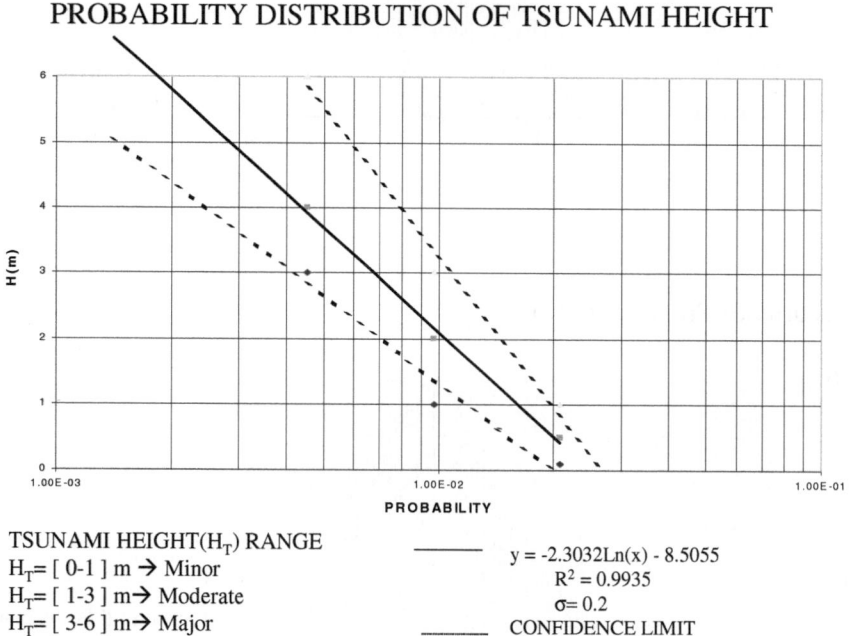

PROBABILITY DISTRIBUTION OF TSUNAMI HEIGHT

TSUNAMI HEIGHT(H_T) RANGE
H_T= [0-1] m → Minor
H_T= [1-3] m→ Moderate
H_T= [3-6] m→ Major

——— $y = -2.3032Ln(x) - 8.5055$
$R^2 = 0.9935$
σ= 0.2
--------- CONFIDENCE LIMIT

Figure 2. Tsunami heights in İstanbul and the adjacent coasts of Marmara in terms of probability of occurrences of the mean values.

TABLE 6. Simulation of failure function for 100 years.

Cases	Tsunami not included	Tsunami risk included
Fitted distribution of g	Gumbel	Gumbel
Distribution parameters	Mode: 1.75 Scale: 0.77	Mode: 1.71 Scale: 0.8
Average (μ_g m.)	1.1	1.08
Minimum of range	-44.2	-44.3
Maximum of range	3.68	3.64

Occurence Probability (%)

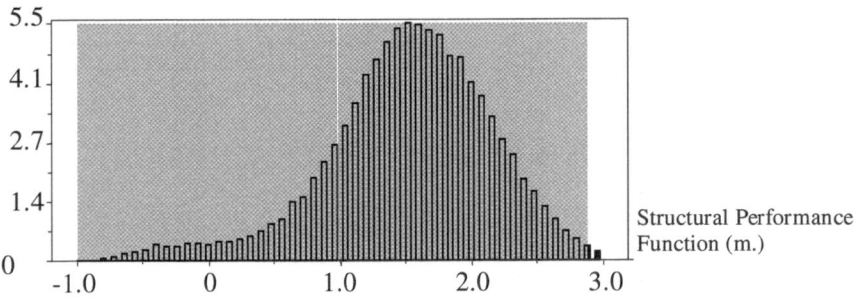

(a)

Occurrence Probability (%)

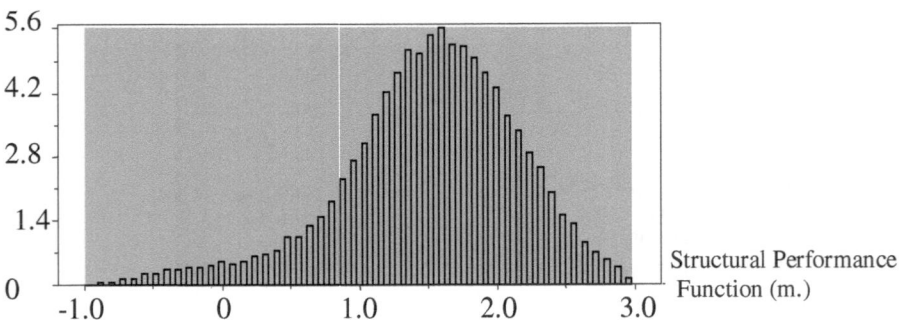

(b)

Figure 3. Distribution of the Van der Meer plunging performance function at the limit state in 100 years. (a) Case I: Tsunami risk not considered (b) Case II: The effect of tsunami risk.

4. Conclusions

The reliability-based structural risk assessment (REBAD) model serves as a basis for risk identification and risk response in coastal projects by integrating time, cost and risk information. In the model, the resistance and load parameters are contemplated as random variables and the consequences of structural failure are reflected in terms of probability distributions by using SORM. The second order estimate to the failure probability is obtained by approaching the joint failure surface by the set bounded by hyperplanes at the design point on the joint failure surface closest to the origin of the coordinate system.

REBAD that can be efficiently utilized for risk assessment of coastal projects in project management, permit the coastal engineer to inspect the sensitivity of structural design to various parameters and contributes a quantitative foundation for comparing design alternatives with assorted damage levels. It provides a valuable tool in the design of coastal structures, which are characterised by large failure consequences and substantial capital expenditures.

The model could incorporate the uncertainties inherent in tsunami, storm surge and wave data to reliability based design of rubble mound breakwaters by using Van der Meer performance function. The tidal range of sites and the storm surge were randomly generated by simulation. Afterwards, the failure mode probability was predicted by the parabolic limit state surface having the identical curvature at the design point with the higher degree failure surface.

The reliability method had advantages when compared to the deterministic practice, since the random behaviour of structural performance could be estimated at the planning stages. Therefore the new reliability approach, which can be applied within few minutes of CPU time in portable computers, was recommended for the design of breakwaters.

Tsunami risk should be included in the reliability-based model, since it increased the failure risk as a risk parameter in this case study. Especially in places with great seismic activity, where the magnitude of the tsunami and its occurrence probability is high, tsunami gains vital importance. Such a structural risk assessment, carried out by using reliability-based models, may be used as a successful tool in emergency preparedness and response to natural hazards, as an early risk mitigation mechanism for important coastal projects, such as nuclear power plants.

References

1. Balas, C.E. and Balas L. (2002) Risk Assessment of Some Revetments in South West Wales, UK, ASCE *Journal of Waterway, Port, Coastal and Ocean Engineering, American Society of Civil Engineers,* **61** (In print).
2. Balas, C.E. and A. Ergin (2002) Reliability-Based Risk assessment in Coastal Projects: Case Study in Turkey, ASCE *Journal of Waterway, Port, Coastal and Ocean Engineering, American Society of Civil Engineers,* **128,** No. 2, 52-61.
3. Balas, C.E. and Koç L. (2002) Risk Assessment of Vertical Wall Breakwaters- A Case Study in Turkey, *China Ocean Engineering,* **15,** No: 4, 453-466.
4. Balas, C.E., A.T. Williams, S.L. Simmons and A. Ergin (2001), "A Statistical Riverine Litter Propagation Model, *Marine Pollution Bulletin,* **42,** No. 11, 1169-1176.
5. Balas, C. E. and A. Ergin (2000) A Sensitivity Study for the Second Order Reliability-Based Design Model of Rubble

Mound Breakwaters, *Coastal Engineering Journal*,**42**, No.1, 57-86.

6. Ergin, A. and Özhan, E. (1986). *Wave Hindcasting Studies and Determination of Design Wave Parameters for 15 Sea Regions*, T. Report No: 35, Coastal Engineering Research Center, Civil Eng. Department, Middle East Technical University, Ankara, Turkey.

7. Goda, Y. (2000). *Random Seas and Design of Maritime Structures*, World Scientific Publications, London

8. Farreras, S.F. (2000). Post tsunami field surveys procedures: An outline, *Natural Hazards,* Kluwer Academic Publishers, Dordrecht, **21**: 207-214.

9. Altinok, Y. and Ersoy, S. (2000). Tsunamis observed on and near the Turkish coast, *Natural Hazards,* Kluwer Academic Publishers, Dordrecht, **21**: 185-205.

10. Sieberg, A. (1927). *Geologische, Physikalishe und Angewandte Erdbebenkunde*, Verlag von Gustav Fisher, Jena.

11. Ambraseys, N.N. (1962). Data for the investigation of the seismic sea-waves in the Eastern Mediterranean, Bulletin of Seismological Society of America, **52** (4), 895-913.

12. Imamura, A. (1942). History of Japanese tsunamis, *Kayo-No-Kagaku (Oceanography)*, **2**, 74-80 (in Japanese).

13. Imamura, A. (1949). *Journal of Seismological Society of Japan*, **2**, 23-28 (in Japanese).

14. Iiada, K. (1956). Earthquakes accompanied by tsunamis occurring under the sea of the islands of Japan, *Journal of Earth Sciences,* Nagoya University, **4**, 1-43.

15. Iiada, K. (1970). The generation of tsunamis and the focal mechanism of earthquakes, in Adams, W.M. (ed.), *Tsunamis in the Pacific Ocean* ,East-Western Centre Press, Honolulu, pp. 3-18.

16. Abe, K. (1979). Size of great earthquakes of 1837-1974 inferred from tsunami data, *Journal of Geophysical Research*, **84**, B4, 1561-1568.

17. Shuto N. and Matsutomi H. (1995). Field surveys of the 1993 Hokkaido-Nansei-Oki earthquake tsunami, *Pure and Applied Geophysics,* **144**, 3-4, 649-663.

VULNERABILITY ASSESSMENT AS A TOOL FOR HAZARD MITIGATION

ALLAN T. WILLIAMS [1,2] & RICARDO A. ALVAREZ[2]
[1] *Applied Science Department, University of Glamorgan, Pontypridd, Wales, UK.*
[2] *International Hurricane Centre, Centre for Engineering and Applied Sciences*
Florida International University, 10555 West Flagler Street, Miami, 33174, USA

Abstract

In emergency management, hazards are considered as sources of damage and damage reduction is the core of *hazard mitigation*, defined as the cost-effective measures taken to reduce the potential for damage on a community from the hazard impact. The equation seems simple: *HAZARD (SOURCE OF DAMAGE) +DAMAGE REDUCTION ALTERNATIVES = MITIGATION.* However, important questions arise when this definition occurs within the context of annual or historical damage from hazard impact. If the relationship is simple, why are damages, often repetitive, mounting as hazards strike vulnerable communities/specific facilities? *Knowledge gaps* exist regarding the causal relationship between hazards and the damage that results from their impact on the realm of human activity. These gaps reflect a general lack of understanding about the sequence of events that lead to actual damage. In the simplest terms, humanity is generally ignorant about its *vulnerability* to the adverse effects of hazards. Consequently, vast segments of human society continue to engage in building structures and facilities, in developing infrastructure, and in all the wide range of human activity seemingly without utilizing the assessment of its vulnerability as a tool to reduce the potential for damage from the impact of hazards. Truly effective mitigation – *hazard damage reduction* - must be based on a clear understanding of the causes of damage. This knowledge is gained by applying the methodology of *vulnerability assessment*. The methodology is applicable regardless of the specific types of hazards that may strike a community or facility. Assessment takes place at three levels: *Hazard identification* defines the magnitudes and probabilities of the hazard that threatens anthropogenic interests; *vulnerability assessment* characterises the population exposed to the hazard and the damage/injuries resulting; *risk analyses* incorporates the probability of damage/injury. As an emergency management tool, vulnerability assessment is a sound foundation for hazard mitigation. Vulnerability assessment and hazard mitigation must be essential components in the practice of any anthropogenic activity in a hazardous area.

A. C. Yalçıner, E. Pelinovsky, E. Okal, C. E. Synolakis (eds.)
Submarine Landslides and Tsunamis 303-313.
©2003 *Kluwer Academic Publishers.*

1. Preamble

Most weeks, the media brings news of disasters and catastrophes, causing death, injury, immeasurable human suffering and considerable damage to the built and natural environments. Poverty, war, and a coastal migration of people, together with poor infrastructures, make many world areas unprepared for these catastrophes. With each new disaster the cycle of blame allocation, finger pointing, calls for action and analysis of causes of damage, of lessons learned, not learned, begins again. A common thread in all this is the endless capacity of humankind to be surprised by the power inherent in nature, or its ability to render the best emergency plans quite ineffective in reducing potential damages.

The United Nations declared the period 1990-2000 as the International Decade for Natural Disaster Reduction (IDNDR), stressing that solutions should be supported at national level, but the fulcrum for all consequent mitigation measures should be implemented at the local level. In this decade, 82 tsunamis were reported, eleven causing >4,600 deaths and more than $1 billion damages [1]. Technology alone will not protect inhabitants in any potential tsunami risk area; awareness and the reading of nature's signals must be drilled into inhabitants in order to create a cultural mitigation.

2. Introduction

All tsunamis are potentially dangerous; although damaging ones are rare the potential for death and property destruction is huge. Tsunami hazards occur as a major event *circa* once every 100 years in the Atlantic. The 1755 Lisbon earthquake spawned in the Azores-Gibraltar fracture zone, generated waves that reached a height > 6m., off the southwest UK coastline, and *circa* 4m. for the USA east coast (Maul, *pers. comm*). In 1867, the Virgin Islands tsunami generated by an earthquake in the Anegada Trough caused huge damages in Puerto Rico and the Virgin Islands. In the Pacific 'Ring of Fire', some five tsunami events take place annually, but only one is large enough to be observed [2]. The 7.3 Richter scale earthquake in the Aleutian Islands in 1946 generated a 17m. tsunami wave height in Hawaii and $26 million in damages. The great Alaskan earthquake at Prince William Sound in 1964 measured 9.2 on the Richter scale and caused > $84 million damages in Alaska and 123 deaths. At Crescent City, Ca, the wave reached 6.3m. in height resulting in > $7 million damages [2]. The Krakatoa tsunami (1883) caused > 36,000 casualties, the Chilean tsunami (1868), >25,000. Since 1946, six tsunamis have caused $0.5 billion damages and killed 350 people in Hawaii, Alaska and the west coast of the USA [3]. Drowning is the main cause of death, however flooding; polluted water supplies and damaged gas lines all make a contribution. Turkey experienced a minor tsunami on the 17th August 1999 due to the Kocaeli earthquake in Izmit Bay. Minor damage and a few deaths occurred but these paled into insignificance when compared to the utter devastation caused by the earthquake to people (>15,000 deaths), homes (>60,000 made homeless) and the destruction of large-scale industries in a major economic region of Turkey [4].

Currently, a controversy exists re the possibility of the lateral collapse of the Cumbre Vieja Volcano in the Canary islands which would create the sub-aqueous 'La Palma slide' involving >500cubic km. of material. It has been postulated that this would produce a tsunami wave of some 50m. in height, that would cross the Atlantic and shoal at the eastern seaboard of the USA. A tsunami is *circa* a 100km wave and this tsunami would generate some 306,250,000,000 J/m energy per unit metre of wave front, or a force of 306,250,000,000 Newtons. If this scenario ever takes place, and it is very questionable, current mitigation measures would be miniscule and ineffective.

3. Hazard and risk

Assessments of hazard and risk are essential pre-requisites, indeed a required foundation for effective hazard mitigation, to the formulation of policies for the reduction of potential damage and managing risks for all situations. Both draw upon experience, the application of common sense, as well as data interpretation. Additionally, they are concerned with the future i.e. what has not happened, and always involve decision-making. Isolated measurements of risk are not very helpful when decisions are made for managing or developing policies for controlling hazards. Generally, these terms are interchangeable, although specifically a *hazard* is defined as a set of circumstances that could lead to harm i.e. death, injury or an illness of a person. The *risk* of such an event happening can be considered as the probability that it will occur as a result of exposure to a defined amount of hazard. Therefore, risk analysis is concerned with chance, consequences and context [5], whereas risk management is undertaken in order to reduce the adverse events identified by risk analysis. The *rate of incidence* (frequency of recurrence) can be viewed as the expected number of events that occur for this defined hazard amount. Probabilities and rates obey different mathematical laws, but if the events are independent and probabilities small, the two values are basically the same. Risks can vary from negligible - an adverse event occurring at a frequency of one per million plus e.g. an asteroid hitting the earth, to high - fairly regular events occurring at a rate of greater than one in a hundred, e.g. hurricanes, tsunamis. For example, with respect to the former, the chances of a category 1 hurricane striking a site specific spot in Florida is *circa* 3 per annum; a category 5, once in 5,000 years. For tsunamis, a damaging one can be expected in Hawaii/Alaska, once every seven years.

Risk reduction emphasises the vulnerability aspect of natural disasters and the differing perceptions associated with risk, as within social systems different cultural values exist. Currently, an intellectual vacuum seems to exist with respect to a theoretical framework between risk and disaster management. Perhaps in the 21st Century, the concept of impact thresholds should be stressed [6]. This would examine links between critical biophysical change (e.g. as a result of flooding associated with a tsunami, hurricane), and socio-economic (behavioral) impact. This would involve management, geography and risk, at a local or regional scale. In this instance, critical would refer to a local/regional scale.

4. Damage

The components of damage resulting from a hazard occurrence (tsunami, hurricane, typhoon etc.) can be sub-divided into direct, indirect and consequential, and the types classified as physical, structural, environmental, political, socio-economic. Direct damage can be quantified in a relatively easy manner, e.g. the replacement cost of houses, businesses demolished. The indirect and especially the consequential damages are much harder to assess. In the wake of many natural disasters, drunkenness, suicides, bankruptcy, psychological traumas - often with a variable time lag, occur which add to the indirect costs that wipe out the coherent social fabric of a community and seem incalculable on any monetary basis.No current mechanism is able to systematically identify and evaluate these latter factors.

5. Vulnerability

The etymology of the word, from the Latin noun *vulnus* = wound or the verb *vulnerare* = to wound, certainly evokes an image of pain, of suffering, i.e. of damage. Vulnerability *has* many meanings and is a dynamic natural process involving change. In order to be vulnerable, there must be a source of the damage i.e. a hazard and also a damage receptor, e.g. a community, environment. It can be defined as: the capacity loss of a system after a disturbance to return to its former dynamic equilibrium; the interaction of human activity with a hazard; interference of human activity with natural processes; incapacity of human activity to confront the consequences of impact from a hazard. There always has to be a two-way exchange because local factors including some under anthropogenic control, affect an area's vulnerability and it is a dynamic process as it continuously changes in response to population and urban development [7].

There are two types of vulnerability [8]:

•*Absolute*, which is a function of location, global processes and is beyond human control e.g. the siting of Japan, Hawaii in relation to tsunamis; tropical cyclones in Florida or the Caribbean; or earthquakes in California or Turkey.

•*Relative*, which results from specific factors present in any given location/site that may affect or modify how the hazard impacts the area. Some degree of control together with modifiers exists, and local factors would include, population density, culture, demographics, infrastructure (housing, communication), tourism, as well as geomorphological, ecological and other physical features.

The recognition that a region, community, facility, or site-specific factors exist, that can to some degree modify the impact of a hazard on the receptor community is of importance. It provides an opportunity for human society to exercise some control over the actual impact of a hazard to the extent that such local factors or modifiers can be analysed and understood. Relative vulnerability allows for the use of vulnerability assessment as a method leading to the identification of mitigation alternatives through an understanding of the causes of damage, the components of a hazard, and of the sequence of events that lead from impact to resulting damage. The assessment of relative vulnerability supports the practice of hazard mitigation.

6. Vulnerability assessment

The considerations examined above help establish that vulnerability assessment is a source of knowledge, a tool and a method to acquire additional knowledge about the causal relationship between a hazard and damage as well as about the mode of damage. To be useful as a tool for knowledge acquisition, the methodology itself needs to be logical, comprehensive and, above all, understood by those who can use it in the practice of the professions or duties on a daily basis.

A sound vulnerability assessment starts by charting the value at risk, meaning the actual monetary cost of replacing or repairing that, which may be damaged by the impact of a hazard. This includes the value of the human function not limited only to the economic losses associated with damage caused by a hazard, but also to the cost of interruption of any element of human function including that of governance. The methodology then assesses the hazard impact in terms of:

•Space, i.e. the geographic area that may suffer the direct strike from a hazard.

•Time, referring to the direct duration of a hazard. This could be seconds, as in the case of earthquakes, to hours as with tropical cyclones, to days and even years for drought and other extreme natural events.

•Intensity, meaning the actual category or magnitude of each hazard event as measured by pertinent ad-hoc scales, such as the Saffir-Simpson scale for hurricanes, or the Modified Mercalli scale for earthquakes.

•Frequency, which goes to the issue of what is the annual probability that a specific hazard may impact a given community.

This assessment methodology also identifies specific local factors, both natural and anthropogenic, that may either contribute to an exacerbation of damage or modify the actual impact on a location specific basis. The objective of this methodology is to provide analytical criteria, which may be used to obtain a realistic evaluation of the type and cost of damage that could result from the impact of a hazard on a location specific basis. Application of this method should provide a foundation for effective mitigation. From emergency management viewpoints, an understanding of the relative vulnerability of a region is absolutely essential in order to refine the response mechanisms to be implemented. Proper hazard mitigation can only be analysed and implemented if based upon a fundament of vulnerability analyses at various (local, regional, country) levels. Analysis must be based on a methodology that includes hazard assessment from a catholic spectrum of impacts (e.g. physical, structural, ecological, socio-economic). The level of understanding, or lack thereof, can directly affect preparedness, mitigation and response, as emergency plan components, especially with regard to location specific factors that can affect the relative vulnerability. This is a function of the level of resolution of the analysis. Methodological updating is another core essential.

Therefore, vulnerability assessment is an attempt to predict how different property types and populations will be affected by the hazard in question. As such, it comprises a fundamental data source inventory upon which to base any emergency response actions. A standard application is that of damage loss and these are termed deterministic as distinct from probabilistic assessments where probabilities are assigned to a complete scenario of possible events. Deterministic assessments can predict the demand upon emergency services, insurance loss, assistance etc. for any large-scale disaster. Housing areas, schools, hospitals, population, age ethnicity, income, health etc. all should be included in relevant databases and much of this information can come from census returns, tax assessments etc.

Many workers have developed methodologies of estimating the societal impacts of natural hazards, e.g. Perkins [9]. Building inventories should include type (wood, steel, reinforced concrete, etc.), location, age, are all-important facets of the inventory. A region's infrastructure takes into consideration roads, sewerage, water supplies, gas and electric supplies, bridges, and details are vitally important for any mitigation. Details of the economic vulnerability of the infrastructure are very important to any successful mitigation measures.

Mintzberg & Waters [10] have described four strategic management planning types: i) *Deliberate*, ii) *Imposed* iii*) Umbrella* and iv) *Emergent.* Of these, the latter two appear to be applicable to tsunami mitigation research, i.e.

iii).'*Umbrella type'*: applicable where elements of the environment are uncontrollable and unpredictable. Only general guidelines for behaviour can be set in such context i.e. overall boundaries are defined within which some parameters can be manoeuvred. This strategy requires the maintenance of a delicate balance between pro-action and reaction.

iv). *'Emergent type'*: this is appropriate where the environment is even more unstable or complex to comprehend. Such a system requires open flexible and responsive management styles.

7. Management and Mitigation

Tsunamis cannot be prevented or predicted and warnings seem unable to prevent destruction of boats, housing, or anything that lies in the path of the runup, but areas at risk can be identified and stringent controls put into place. For example, avoidance of potential runup areas for buildings, placing of potential inundation areas under a floodplain zoning, constructing breakwaters or wave-energy attenuating structures at harbour entrances, planting tree belts between shorelines and areas needing protection, having adequate warning systems (real time) in place, setting in place sound construction and building elevation standards and have streets/homes aligned perpendicular to the wave advance, public education campaigns, etc.

For all the above, when a tsunami or any other natural hazard breaks on a densely populated area, emergency plans are often inadequate and inhabitants frequently seem surprised at the results. *Reduction of the potential for damage is the key objective of mitigation.* Mitigation is those actions taken singly or in combination, which attempt to rectify impacts, associated with a particular activity i.e. a natural hazard. The goal '*should be to substantially increase public awareness of natural hazard risk so that the public demands safer communities in which to live and work'* (Jamieson and Drury; [p257; 11]).

The Puerto Rico Tsunami Warning and Mitigation programme can be taken as an exemplar of the above [12]. It emphasises the preparation and supply of tsunami flooding and evacuation maps together with awareness raising of the potentially affected populations. The latter will:

•Hold regional conferences (with representatives from state/regional/local emergency managers, schools, hospitals, hotels, and industries) for potentially tsunami hazard areas.

•Place tsunami hazard signs in exposed coastal locations urging people to seek higher ground if shaking ground can be felt.

•Carry out evacuation exercises.

•Utilise the WEB to give tsunami information, flood maps, information.

•Produce a tsunami video from the Caribbean viewpoint.

In addition, a network of seismic networks termed the Puerto Rican Seismic Network, (PRSN) will be centred on the University of Puerto Rico and part funded by the US Federal Emergency Management Agency (FEMA). Partnership in the National Tsunami Hazard Mitigation Programme (with the states of Alaska, Washington, Oregon and Hawaii), is envisaged. The Caribbean is unique in that as well as having tsunami potential via earthquakes, slumping and volcanic activity it may also have seismic induced waves as a result of hurricanes or low pressures.

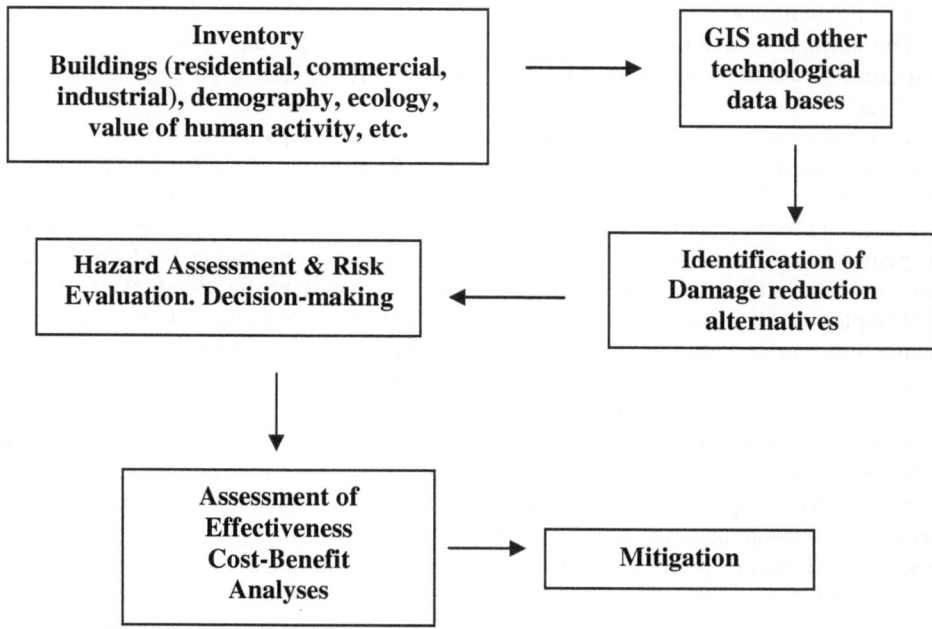

Figure 1. Mitigation flow chart.

Change in coastal systems generates through interaction between the objective and subjective variables that constitute the environment. Huge changes in coastal systems, occur as a result of large-scale natural disasters, such as tsunamis, hurricanes. This change occurs via interaction between objective variables (quantifiable measurements i.e. the scale of the tsunami, hurricane, volcanic eruption etc.), and subjective variables (set within the complex of socio-economic and cultural factors). The latter are much more difficult to quantify. Effective mitigation strategy depends upon a detailed management planning procedure concerning structured data collection pre a natural disaster plus the availability of correct quality information during the event, so that clear guidelines are set (Figure 1). Emphasis should be placed on the gathering of reliable, sufficient, impartial, consistent, comprehensive information that can be of predictive value. Data must be organised into a logical format particularly where problems are complex, as in natural disaster areas [13]. It is an axiom that management of nature/anthropogenic interactions is always concerned with, *'mutual interactions and feedback based on power differentials, conflicting values and competing interests and expectations'* (Boehmer-Christiansen, p84; [14]).

Long-term behavioural characteristics to natural disasters may be identified/responded to via a strategic management approach, as policy making has an inexact and non-scientific nature that is characterised by considerable uncertainty and ambiguity [15, 16]. This is evidenced by the 'surprise factor' that occurs in virtually all such natural disasters, epitomised by the, '*it will not happen to me*' attitude, or the cognitive dissonance factor of people blocking out unwelcome reality and believing only what they want to believe. Therefore, management mitigation decisions should be based on models that have:

- Evidence obtained from environmental monitoring
- A set of clear objectives
- Evaluation of strategic options.

The ultimate tsunami mitigation strategy is to keep people and critical facilities away from areas prone to flood. Bernard [1], postulated three effective paradigms that would save lives:

1.Develop high-resolution tsunami inundation hazard maps to identify areas liable to flood. Numerical models now exist that can simulate tsunami behaviour and estimate the extent of potential flooded areas. These should be locally based and mapped ((hardcopy and electronic format) on scales of 1:2000 or less. GIS systems now make this task routine, but greater precision means greater costs. Building foundations should be deep enough to reduce erosion and scour, and buildings should be elevated above the flood levels

2.Implement/maintain a community wide awareness/educational programme on tsunami dangers to provide community understanding and commitment. Education, is frequently commented upon as a panacea for virtually all matters, but it is crucial in hazard mitigation strategies and stakeholder participation is a mandatory aim. It would include evacuation procedures, practice drills, video presentations etc. as disaster mitigation should be the responsibility of all citizens. This should be carried out in the communities. To date, it appears that people still need to increase awareness, preparation and response to warnings, but awareness is '*the most cost effective way to create a tsunami-resistant community*' (Bernard, p60; [1]). A lack of impact expectation, minimal preparation, and confusion still appears to be the norm in many areas – see below.

3.Have an *efficient* early warning system in place to alert coastal inhabitants about the potential forth-coming danger. Many false alarms have been triggered off by regional systems as currently many do lack precise accuracy. A May 1986, a false alarm in Hawaii cost an estimated $30-60 million in lost business revenue and it also undermined credibility

The above points are exemplified in the following two case studies. On 12 July 1993, a tsunami struck the village of Aonae, Japan. Eighty five percent of a population of *circa* 1,400 people were saved simply by moving to high ground at the onset of the earthquake [17]. The 17th July 1998 tsunami at Papua New Guinea (PNG) generated from an earthquake epicentre 12km. offshore (magnitude of 7.1), had a wave of *circa* 10m in height, which affected an area some 30km wide along the northern coast of PNG where the highest land elevation was some 3m. Three villages arranged alongside the Sissano lagoon were devastated and out of a total population of some 10,000, over 3,000 died [18]. Dengler and Preuss [19], concluded that more than half the population of 2,730 people inhabiting the village of Warupu survived as a result of knowledge about tsunami behaviour. Awareness saved many, but many died due to no vertical elevation structures.

Demonstration micro-projects, as carried out by the European Community Humanitarian Office (ECHO), have stressed preparedness via the Disaster Preparedness ECHO (DIPECHO) programme. To be effective this needs to have up to date accurate information and the means of disseminating that information quickly through the media.

TABLE 1. *A Hazard Mitigation Summary*

HAZARD	A source of danger that may cause damage.
MITIGATION	To soften, mollify, make less harsh.
HAZARD MITIGATION	Cost-effective measures to reduce the potential for damage from hazards.
H AZARD MITIGATION ACTIONS	a) Acts upon the hazard – e.g. fire fighting. b) Keeps the hazard away – e.g. flood control. c) May interact with a hazard – e.g. hurricane shutters. d) Keeps people away i.e. relocate – e.g. relocation. With respect to tsunamis, b) and d) are the main actions.
LONG TERM HAZARD MITIGATION	a) Strengthening building codes. b) Strict zoning regulations. c) Responsive development. d) Education/Awareness.
WHEN TO MITIGATE	a) During the design phase of new buildings. b) During the restoration effort post a disaster. c) At any time as a retrofit. d) During the daily practice of a profession/job. e) When planning for development or re-development of a community.
VULNERABILITY ANALYSIS	a) Sets the foundation for effective mitigation. b) Commences with hazard assessment. c) Documents time, space and frequency components of the hazard. d) Reviews the physical, social and economic aspects.
ACTIONS	a) Macro (Regional) mitigation b) Micro (Site specific) mitigation. c) Damage function (benchmark) d) Cost-benefit analysis. e) Environmental issues.

When a natural hazard strikes, the 'well oiled machinery', rarely functions smoothly with the best-laid plans all tending to stutter along, but summary checklists, as depicted in Table 1, can help the process. Better interdisciplinary, inter-agency and sectoral co-operation should be the norm, but frequently a lack of co-ordination and leadership prevails. Governmental support via legislation if necessary should be a common denominator for natural disaster prone areas. For example, in Turkey, there exists no national co-ordinating agency for disaster management. The nearest that currently exists is termed 'The National Preparedness Plan; law number 7269.' Additionally, the development of regional, national and international incentives should be enhanced in order to encourage localities, regions, and countries to practice mitigation in order to reduce the adverse consequences of any disaster. The creation of natural disaster research, education centres and/or schools would be an extremely advantageous measure for any state or country. The syllabus should concentrate on remote sensing, GIS, sound data acquisition/transfer especially for low lying, high risk, flood prone areas, geomorphology, erosion trends, socio-economic characteristics, including valuations, i.e. Willingness to Pay and Contingency Evaluation, sea level changes, ecological, hydrological, meteorological, studies etc.

8. Conclusions

The continued development of coastal areas plus population growth ensures a higher vulnerability and loss probability in the event of an impact from a natural hazard. Real time warning, mapping of inundation and wave run up limits and public education/awareness campaigns are the main key tsunami hazard mitigation concerns. It is of paramount importance that people understand the vulnerability of the natural and human systems, together with the measures that must be taken to reduce the potential for damage. Public Disaster plans have to be developed which must have the support of all stakeholders. Hazard mitigation relates to any cost-effective measure undertaken to reduce the potential for damage from a hazard, but the limits of cost/benefit analyses are well known. The hazard can be a hurricane, tsunami, earthquake, etc., but mitigation matters remain fairly constant irrespective of the natural disaster. Hazard identification is the fundament re disaster mitigation measures. The base line for these studies should be the local area and integration of mitigation planning into local decision process is imperative, but many communities rarely use formal risk analyses. The growth of GIS, and usage of other technologies such as LIDAR (LIght Detection And Ranging) and computer-based simulation and visualization, now enables planners and scientists to utilise probabilistic risk analyses to predict impacts and assess cost-benefits of various management strategies. However, there is limited knowledge of the probabilities/magnitudes of many events. This is often coupled with poor regional building codes along with scant Legislature and/or Institutional implementation of mitigation measures; the existence of communication gaps between local vested interests and pure science based researchers who couch hazard assessments in a highly technical language. The preservation of wetlands, currently under threat in most areas of the world, should be done, as these act as buffers to flood waters. This infers that there is a long pathway to travel before reaching the optimum mitigation strategy for any potential natural hazard. Goethe commented that, *'nature understands no jesting. She is always right, and the errors and faults are always those of man.'* This is extremely applicable to tsunami mitigation.

Acknowledgements

We wish to thank Prof. George Maul, Florida Institute of Technology, Miami, USA and Prof. Aurelio Mercado-Irizarry, Dept of Marine Sciences, University of Puerto Rico, Mayaguez, PR, for their detailed advice and insight regarding tsunamis in general. Many meaningful discussions were held with Mr. Scott Caput, International Hurricane Centre, Florida International University, Miami, USA, and we thank him. However, the views expressed are the authors and do not necessarily reflect official university policy. ATW wishes to acknowledge financial help from the Andrew W. Mellon Foundation.

References

1. Barnard, E. (1999). Tsunami. In, Jon Ingleton, (ed.), *Natural Disaster Management*, Tudor Rose Ltd. Leicester, UK, 58-60.
2. Lander, J.F., Lockeridge, P.A and M.J. Kozuch (1993). *Tsunamis Affecting the West Coast of the United States, 1006-1992, Key to Geophysical Records*, Documentation, N0 29. NOAA, National Geophysical Data Centre, Boulder, Colorado, USA.
3. www.redcross.org/services/disaster/keepsafe/tsunami/html
4. Yalciner, A.C., Synolakis, C.E., Borrero, J., Altinok, Y., Watts, P., Imamura, F., Kuran, U., Ersoy, S., Kanoglu, U., and S. Tinti, 1999. Tsunami Generation in Izmit Bay by the 1999 Izmit Earthquake. *Proceedings of ITU-IAHS International Conference on the Kocaeli Earthquake*, Istanbul Technical University, Turkey, December 1999, 217-220.
5. Elms, D.G. (1992). Risk Assessment. In, D Blockley, (ed.), *Engineering Safety*, MacGraw-Hill International Series in Civil Engineering, London, 28-46.
6. Jones, R.N. 2001. An Environmental Risk Assessment/Management Framework for Climate Change Impact Assessments. *Natural Hazards*. 23 (2-3), 197-230.
7. Mitchell, James K., (ed; 1999) *Crucibles of Hazard: Mega-Cities and Disasters in Transition,* United Nations University Press, Tokyo, New York, Paris.
8. Alvarez, Ricardo. A (1999). Tropical Cyclone. In, Jon Ingleton, (ed.), *Natural Disaster Management,* Tudor Rose Ltd. Leicester, UK, 34-36.
9. Perkins. J.B. (1992). *Estimates of Uninhabitable Dwelling Units in Future Earthquakes Affecting the San Francisco Bay Region.* Oakland, California: Association of Bay Area Governments, San Francisco, Ca, USA
10. Mintzberg, H. & Waters, J.A. (1989). Of Strategies, Deliberate and Emergent. In, D, Asch and C. Bowman, (eds.), *Readings in Strategic Management*, Macmillan, London, in association with the Open University, 37-56.
11. Jamieson, G. and Drury, C. (1997). Hurricane Mitigation Efforts at the US Federal Emergency Management Agency. In, H.F. Diaz and R.S. Pulwarty, (eds.), *Hurricanes: climate and socio-economic impacts*, Springer-Verlag, Berlin.
12. www.poseidon.uprm.edu/welcome.html
13. Taylor, J.W. (1961). *How to Create Ideas*. Prentice Hall, Inglewood Cliffs, New Jersey, USA.
14. Boehmer-Christiansen, S. (1994). Politics and Environmental Management. *Journal of Environmental Planning and Management*, **37 (1)**, 69-85.
15. Donaldson, G. and Lorsch, J.W. (1983). *Decision Making at the Top*. Basic Books, New York, USA.
16. Lentz, R.T. and Lyles, M.A. (1989). Paralysis by Analysis: is your Planning System Becoming too Rational? In, D. Asch and C. Bowman, (eds.), *Readings in Strategic Management*, Macmillan, London, in association with the Open University, 57-70.
17. Barnard, E. (1998). Program Aims to Reduce Impact of Tsunamis on Pacific States. *Eos, AGU*, 70(22), 258-263.
18. www.geocities.com/CapeCanaveral/Station/Lab/1029TsunamiPNG.html
19. Dengler, L. and Preuss, J. (1999). Reconnaissance Report on the Papua New Guinea Tsunami of 17July, 1998, in, *EERI Special Earthquake Report* January 1-8.

References

PRODUCING TSUNAMI INUNDATION MAPS: THE CALIFORNIA EXPERIENCE

J. BORRERO[1], A.C. YALÇINER[2], U. KANOGLU[3], V. TITOV[4],
D. McCARTHY[5], C.E. SYNOLAKIS[1]

[1]University of Southern California, School of Engineering, Dept. of Civil and Environmental Engineering, *Los Angeles, California , 90089-2531, CA, USA*
[2] Middle East Technical University, Dept. of Civil Engineering, Ocean Engineering Research Center, *06531, Ankara, Turkey*
[3] Middle East Technical University, Dept. of Engineering Sciences, *06531, Ankara, Turkey*
[4] Pacific Marine Environmental Laboratory, NOAA, *7600 Sand Point Way, NE, Seattle, Washington, 98115, USA*
[5] California Seismic Safety Commission, *1755 Creekside Oaks Drive, Suite 100, Sacramento, California 95833, USA*

Abstract
More than 20 tsunami events have impacted the State of California in the past two centuries. While some earlier 19th century reports are subject to interpretation, there is little question that offshore seismic sources exist and could trigger tsunamis directly or through coseismic submarine offshore landslides or slumps. Given the intense coastal land use and recreational activities along the coast of California, even a small hazard may pose high risk. California presents nontrivial challenges for assessing tsunami hazards, including a short historic record and the possibility of nearshore events with less than $20min$ propagation times to the target coastlines. Here we present a brief history of earlier reports to assess tsunami hazards in the State, and our methodology for developing the first generation inundation maps. Our results are based on worst case scenario events and suggest inundation heights up to $13m$. These maps are only to be used for emergency preparedness and evacuation planning.

1. Introduction

Inundation maps are depictions of coastal areas that identify regions, population and facilities at risk from tsunami attack, and are thus helpful for planning emergency preparedness and response. They are necessary tools for allocation resources for disaster mitigation. The preparation of inundation maps involves both the assessment of the local geologic hazards, their interpretation in terms of tsunami initial conditions, and the calculation of the resulting potential coastal inundation. Inundation maps are now under preparation for most coastal areas of the US Pacific states, most coastal areas of Japan, and several other vulnerable areas around the world. Even using these state--of--the--art inundation prediction

A. C. Yalçiner, E. Pelinovsky, E. Okal, C. E. Synolakis (eds.)
Submarine Landslides and Tsunamis 315-326.
©2003 Kluwer Academic Publishers.

tools [1], California presents unique challenges in assessing tsunami hazards, among them,

* One, there is an extremely short historic record of tsunamis in the State. Whereas in some areas in the Pacific 1000 year long records exist, in California there are none known before the 19th century. Several tsunamis have been reported since 1800, but in most cases the information is not sufficient for reasonable inferences of the associated inundation.

* Two, most of the geologic work in the State has concentrated on identifying the risks associated with onshore faults. Even at the time of writing, there is scant and mostly unpublished information on offshore faults or landslide and slump scars suggestive of past submarine mass failures.

* Three, earlier estimates of tsunami hazards had relied almost entirely on farfield sources and had used pre 1980's inundation mapping technology. This practice had created the impression among policy planners and the general public that the tsunami hazard was small.

* Four, nearshore seismic events may trigger tsunamis arriving within less than $20min$ from generation, allowing little time for evacuation.

* Five, the coastal population density is the highest among the five Pacific States; even a small tsunami arriving to the southern California beaches on a summer Sunday afternoon with tens of thousands of people on the beach poses nontrivial risks whose mitigation needs to be carefully planned for.

2. Existing analyses of tsunami hazards in California

The most comprehensive calculation of tsunami hazards for California was [2] and [3]. Both studies described the hazard in Southern California from farfield events. [4] also focused on farfield events, but also considered several local events off Los Angeles. [5] analyzed the 1927 Lompoc earthquake and associated local hazards. In a seminal review, [6] performed a systematic analysis of all historic and possible tsunami hazards in California and they qualitatively calculated the tsunami hazard in California as high along the coast from Crescent City to Cape Mendocino, moderate south of the Cape to north of Monterey, high, south of Monterey to Palos Verdes, and moderate south of Palos Verdes to San Diego. [7] revisited the [6] estimates and identified the need for modelling from nearshore events. As an example, they considered a hypothetical fault rupture along the San Clemente fault, south west off Los Angeles. They found that results using the older pre 1980 methodology were as much as 50% lower than results using current inundation models. When performed with the new generation of inundation models, runup estimates were occasionally up to 100% higher than what the earlier calculations suggested, depending on the nearshore topography.

[8, 9] studied nearshore tectonic, landslide and slump sources in East Santa Barbara channel and produced runup estimates ranging from $2m$ to $13m$. Unpublished results suggest that there is tentative sedimentologic evidence to also associate a landslide with the 1812 event. [9] found that purely tectonic sources could generate tsunamis with $2m$ runup, while combination of tectonic sources and submarine mass movements could generate extreme runup of up to $20m$, in one location. They observed narrow runup peaks and warned that "*A wave of this size anywhere along the populated shores of southern California would be devastating, and further mapping work is urgently needed to quantify this possibility*". Clearly without further marine geologic investigations it is not easy to assess the relative likelihood or repeat interval of the events.

The bathymetry off Santa Barbara is shown in Figure 1, as first presented by [10]. The features are suggestive even to non-marine geologists of landslide scarps. Figure 2 shows the [9] runup estimates based on worst case scenarios.

The estimates for the leading wave heights for landslide generated waves off Palos Verdes, a peninsula separating Santa Monica and San Pedro Bays in Los Angeles County is provided by [11]. Their estimates ranged from 10m to 40m depending on the initiation depth. [1] used a combination FD solution and analytic solution of the linear shallow water wave equations -see [12]- to calculate tsunami propagation, except in the Santa Monica and San Diego bays where they used an FE solution to resolve possible local resonance effects. They argued that the only reliable data for defining source characteristics at that time were from the 1964 Alaskan and the 1960 Chilean earthquakes. Based on these data, they approximated the initial ground deformation by a hypothetical uplift mass of ellipsoidal shape, about 600 miles long with an aspect ratio of 1:5 and maximum vertical uplift of 8m to 10m. They then divided the Aleutian trench into segments and calculated the wave evolution from each segment, and repeated the procedure for tsunamis from the Peru--Chile trench to derived 100-year (R_{100}) and 500--year (R_{500}) tsunami runup heights. [13] and [14] argued that the 100-year hazard in California is dominated by distant events, hence the Houston and Garcia analysis is probably adequate. They also argued that the 500-year hazard is dominated by local offshore landslide events, hence they revised [2]'s estimates, as shown in Table 1.

TABLE 1. Revised predictions for tsunami runup heights in Southern California.

Location in Southern California	R_{100}	R_{500}
Port of Los Angeles/Port of Long Beach	2.4m	4.6m
Port Huaneme	3.4m	6.4m
Santa Barbara	4.5m	10m

[4] was a seminal work on tsunami hazard potential in Southern California. [4] relied on [3] results for farfield tsunamis and then used seismological data to make predictions for nearshore events using several empirical formulae developed in Japan. These formulae had been extensively used before 1992; since then, high quality runup data from the 1992-2002 event for many areas around the Pacific suggest that these formulas are may only be applicable in Japan and that they can substantially underpredict the runup elsewhere. [4] used the Japanese data to argue that a local seafloor earthquake having a magnitude 7.5 and a hypocentral depth of 4km to 14km could produce a tsunami accompanied by a runup height of 4m to 6m. In 1985, a 6m tsunami may have appeared a marginal hazard, even though the tsunami height in the 1964 Alaskan tsunami in Crescent City which killed 11 people was about 6.2m, while the runup height was 3.8m. The 1992-2002 post-event field surveys have shown that even a 4m tsunami can cause extensive damage and flooding in flat coastlines, such as those in Santa Monica bay or in Orange and San Diego counties. Note that the more recent estimates in table 2 are compatible with McCulloch's results for tectonic tsunamis.

Perhaps the most serious implication of [4]'s assessments is his conclusion that landslide generated waves off Palos Verdes (PV) would be small. This conclusion was based on an arithmetic error and his 0.14m value for the PV slide should had been *14m*, again consistent with newer computations. [11] provide estimates for the same slide ranging from 10m to 40m. The PV slide that was studied by McCulloch and others is shown in Figure 3, as mapped by [15].

318

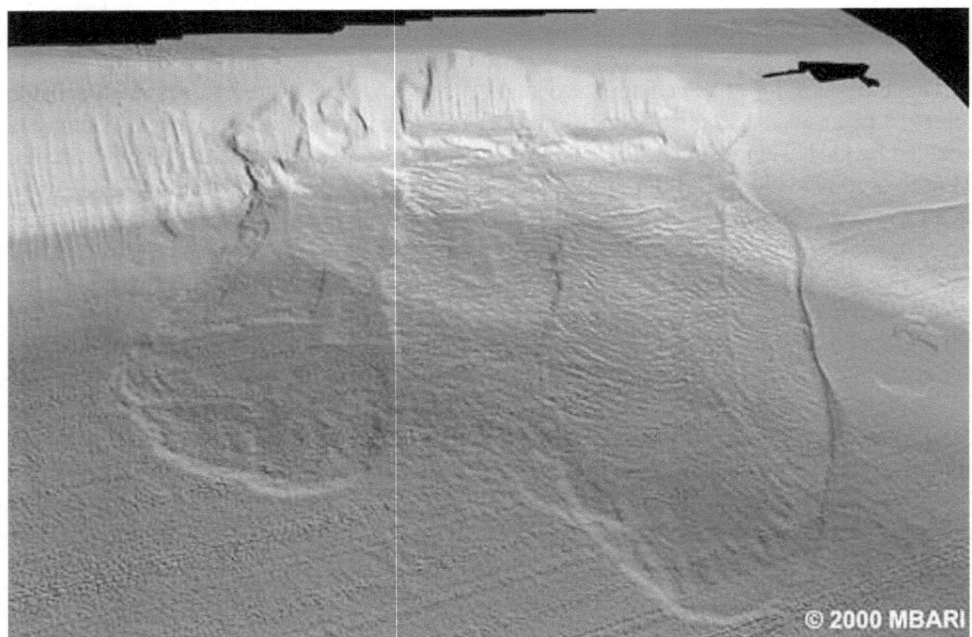

Figure 1. The bathymetry off Santa Barbara, California (after [10]). Two paleolandslide scarps are seen as presented in USC's NSF workshop on landslide hazards.

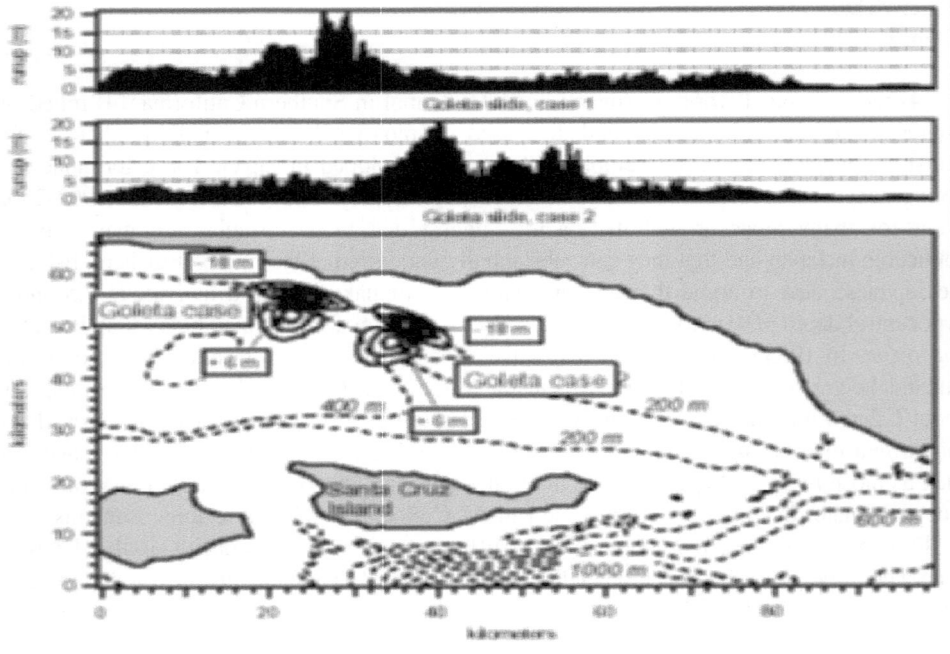

Figure 2. The worst case scenario runup estimates for Santa Barbara County [9].

3. Developing inundation maps for the State of California

In 1996, the Tsunami Hazard Mitigation Federal/State Working Group prepared a report to the US Congress recommending the preparation of inundation maps for the five States, Alaska, California, Hawaii, Oregon and Washington. The report led to mobilization of significant federal resources for tsunami hazards mitigation, and to the establishment of the US National Tsunami Hazard Mitigation Program (NTHMP) program which provides resources in all five states for mitigating tsunami hazards. The NTHMP was the focus of a program review during the ITS held on August 5-7, 2001 in Seattle [16]. As early as 1997, the California's Coastal Region Administrator Rich Eisner of the Governor's Office of Emergency Services (OES), through a series of workshops and publications informed local governments and emergency agencies of the plans to address tsunami hazards and presented the NTHMP. OES solicited input as to the levels of hazards to be represented on the maps, as the short length of the historic record did not permit a comprehensive probabilistic hazard assessment. It was thus decided that the maps would include realistic worst case scenarios to be identified further in the mapping process. In 1998, as funding became available for the State, OES contracted to University of Southern California (USC)'s Tsunami Research Program the development of the first generation of inundation maps for the State.

The State of California has the most densely populated coastlines among all five States in the NTHMP. The State had to utilize the same limited resources as the other four but assess offshore tsunami hazards over a much longer coastline. A comprehensive tsunami hazards evaluation of involves both the probabilistic hazard assessment of different farfield and nearfield, onshore and offshore sources and the hydrodynamic computation of the tsunami evolution from the source to the target coastline. Given the level of funding, this was not feasible, and this presented another challenge for California. Given the quantitative agreement between model results and measurements for the 1964 tsunami of the work of [2], it was decided to focus on nearshore tsunami hazards, which had not been modelled before 1999. If inundation predictions from nearshore events proved smaller than twice the farfield tsunami results of Houston and Garcia, then farfield sources would have to be considered as well. Early results suggested that for the areas studied, nearshore sources produced higher inundation heights that were twice the 100--year values of Houston and Garcia, hence only nearshore sources were considered. Revised estimates for selected locales for the 100 and 500 year tsunamis are provided in Table 1.

The State was also faced with the decision of choosing its mapping priorities. By considering the geographic distribution of population centers, the State opted to perform modeling of the Santa Barbara and San Francisco coastlines in year one, of Los Angeles and San Diego in year two, and of Monterey Bay in year three. The next decision was the resolution of the numerical grids to be used in the maps. While technology existed for high resolution maps with grids as small as $5m$ this would have resulted in a relatively small spatial coverage with large grids and uneconomic computations. It was opted to produce maps at $125m$ resolution, based on [17] who had argued that dense grid may improve numerical accuracy but do not improve realism if the bathymetric/topographic sets are not of similar resolution; in California, the sets varied in resolution between $50m$ and $150m$. Also, given the uncertainties in locating and understanding source mechanisms, results with higher resolution would be misleading. With a few exceptions, most state maps conformed to these resolution criteria.

Figure 3. *The bathymetry off Los Angeles, California. The Palos Verdes peninsula is seen on the left separating Santa Monica (left) and San Pedro bays (right). Various scarps suggestive of paleotsunami slides are shown including the PV debris avalanche.*

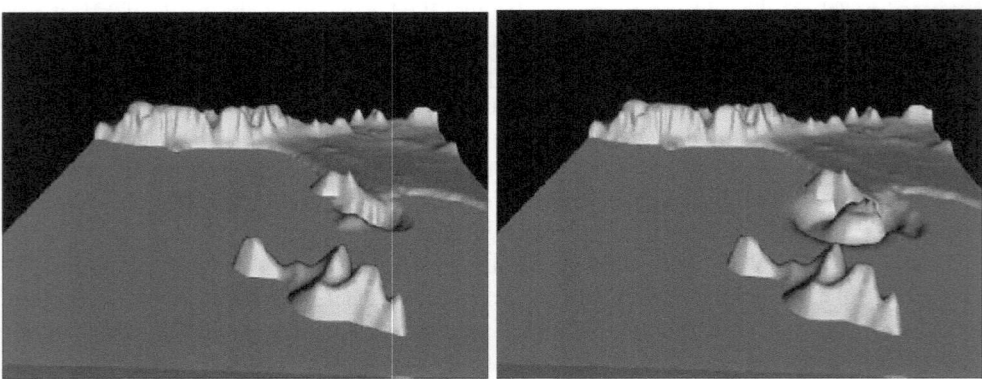

Figure 4. *Two instances of the wave evolution off Palos Verdes to simulate the PV debris avalanche. Calculations of [13], animation images by Salim Pamukcu.*

The next question was whether to provide emergency planners with inundation results at different levels of risk. For example, one suggestion was to include a low and a high risk lines on the inundation maps, another to provide separate lines for nearfield and farfield events.

On discussing these issues with emergency preparedness professionals across the State, it was felt that a single line representing a worst case scenario was preferable, for it simplified the preparedness response of city officials and it better informed the general public. Further, without a probabilistic hazards assessment it was difficult to rank the relative risk from different scenarios. Lines identifying risk zones for nearfield and farfield events could also prove cumbersome and confusing for the public. It was therefore decided to consider for every locale in the region under consideration, the worst case nearshore event that was plausible based on the available historic earthquake and tsunami information. The inundation mapping effort first identified offshore faults and offshore landslide and slump hazards. Difficulties encountered included the lack of detailed high resolution marine surveys over all target coastlines. With the exception of marine surveys undertaken by the USGS off Santa Monica Bay (see Figure 3) and of the Monterey Bay Acquarium Marine Institute (MBARI) off Santa Barbara (see Figure 1) and Monterey Bay, high resolution surveys are not available for other parts of the State, if indeed they do exist at all. Hence, and given that onshore earthquakes can trigger submarine landslides, in regions where marine geology data did not exist, steep submarine soft sediment slopes were considered as possible sources. Offshore faults and slide-prone areas were then used to develop initial tsunami waves as discussed in [9] while the methodology for calculating landslide generated waves is in [18]. Then, the inundation model MOST was used to obtain inundation heights and penetration distances along the target coastline. The evolution of MOST is described in [17, 19, 20, 21].

The inundation predictions for any given event are highly bathymetry and topography dependent and vary substantially along the coast. Since the location of the source is seldom accurately known, the source was moved around within the range of uncertainty. Along California's flat coastlines, this relocation of the tsunami sources resulted in relocation of the maximum along the coast. When asked, emergency planners preferred to have a single value for each region identifying the maximum elevation that tsunami waves from the different local offshore sources would attain. This practice would simplify the communication of the risk to the public and it would provide a information that was easy to remember and implement in regional emergency preparedness. For example, a region could plan for tsunami evacuation areas above a certain minimum elevation across its jurisdiction. Hence, in the development of the maps, sources were relocated along the coast and the highest inundation value among different runs identified. Interestingly, in the areas studied there were no areas that consistently experienced higher runup that adjacent locales. [14] found that most low lying coastal areas could experience the high runup, if the source was relocated in an appropriate direction, within the uncertainties of defining the source. Thus the inundation maps for California do not represent the inundation from any particular event or characteristic earthquake, but the locus of maximum penetration distances from relocating worst case scenario events.

One example of the maps is shown in Figure 5 which presents both a photograph with a view of the west coast of the City of San Francisco, south of the Golden Gate Bridge and the corresponding inundation map, as drawn by the OES with data supplied by USC.

Figure 6 shows prediction of wave runup heights from the Palos Verdes debris avalanche as shown in Figure 4. It should be emphasized that the assumption is that the entire avalanche takes place all at once, based on the slide characteristics used by [11] and [15]. This is the same assumption as the one used in calculating the impact of the Goleta slides off Santa Barbara. Most of the extreme runup is located around the PV cliffs, yet there is significant impact in the ports of Los Angeles (POLA) and Long Beach (POLB) in San Pedro Bay. The OES did an approximate calculation to compare the impact from this particular tsunami scenario to a hypothetical OES scenario for flooding from a dam break. About 75,000 residents are exposed to the tsunami scenario with projected losses of 4.5 billion US dollars, versus 15,000 residents

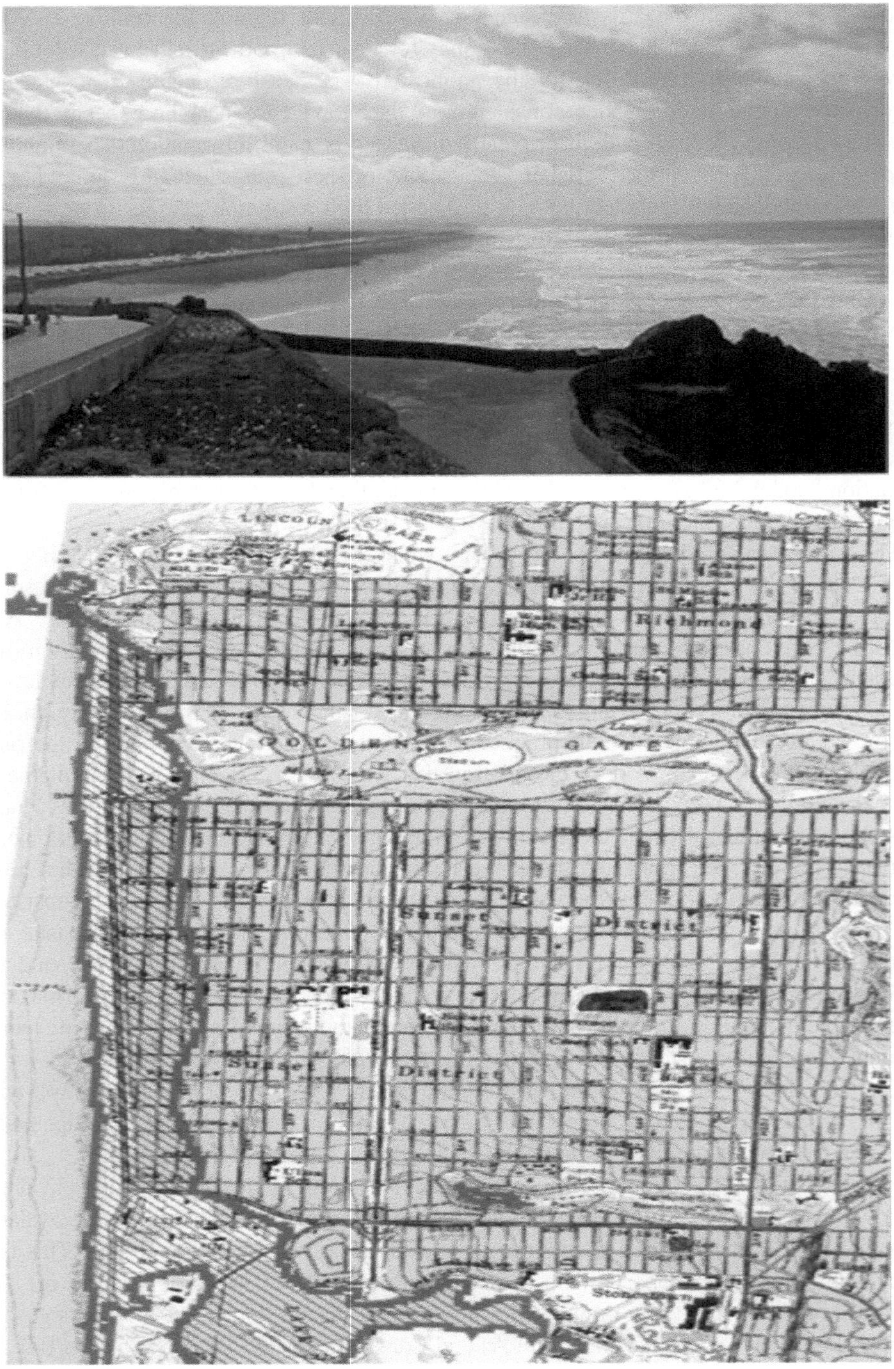

Figure 5. *A view of the coastline and the corresponding inundation maps for the City of San Francisco, California.*

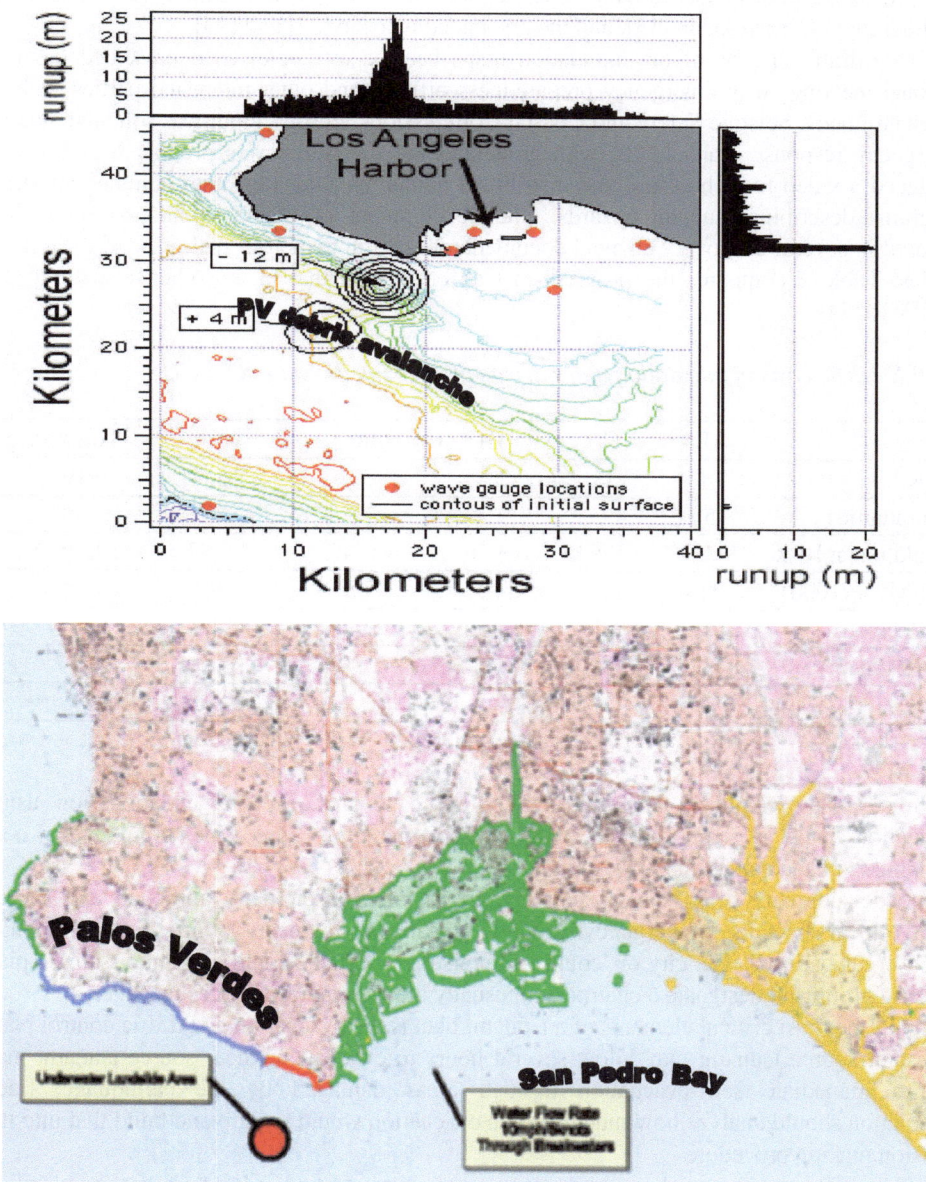

Figure 6. Predicted inundation from the PV debris avalanche. OES Estimates of the affected region comparing the impact of a tsunami attack (green) and of a scenario dam break (yellow).

exposed to the dam break flooding scenario with losses of up to 1 billion US dollars. The methodology is described in [13], and the associated maps in [22] and [14].

Once draft versions of the inundation maps became available, OES presented them in regional meetings with emergency preparedness officers and other interested parties such as the State Lands, Seismic Safety and Coastal Commissions. Further input was solicited, and an emergency response manual [23] with guidelines for mitigation was produced. OES also produced a video for school use and distributed numerous copies of other commercial video programs describing tsunami hazards. The development of State's inundation maps was featured in several Discovery channel documentaries and in numerous national and local news stories. Table 2 compares the progress in California with that in the other 4 states in the NTHMP [24].

TABLE 2. The costs of inundation maps across the five Pacific states of the US.

	AK	CA	HI	OR	WA	Total/Avrg
Maps	1	5	2	6	5	19
Communities	5	42	9	7	25	88
Population at Risk	9608	857915	66916	41743	44383	1020565
Cost/Map(x1000)	414	82	112	65	79	97
Cost/person	$43	$0.48	$3.36	$9.4	$8.9	-
Months/map	42	8.4	15	9	10.8	11.7

AK : Alaska, CA : California, HI : Hawaii, OR : Oregon, WA : Washington

Since many engineers are occasionally asked to make recommendations on using inundation maps, it is appropriate to end with some of the guidelines from [23]. Local Planning Guidance on Tsunami Response

* The development of a tsunami plan requires a multi-disciplinary approach and should involve local specialists (emergency responders, planners, engineers, utilities and community based organizations.) The city or county administrative office should appoint a tsunami plan working group and designate a chairperson, usually the emergency services manager.

* One of the most critical elements of a tsunami plan is the evacuation and traffic control plan. A distant source tsunami may allow several hours to evacuate. A near source tsunami may require immediate self--evacuation through areas damaged by the earthquake. Each jurisdiction should analyze how much time an evacuation would require and build that into the decision making procedure.

* Inundation projections and resulting planning maps are to be used for emergency planning purposes only. They are not based on a specific earthquake and tsunami. Areas actually inundated by a specific tsunami can vary from those predicted. The inundation maps are not a prediction of the performance in an earthquake or tsunami of any structure within or outside the projected inundation area.

* Elements to consider in developing an evacuation plan are to locate optimum evacuation routes. The primary objective is to move up and inland away from the coast.

4. Summary

California for the first time ever now has inundation maps covering a significant portion of the State. The maps were developed based on offshore tsunami sources, including both tectonic motions and mass movements. The mapped inundation line is based on runup computations developed by relocating worst case offshore sources that trigger tsunamis. In most mapped areas the projected inundation extends from 12 to 14m. The maps are for emergency preparedness and evacuation planning only.

Acknowledgements

We gratefully acknowledge the support of NOAA-PMEL and in particular of Drs. Eddie Bernard, Frank Gonzalez and Vasiliy Titov. Most MOST simulations used in these maps used Dr. Titov's codes, others used TUNAMI-N2 as copyrighted by Professors Yalciner, Imamura and Synolakis. Rich Eisner of the Office of Emergency Services provided valuable input and co authored earlier incarnations of this work. While some preliminary assessments used the PNG initial condition a version of which was provided by Dr. Phil Watts of Applied Fluids Engineering, all maps were produced with initial conditions calculated as appropriate for the local conditions.

References

1. Synolakis, C.E., Liu, P.L -F., Yeh, H., Carrier, G., (1997), Tsunamigenic Seafloor Deformations, *Science*, 278, 598-600.
2. Houston, J.R., and Garcia, A.W. (1974) Type 16 Flood Insurance Study, USACE, WES Report H-74-3.
3. Houston, J.R. (1980), Type 19 Flood Insurance Study, USACE, WES Report HL-80-18.
4. McCulloch, D.S., (1985), Evaluating Tsunami Potential, USGS Prof. Paper 1360, 374-413.
5. Satake K., and Somerville, P. (1992) Location and size of the 1927 Lompoc, California earthquake from tsunami data, Bulletin of the Seismological Society of America, 82, 1710-1725.
6. McCarthy, R.J., Bernard, E.N., Legg, M.R., (1993) Coastal Zone´93, Proc. Amer. Shore and Beach Preserv. Assoc., ASCE, New Orleans, Louisiana, 2812-2828.
7. Synolakis, C.E, McCarthy, D., Titov, V.V., Borrero, J., (1997) Evaluating the tsunami risk in California, in California and the World Oceans 97, Proc.ASCE, San Diego, California.
8. Borrero, C., Kanoglu, U., Synolakis, C.E., (1999), Tsunami Generation Mechanisms Along the California Coast and the Inundation Mapping Effort, EOS, Bulletin of the American Geophysical Union, 80, San Francisco, California (Abstract).
9. Borrero, J.C., Dolan, J.F., and Synolakis, C.E. (2001), Tsunamis within the Eastern Santa Barbara Channel, Geophysical Research Letters, 28, 643-646.
10. Greene, G. H., Maher, N., Pauli, C.K., (2000), Landslides off Santa Barbara, California, EOS, Trans.A GU, 81, F750.
11. Locat, J., Locat, P., Lee, H. J., Imaran, J., (2002) Numerical analysis of the mobility of the Palos Verdes debris avalanche and its implication for the generation of tsunamis. Was to appear in of the NSF Workshop on Underwater Landslide Prediction, and manuscript accepted in final form, current status unknown.)
12. Synolakis, C.E. (2003) Tsunami and Seiche, in *Earthquake Engineering Handbook*, W-F Chen and C Scawthorn (editors), CRC Press, 9-1 to 9-90.
13. Borrero, J.C., (2002) Analysis of the Tsunami Hazards in Southern California PhD Thesis, Department of Civil Engineering, University of Southern California, Los Angeles, California, 262pp.
14. Synolakis, C.E., Borrero, J., Eisner, R., (2002b) Developing inundation maps for Southern California, in Solutions to Coastal Disasters, Edited by Ewing L., and Wallendorf L., Proc. ASCE., Reston Virginia, 848-862.
15. Bohannon, R.G. and Gardner, J.V., (2001) Submarine Landslides of San Pedro valley, Was to appear in of the NSF Workshop on Underwater Landslide Prediction and manuscript accepted in final form, current status unknown.)

16. Bernard, E.N., (2001) The US National Tsunami hazard Mitigation Program Summary, Proceedings of the 2001 ITS, Review Session, Number R-1, 9-16, Seattle, Washington.

17. Titov, V. V. and Synolakis, C.E., (1997), Extreme inundation flows during the Hokkaido- Nansei-Oki tsunami, Geophysical Research Letters 24, 1315-1318.

18. Synolakis, C.E., Bardet,J.P., Borrero, J.C., Davies,H., Okal, E., Silver, E., Sweet, S., Tappin,D., (2002a) Slump origin of the 1998 Papua New Guinea tsunami, *Proceedings of the Royal Society*, 458, 763-769.

19. Titov, V.V. (1997) Numerical modeling of long wave runup, Ph.D. thesis, University of Southern California, Los Angeles, California. 130pp.,

20. Titov, V.V. and Synolakis, C.E. (1995), Modeling of Breakingand Nonbreaking Long Wave; Evolution and Runup Using VTSC-2. Journal of Waterway Port, Coastal and Ocean Engineering 121 (5), 308-316.

21. Titov, V.V.and .Synolakis, C.E., (1998) Numerical Modelling of Tidal Wave Runup, Journal of Waterway, Port, Coastal and Ocean Engineering, 124, 157-171.

22. Eisner, R.K., Borrero, J.C., Synolakis, C.E., (2001) Inundation maps for the State of California, Proceedings of the 2001 ITS, Review Session, Number R-4, 55-69, Seattle, Washington.

23. OES, (2002), Emergency Response Manual published by Offshore Emergency Services

24. Gonzalez, F, Titov V.V., Mofjeld, H.O.1, Venturato, A.J., Newman, J.C. (2001) The NTHMP Inundation Mapping Program, ITS 2001 Proceedings, NHTMP Review Session, Number R-2 29, NOAA-PMEL.

SUBJECT INDEX